Student Interactive Workbook
for Starr and McMillan's

HUMAN BIOLOGY

Sixth Edition

Shelley Penrod
North Harris College

Ann Maxwell
Angelo State University

Jane B. Taylor
Northern Virginia Community College

John D. Jackson
North Hennepin Community College

THOMSON

BROOKS/COLE

Australia • Canada • Mexico • Singapore • Spain • United Kingdom • United States

THOMSON
™
BROOKS/COLE

Student Interactive Workbook for Starr and McMillan's Human Biology
Sixth Edition
Shelley Penrod, Ann Maxwell, Jane B. Taylor, John D. Jackson

Biology Publisher: Jack C. Carey
Senior Assistant Editor: Suzannah A. Alexander
Editorial Assistant: Chris Ziemba
Technology Project Manager: Keli Sato Amann
Marketing Manager: Ann Caven
Marketing Assistant: Leyla Jowza
Advertising Project Manager: Nathaniel Bergson-Michelson
Project Manager, Editorial Production: Suzanne Barnecut

Print/Media Buyer: Lisa Claudeanos
Permissions Editor: Joohee Lee
Production Service: Matrix Productions
Copy Editor: Anna Trabucco
Cover Image: David Madison/Getty Images
Cover Printer: Thomson/West
Compositor: Interactive Composition Corporation
Printer: Thomson/West

Printed in the United States of America
1 2 3 4 5 6 7 08 07 06 05 04

For more information about our products,
contact us at:
Thomson Learning Academic Resource Center
1-800-423-0563
For permission to use material from this text or product,
submit a request online at **http://www.thomsonrights.com.**
Any additional questions about permissions can be submitted
by email to **thomsonrights@thomson.com.**

ISBN 0-534-99785-6

Thomson Higher Education
10 Davis Drive
Belmont, CA 94002-3098
USA

Asia (including India)
Thomson Learning
5 Shenton Way
#01-01 UIC Building
Singapore 068808

Australia/New Zealand
Thomson Learning Australia
102 Dodds Street
Southbank, Victoria 3006
Australia

Canada
Thomson Nelson
1120 Birchmount Road
Toronto, Ontario M1K 5G4
Canada

UK/Europe/Middle East/Africa
Thomson Learning
High Holborn House
50–51 Bedford Road
London WC1R 4LR
United Kingdom

Latin America
Thomson Learning
Seneca, 53
Colonia Polanco
11560 Mexico
D.F. Mexico

Spain (including Portugal)
Thomson Paraninfo
Calle Magallanes, 25
28015 Madrid, Spain

CONTENTS

Credits

This page constitutes an extension of the copyright page. We have made every effort to trace the ownership of all copyrighted material and to secure permission from copyright holders. In the event of any question arising as to the use of any material, we will be pleased to make the necessary corrections in future printings. Thanks are due to the following authors, publishers, and agents for permission to use the material indicated.

Chapter 3
p.36 (bottom): Art by Raychel Ciemma and American Composition and Graphics
p.39: Art by Raychel Ciemma and American Composition and Graphics

Chapter 4
p.60: Art by Raychel Ciemma
p.63: Art by L. Calver
p.65: Art by Robert Demarest

Chapter 5
p.76: Art by Joel Ito
p.76: Micrograph Ed Reschke
p.77: Art by Robert Demarest
p.79: Art by Raychel Ciemma

Chapter 6
p.89: Art by Raychel Ciemma
p.90: Art by Robert Demarest with Gary Head/micrograph, © Don Fawcett/Visuals Unlimited, from D. W. Fawcett, *The Cell*, Philadelphia; W. B. Saunders Co., 1966
p.93: Art by Kevin Sommerville

Chapter 7
p.99: Art by Kevin Sommerville

Chapter 8
p.117: Art by Lisa Starr

Chapter 9
p.126: Art by Raychel Ciemma
p.127: Art by Kevin Sommerville
p.132: Art by Raychel Ciemma
p.135: Art by Robert Demarest, based on *Basic Human Anatomy*, by A. Spence, Benjamin-Cummings, 1982.

Chapter 10
p.142: Art by Rachel Ciemma

Chapter 11
p.155: Art by Kevin Sommerville
p.159: Art by Lisa Starr

Chapter 12
p.170: Art by Robert Demarest
p.171: Art by Robert Demarest

Chapter 13
p.181: Art by Raychel Ciemma
p.185: Art by Robert Demarest
p.186: Art by Robert Demarest
p.187: Art by Robert Demarest
p.190: Art by Robert Demarest
p.190: Manfred Kage/Peter Arnold, Inc.
p.193: Yokochi and J. Rohen, *Photographic Anatomy of the Human Body*, 2/e. Igaku-Shoin, Ltd., 1979.

Chapter 14
p.205: Art by Robert Demarest
p.208: Art by Robert Demarest

Chapter 15
p.217: Art by Kevin Sommerville

Chapter 16
p.231: Art by Raychel Ciemma
p.232: Art by Raychel Ciemma
p.235: Art by Raychel Ciemma
p.237: Art by Robert Demarest and Precision Graphics

Chapter 17
p.250: Art by Raychel Ciemma
p.254: Art by Raychel Ciemma/photographs from Lennart Nilsson, *A Child is Born*, © 1966, 1977 Dell Publishing Company, Inc.

Chapter 19
p.277: Art by Raychel Ciemma
p.282: Art by Raychel Ciemma

Chapter 21
p.303: Art by Precision Graphics and Gary Head

Chapter 22
p.315: Gary Head and Lisa Starr
p.322: Art by Lisa Starr

Chapter 24
p.344: Art after *Evolving*, by F. Ayala and J. Valentine, Benjamin-Cummings, 1979

PREFACE

Tell me and I will forget, show me and I might remember, involve me and I will understand.
                    —Chinese Proverb

The proverb outlines three levels of learning, each successively more effective than the method preceding it. The writer of the proverb understood that humans learn most efficiently when they *involve* themselves in the material to be learned. This student workbook is like a tutor; when properly used it increases the efficiency of your study periods. The interactive exercises actively involve you in the most important terms and central ideas of your text. Specific tasks ask you to recall key concepts and terms and apply them to life; they test your understanding of the facts and indicate items to reexamine or clarify. Your performance on these tasks provides an estimate of your next test score based on specific material. Most important, though, this biology student workbook and text together help you make informed decisions about matters that affect your own well-being and that of your environment. In the years to come, human survival on planet Earth will demand administrative and managerial decisions based on an informed biological background.

HOW TO USE THIS STUDENT WORKBOOK

Following this preface, you will find an outline that will show you how the student workbook is organized and will help you use it efficiently. Each chapter begins with a title and an outline of the level one and level two headings in that chapter. The Interactive Exercises follow, wherein each chapter is divided into sections of one or more of the main (1-level) headings that are labeled 1.1, 1.2, and so on. *For easy reference to an answer or definition, each question and term in this unique student workbook is accompanied by the appropriate text page(s) in the form:* [p.352]. The Interactive Exercises begin with a list of Selected Words (other than boldfaced terms) chosen by the authors as those that are most likely to enhance understanding. In the text chapters, the selected words appear in italics, quotation marks,

or roman type. This is followed by a list of Bold-faced, Page-Referenced Terms that appear in the text. These terms are essential to understanding each student workbook section of a particular chapter. Space is provided by each term for you to formulate a definition in your own words. Next is a series of different types of exercises that may include completion, short answer, true/false, fill-in-the-blanks, matching, choice, dichotomous choice, label and match, problems, labeling, sequencing, multiple choice, and completion of tables.

A Self-Quiz immediately follows the Interactive Exercises. This quiz is composed primarily of multiple-choice questions although sometimes we present another examination device or some combination of devices. Any wrong answers in the Self-Quiz indicate portions of the text you need to reexamine. A series of Chapter Objectives/Review Questions follows each Self-Quiz. These are tasks that you should be able to accomplish if you have understood the assigned reading in the text. Some objectives require you to compose a short answer or long essay; others may require a sketch or supplying correct words.

Following the Chapter Objectives/Review Questions section are the Media Menu Review Questions. These questions test your understanding of major points from the Media Menu InfoTrac College Edition articles found at the end of each text chapter.

The final part of each chapter is Integrating and Applying Key Concepts. It invites you to try your hand at applying major concepts to situations in which there is not necessarily a part answer and so none is provided in the chapter answer section. Your text generally will provide enough clues to get you started on an answer, but this section is intended to stimulate your thought and provoke group discussions.

A person's mind, once stretched by a new idea, can never return to its original dimension.
                    —Oliver Wendell Holmes

STRUCTURE OF THIS STUDENT WORKBOOK

The outline below shows how each chapter in this student workbook is organized.

Chapter Number ──────────→

3

Chapter Title ──────────→

CELLS AND HOW THEY WORK

Chapter Outline ──────────→

Impacts, Issues: When Mitochondria Spin Their Wheels

CELLS: ORGANIZED FOR LIFE
 All cells are alike in three ways
 There are two basic kinds of cells
 Why are cells small?
 The structure of a cell's membranes reflects
  their function

THE PARTS OF A EUKARYOTIC CELL

THE PLASMA MEMBRANE: A LIPID BILAYER
 The plasma membrane is a mix of lipids and
  proteins
 Membrane proteins carry out most membrane
  functions

SUGAR WARS
 The carbo corps
 Carbo culprits, carbo cops
 Cancer in the crosshairs

THE NUCLEUS
 A nuclear envelope encloses the nucleus
 The nucleolus is a "workshop" where cells make
  the subunits of ribosomes
 DNA is organized in chromosomes
 Events that begin in the nucleus continue to
  unfold in the cell cytoplasm

THE ENDOMEMBRANE SYSTEM
 ER: A protein and lipid assembly line
 Golgi bodies: Packing and shipping
 A variety of vesicles

MITOCHONDRIA: THE CELL'S ENERGY
 FACTORIES
 Mitochondria make ATP
 ATP forms in an inner compartment of the
  mitochondrion

MICROSCOPES: WINDOWS INTO THE WORLD
 OF CELLS

THE CYTOSKELETON: CELL SUPPORT AND
 MOVEMENT
 The cytoskeleton provides internal support for
  cells
 Structures that allow cells to move: flagella and
  cilia

MOVING SUBSTANCES ACROSS MEMBRANES
 BY DIFFUSION AND OSMOSIS
 The plasma membrane is "selective"
 Diffusion: A solute moves down a gradient
 Osmosis: How water crosses membranes

OTHER WAYS SUBSTANCES CROSS CELL
 MEMBRANES
 Small solutes cross membranes through transport
  proteins
 Vesicles transport large solutes

REVENGE OF EL TOR

METABOLISM: DOING CELLULAR WORK
 ATP—The cell's energy currency
 There are two main types of metabolic pathways
 A closer look at enzymes

HOW CELLS MAKE ATP
 Cells make ATP in three steps
 Step 1: Glycolysis breaks glucose down to
  pyruvate
 Step 2: The Krebs cycle produces energy-rich
  transport molecules
 Step 3: Electron transport produces a large harvest
  of ATP

SUMMARY OF AEROBIC RESPIRATION

ALTERNATIVE ENERGY SOURCES IN THE BODY
 Carbohydrate breakdown in perspective
 Energy from fats
 Energy from proteins

Interactive Exercises ──────────→ The interactive exercises are divided into numbered sections by titles of main headings and page references. Each section begins with a list of author-selected words that appear in the text chapter in italics, quotation marks, or roman type. This is followed by a list of important boldfaced, page-referenced terms from each section of the chapter. Each section ends with interactive exercises that vary in type and require constant interaction with the important chapter information.

Self-Quiz ──────────→ The self-quiz is a set of multiple-choice questions that sample important blocks of text information.

Chapter Objectives/ ──────────→ Combinations of relative objectives to be met and questions to be
Review Questions answered

Media Menu ──────────→ These questions examine readings from InfoTrac College Edition articles
Review Questions cited in the Media Menu at the end of each text chapter.

Integrating and ──────────→ Applications of text material to questions for which there may be more
Applying Key Concepts than one correct answer.

Answers to Interactive ──────────→ Answers for all interactive exercises can be found at the end of this
Exercises and Self-Quiz student workbook by chapter and title, and the main headings with their page references, followed by answers for the Self-Quiz.

1

LEARNING ABOUT HUMAN BIOLOGY

Interactive Exercises

Note: In the answer section of this book, a specific molecule is most often indicated by its abbreviation. For example, adenosine triphosphate is ATP.

Impacts, Issues: What Am I Doing Here? [p.1]

1.1. THE CHARACTERISTICS OF LIFE [p.2]

For additional practice, use the interactive vocabulary exercises linked with your BiologyNow CD-ROM.

Selected Words: deoxyribonucleic acid [p.2]

In addition to the boldfaced terms, the text features other important terms essential to understanding the assigned material. "Selected Words" is a list of these terms, which appear in the text in italics, in quotation marks, and occasionally in roman type. Latin binomials found in this section are underlined and in roman type to distinguish them from other italicized words.

Boldfaced, Page-Referenced Terms

These terms are important; they were in boldface type in the chapter. Write a definition for each term in your own words without looking at the text. Next, compare your definition with that given in the chapter or in the text glossary. If your definition seems accurate, allow some time to pass and repeat this procedure until you can define each term rather quickly (how fast you answer is a gauge of your learning efficiency).

[p.2] DNA _____

_____

[p.2] cell _____

_____

[p.2] homeostasis _____

_____

Choice

For examples 1–14, choose from the following characteristics of life:

a. DNA
b. taking in and using energy and materials
c. sensing and responding
d. reproduction and growth
e. cell
f. homeostasis

1. _____ An animal eating food [p.2]

2. _____ Short for *deoxyribonucleic acid* [p.2]

3. _____ A molecule that living things share and nonliving things do not [p.2]

4. _____ A driver stops when a traffic light turns red [p.2]

5. _____ The smallest entity that is alive [p.2]

6. _____ Its activities are powered using ATP as an energy source [p.2]

7. _____ Infancy, childhood, adolescence, adulthood [p.2]

8. _____ Energy and molecules from protein in a meal are used to build muscle tissue in the human eating the meal [p.2]

9. _____ Living things increase in size, mature, and produce more of their kind [p.2]

10. _____ Maintains the internal environment within life-supporting ranges [p.2]

11. _____ Regulates chemical and physical conditions essential for the survival of an organism [p.2]

12. _____ Contains the instructions for using nonliving substances such as carbon to build and operate the organism [p.2]

13. _____ A dog comes when its master whistles [p.2]

14. _____ Means "staying the same" [p.2]

1.2. WHERE DO WE FIT IN THE NATURAL WORLD? [p.3]

Boldfaced, Page-Referenced Terms

[p.3] evolution _____

_____

[p.3] cultural evolution _____

_____

Hierarchical Relationships

Fill in the blanks to represent the place of humans in the natural world. (Figure 1.2 in the text may be helpful.) Choose from the following words:

humans   mammals   vertebrates   primates   animals

(broadest group) 1. _____ kingdom including humans and millions of other species [p.3]

2. _____ animals with backbones [p.3]

3. _____ vertebrates with fur and mammary glands [p.3]

4. _____ humans, apes, and closely related mammals [p.3]

(narrowest group) 5. _____ primates with great manual dexterity and a brain providing verbal skills and analytical ability [p.3]

Listing

6. Using text Figure 1.2 as a guide, list the four nonbacterial kingdoms of life. [p.3]

_____ _____ _____ _____

7. In a five-kingdom system, all of the bacteria are placed in one kingdom. In a six-kingdom system, they are split into two kingdoms. List these two kingdoms. [p.3]

_____ _____

1.3. LIFE'S ORGANIZATION [pp.4–5]

Boldfaced, Page-Referenced Term

[p.4] biosphere _____

Matching

Choose the most appropriate description for each term.

1. _____ organ system [p.4]

2. _____ cell [p.4]

3. _____ community [p.4]

4. _____ atoms and molecules [p.4]

5. _____ ecosystem [p.4]

6. _____ population [p.4]

7. _____ tissue [p.4]

8. _____ biosphere [p.4]

9. _____ complex organism [p.4]

10. _____ organ [p.4]

A. Different tissues combined together to carry out a function in an organism
B. All parts of Earth's waters, crust, and atmosphere in which organisms live
C. The smallest living unit
D. Different organs working together in a coordinated manner
E. The populations of all species occupying the same area
F. Nonliving materials from which cells are built
G. A community and its surrounding environment
H. An individual composed of coordinated organ systems
I. A group of individuals of the same kind, such as all of Earth's humans
J. A group of cells organized to carry out a specific function, e.g., epithelium

Sequence

Arrange the following levels of organization in nature in the correct hierarchic order. Find the letter of the simplest level and write it next to the number 11; write the letter of the most complex level next to the number 21; and so forth. Text Figure 1.4 may be helpful.

11. _____   A. Organ system [p.4]

12. _____   B. Biosphere [p.4]

13. _____   C. Organ [p.4]

14. _____   D. Ecosystem [p.4]

15. _____   E. Atom [p.4]

16. _____   F. Multicellular organism [p.4]

17. _____   G. Population [p.4]

18. _____   H. Community [p.4]

19. _____   I. Cell [p.4]

20. _____   J. Tissue [p.4]

21. _____   K. Molecule [p.4]

Relationships

Determine the word that fits into the sentence on the left. Place that word into the sequence on the right.

22. Energy enters the biosphere from the _____ [p.4]. (22) _____

23. The plants and other organisms that use this energy to make food molecules by photosynthesis are called _____ [p.4].   (23) _____ (25) _____

24. Animals that eat plants or other animals to obtain energy and molecules are called _____ [p.4].   (24) _____

25. Bacteria and fungi obtain the materials and energy they need by _____ [p.4] the remains of organisms, thus recycling substances back to producers.

26. Because of these interconnections, ecosystems can be thought of as a _____ [p.5] of life in which events in one part impact the entire ecosystem.

1.4. SCIENCE AS A WAY OF LEARNING ABOUT THE NATURAL WORLD [pp.6–7]

1.5. SCIENCE IN ACTION: CANCER, BROCCOLI, AND MIGHTY MICE [p.8]

Selected Words: "science" [p.6], "if–then" process [p.6], *logic* [p.6]

Boldfaced, Page-Referenced Terms

[p.6] scientific method _____

_____

[p.6] hypothesis _____

_____

[p.6] prediction _____

[p.7] experiment _____

[p.7] controlled experiment _____

[p.7] control group _____

[p.7] variable _____

Complete the Table

1. Complete the following table of concepts important to understanding the scientific method of problem solving. Choose from *experiment, variable, prediction, control group,* and *hypothesis.*

Concept	Definition
a. [p.6]	An educated guess about what the answer to a question or the solution to a problem might be
b. [p.6]	A statement of what one should be able to observe about a problem if a hypothesis is valid; the "if–then" process
c. [p.7]	A test designed to allow the observer to predict what will happen if his or her hypothesis is not wrong or what won't happen if it is wrong
d. [p.7]	The control group is identical to the experimental group except for this key factor under study
e. [p.7]	Used in scientific experiments as a standard to which the experimental group is compared

Sequence

Arrange the following steps of the scientific method in the correct sequence. Find the letter of the first process and write it next to the number 2, and so forth, finishing with the letter of the final process next to the number 7.

2. _____   A. Develop a hypothesis. [p.6]

3. _____   B. Test the accuracy of prediction. [p.6]

4. _____   C. Repeat tests or devise new tests of the hypothesis. [p.6]

5. _____   D. Make a prediction. [p.6]

6. _____   E. Objectively analyze and report the test results and conclusions. [p.6]

7. _____   F. Observe some aspect of the natural world and identify a question to be explored. [p.6]

After arranging the six steps above in the proper order of the scientific method, now arrange the following six events from actual research in the correct order. Find the letter of the first step and write it next to the number 8, and so forth, finishing with the letter of the final process next to the number 13.

Then, to the right of each letter, identify the step. Choose from: *question, hypothesis, prediction, test, repeat testing, conclusion.*

8. _____ _____   A. Data showing that the experimental group of mice developed fewer tumors than the control group support the prediction made about the effects of sulforaphane. [pp.6,8]

9. _____ _____   B. If sulforaphane helps the body fight off cancer, then mice in the experimental group should develop significantly fewer tumors than mice in the control group. [pp.6,8]

10. _____ _____   C. The experiment is repeated, and a further experiment is added to test the effects of sulforaphane in mice with a mutation that makes them especially vulnerable to cancer. [pp.6,8]

11. _____ _____   D. It is expected that the compound sulforaphane, found in the cabbage family, boosts the body's defenses against cancer. [pp.6,8]

12. _____ _____   E. Observations indicate that diets rich in cabbage and other members of this vegetable family are associated with lower incidence of cancer. Researchers ask the question, "Does a substance in these vegetables called sulforaphane help prevent cancer?" [pp.6,8]

13. _____ _____   F. A control group of mice is fed a normal diet. The variable by which the experimental group of mice differs from the control group is the addition of large amounts of sulforaphane to its diet. [pp.6,8]

Completion

A young girl is stung by a bee that has yellow and black stripes. Several days later, a yellow and black striped wasp stings the child. It only takes a few more stings for the girl to conclude that all black and yellow striped insects give painful stings. This is an example of (14) _____ [p.6] reasoning. Later, the girl is playing outside with a younger friend when she sees a yellow and black striped flying insect. She pulls her friend away, explaining that "*if* you bother a yellow and black striped insect, *then* you will probably get a painful sting." This is an example of (15) _____ [p.6].

1.6. SCIENCE IN PERSPECTIVE [p.9]

1.7. CRITICAL THINKING IN SCIENCE AND LIFE [p.10]

1.8. WHAT IS THE TRUTH ABOUT HERBAL FOOD SUPPLEMENTS? [p.11]

Selected Words: cause vs. *related* factors [p.10], *opinion* [p.10]

Boldfaced, Page-Referenced Terms

[p.9] theory _____

[p.10] critical thinking _____

[p.10] fact _____

True/False

In the blank beside each statement below, write a "T" (true) or an "F" (false).

1. _____ A theory in science is simply a person's idea about something. [p.9]
2. _____ Once a theory is accepted by scientists, its accuracy is never challenged again. [p.9]
3. _____ A theory explains a broad range of related phenomena in the natural world. [p.9]
4. _____ With time and better technology, science will be able to investigate and answer any question that humans ask. [p.9]
5. _____ Science requires strict objectivity. [p.9]
6. _____ Science does not involve value judgments. [p.9]
7. _____ Because science provides knowledge through discovery, it is fair to expect the scientific community to guide the use of this knowledge. [p.9]

Completion

A study shows that, as a group, women who jog regularly develop breast cancer less often than women who never jog. This data reveals that jogging is (8) _____ [p.10] to a lower incidence of breast cancer. However, the study did not control other variables besides exercise in the women's lifestyle (diet, stress, etc.), and does not provide enough information to conclude that jogging is a (9) _____ [p.10] of lower cancer rates. The manufacturer of a home jogging machine publishes the results of the study on breast cancer in order to encourage women to buy its product. Checking the accuracy of the manufacturer's advertising and questioning the source are important aspects of (10) _____ [p.10] thinking. The manufacturer also claims that using the jogging machine will result in a happier lifestyle. Because this information cannot be verified, it is an (11) _____ [p.10], not a (12) _____ [p.10].

Choice

Answer the questions using the choices below.

        a. ginkgo biloba   b. ephedra   c. St. John's wort   d. kava   e. Echinacea

13. _____ Which herbal supplements have been related to a health risk? [p.11]

    _____

14. _____ Which herbal supplements have not demonstrated through scientific studies the benefits attributed to them? [p.11]

    _____

    _____

Self-Quiz

Are you ready for the exam? Test yourself on key concepts by taking the additional tests linked with your BiologyNow CD-ROM.

Multiple Choice

_____ 1. Most of human evolution for the past 40,000 years is based on learning and communicating behaviors within a society. What term refers to this type of evolution? [p.3]
   a. organic evolution
   b. cultural evolution
   c. technological evolution
   d. verbal evolution

_____ 2. About 12 to 24 hours after a meal, a person's blood sugar level normally varies from about 60 to 90 mg per 100 ml of blood, although it may attain 130 mg/100 ml after meals high in carbohydrates. The maintenance of blood sugar level within a fairly narrow range despite uneven intake of sugar is due to the body's ability to maintain a state of _____. [p.2]
   a. adaptation
   b. diversity
   c. metabolism
   d. homeostasis

_____ 3. _____ is a change in the body plan and functioning of organisms through successive generations. [p.3]
   a. Homeostasis
   b. Reproduction
   c. Metabolism
   d. Evolution

_____ 4. Webs of life including producers, consumers, and decomposers that function in the physical environment are called _____. [p.5]
   a. populations
   b. biomes
   c. ecosystems
   d. food chains

_____ 5. The information-gathering system used by scientists to gain knowledge about the natural world is _____. [p.6]
   a. the scientific method
   b. inductive reasoning
   c. deductive reasoning
   d. the "if–then" process

_____ 6. The experimental group and the control group are identical except for _____. [p.7]
   a. the results they give in an experiment
   b. the variable being studied
   c. the number of test subjects
   d. the number of experiments performed on each group

_____ 7. Which of these is true of a hypothesis? [pp.7,8,9]
   a. It is the same as a theory.
   b. It is the same as a prediction.
   c. It is simply a guess.
   d. It must be falsifiable.

_____ 8. Scientific theories are based on all of these except _____. [pp.9,10]
   a. systematic observations
   b. relatedness and opinion
   c. "if–then" predictions
   d. repeated controlled experiments

Choice

For questions 9–10, choose from the following:

   a. theory   b. hypothesis
   c. prediction   d. conclusion

9. _____ A _____ is a statement about what one can expect to observe in nature if a hypothesis is correct. [p.6]

10. _____ A _____ is a statement about whether a hypothesis (or theory) should be accepted, rejected, or modified, based on reproducible results of controlled tests. [p.6]

Chapter Objectives/Review Questions

This section lists general and detailed chapter objectives that can be used as review questions. You can make maximum use of these items by writing answers on a separate sheet of paper. Fill in answers where blanks are provided. To check for accuracy, compare your answers with information in the chapter or glossary.

1. Having a special molecule called _____ sets living things apart from the nonliving world. [p.2]
2. In addition to having DNA, what are the characteristics that all living things share? [p.2]
3. To sustain life, body systems must maintain a dynamic state of internal equilibrium called _____. [p.2]
4. A(n) _____ is an organized living unit that can survive and reproduce itself, using DNA instructions and the necessary energy and materials. [p.2]
5. List the five kingdoms of living things. Which is sometimes split into two kingdoms? To which kingdom do humans belong? [p.3]
6. List characteristics that separate humans from their primate relatives. [p.3]
7. Name and arrange the levels of organization in nature, from the least inclusive to the most inclusive (nonliving cellular components to biosphere). [p.4]
8. Explain how the actions of producers, consumers, and decomposers create interdependency among organisms. [pp.4–5]
9. In what two ways is every part of the living world linked to every other part? [p.5]
10. List and explain the six steps that describe what scientists generally do when they proceed with an investigation. [p.6]
11. Correlating specific events to form a general conclusion is _____ reasoning, whereas drawing specific inferences from a working hypothesis is _____ reasoning. [p.6]
12. What is the function of the control group in an experiment? How does the control group differ from the experimental group? What is a variable? [p.7]
13. How does a scientific theory differ from a scientific hypothesis? [p.9]
14. Explain why no theory is an "absolute truth" in science. [p.9]
15. Describe how the methods of science differ from methods using subjective thinking and systems of belief. [p.9]
16. Define "critical thinking" and list ways in which a piece of information is evaluated critically. [p.10]
17. What does the National Institute of Health propose in regard to purported benefits of herbal supplements? [p.11]

Media Menu Review Questions

Questions 1–6 are drawn from the following InfoTrac® College Edition article: "Why Smart People Believe Weird Things." Michael Shermer. *Skeptic*, Summer 2003.

1. We sort through data and select those ideas most _____ what we already believe, and ignore or _____ away those that are discomforting.
2. Which of these factors seem to be strongly correlated to believing weird things?
   a. gender   b. age   c. education   d. none of these
3. Raw intelligence is not as important as the ability to generate _____ of ideas and select from them those that are most likely to _____.
4. People who measure high on _____ locus of control tend to believe that circumstances are beyond their control.
5. The tendency to seek or interpret evidence favorable to already existing beliefs and to ignore evidence unfavorable to those beliefs is called the _____.
6. Science gets around confirmation bias because other people will _____ results of an experiment.

Questions 7–10 are drawn from the following InfoTrac College Edition article: "Speak No Evil." Nell Boyce. *U.S. News and World Report*, June 2002.

7. Although smallpox is difficult to come by, its _____ is freely available on the Internet.
8. The scientific tradition of openness is clashing with the fear that research could be misused to develop _____.
9. Unlike nuclear physics, which is tightly controlled by the government, _____ has marched forward unencumbered by security concerns.
10. Even though scientists treasure freedom of inquiry, many seem ready to go along with reasonable _____.

Integrating and Applying Key Concepts

1. Humans have the ability to maintain body temperature very close to 37°C.
   a. What conditions tend to make body temperature drop?
   b. What measures do you think your body takes to raise body temperature when it drops?
   c. What conditions cause body temperature to rise?
   d. What measures do you think your body takes to lower body temperature when it rises?
2. What sorts of topics do scientists usually regard as not testable by the methods they generally use?

2

MOLECULES OF LIFE

Interactive Exercises

Impacts, Issues: It's Elemental [p.15]

2.1. THE ATOM [p.16]

2.2. SAVING LIVES WITH RADIOISOTOPES [p.17]

For additional practice, use the interactive vocabulary exercises linked with your BiologyNow CD-ROM.

Selected Words: organic [p.15], *hydrocarbon* [p.15], "atomic number" [p.16], "mass number" [p.16], "atomic weight" [p.16], "radioactivity" [p.16]

Boldfaced, Page-Referenced Terms

[p.15] element

[p.15] trace elements

[p.16] atom

[p.16] protons

[p.16] electrons

[p.16] neutrons

[p.16] isotope

[p.16] radioisotope

[p.17] half-life

[p.17] tracer

[p.17] radiation therapy

Matching

1. _____ atom [p.16]
2. _____ atomic number [p.16]
3. _____ electrons [p.16]
4. _____ element [p.15]
5. _____ isotopes [p.16]
6. _____ mass number [p.16]
7. _____ neutrons [p.16]
8. _____ tracer [p.17]
9. _____ trace element [p.15]
10. _____ organic compounds [p.15]
11. _____ half-life [p.17]
12. _____ radioactivity [p.16]
13. _____ protons [p.16]
14. _____ radioisotope [p.16]
15. _____ radiation therapy [p.17]

A. The smallest unit that has the properties of a given element
B. The number of protons in the nucleus of one atom of an element
C. Subatomic particles in the nucleus of an atom that have no charge
D. An element important to the body that represents less than 0.01 percent of body weight
E. Energy emitted from unstable isotopes
F. Positively charged subatomic particles in the nucleus of an atom
G. Molecules built of two or more elements and containing carbon
H. Unstable isotope that stabilizes itself by spontaneously emitting energy and particles
I. Combined number of protons and neutrons in the nucleus of an atom
J. The time it takes for half of a quantity of a radioisotope to decay into a more stable isotope
K. Negatively charged subatomic particles that move rapidly around the nucleus of an atom
L. Use of radioisotopes, such as radium-226 or cobalt-60, to treat some cancers
M. Fundamental form of matter; cannot be broken down into other substances by ordinary processes
N. Forms of an element, the atoms of which contain a different number of neutrons than other forms of the same element
O. A substance with a radioisotope attached so that its movement can be tracked

Complete the Table

16. Complete the following table by entering the name of the element and its symbol in the appropriate spaces. [p.16]

Element	Symbol	Atomic Number	Mass Number
a.		1	1
b.		6	12
c.		7	14
d.		8	16
e.		11	23
f.		17	35

Dichotomous Choice

17. For each element in the chart above, the (atomic number/mass number) is always the same. [p.16]

18. Atoms of an element with different (atomic numbers/mass numbers) are called isotopes. [p.16]

19. The atomic number is the number of (protons/neutrons). [p.16]

20. A neutral atom will have the same number of protons as it does (neutrons/electrons). [p.16]

21. The mass number is the total number of protons and (electrons/neutrons). [p.16]

2.3. WHAT IS A CHEMICAL BOND? [pp.18–19]

2.4. IMPORTANT BONDS IN BIOLOGICAL MOLECULES [pp.20–21]

2.5. ANTIOXIDANTS: FIGHTING FREE RADICALS [p.22]

Selected Words: "shells" [p.18], "orbitals" [p.18], "octet rule" [p.18], inert [p.19], *ionized* [p.20], *single* covalent bond [p.20], *double* covalent bond [p.20], *triple* covalent bond [p.20], *nonpolar* covalent bond [p.21], *polar* covalent bond [p.21], "electronegative" [p.21], *oxidation* [p.22]

Boldfaced, Page-Referenced Terms

[p.18] chemical bond _____

_____

[p.19] molecule _____

_____

[p.19] compounds _____

_____

[p.19] mixture _____

_____

[p.20] ion _____

_____

[p.20] ionic bond _____

_____

[p.20] covalent bond _____

_____

[p.21] hydrogen bond _____

_____

[p.22] free radicals _____

_____

[p.22] antioxidant _____

_____

Short Answer

1. Oxygen gas consists of two atoms of oxygen bonded together. Water consists of two atoms of hydrogen and one atom of oxygen bonded together. Which is/are molecules? Which is/are compounds? Explain. [p.19]

_____

_____

_____

Identification

2. Following the atomic model of helium gas (below), where the number of protons and neutrons are shown in the nucleus and the electrons are arranged in shells, identify the electron configuration of the atoms illustrated by entering appropriate electrons in the form $2e^-$. [p.19]

He

  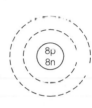

H   C   N   O

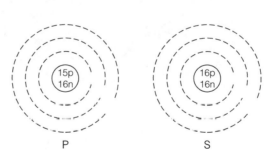

P   S

Short Answer

3. Of all the atoms shown above, which would not interact and form chemical bonds? Explain. [p.19]

_____

_____

_____

4. Of all the atoms shown above, which represent the four most abundant elements in the human body? Why will atoms of these elements form chemical bonds? [pp.18,19]

_____

_____

Identification

5. Following the model of NaCl (sodium chloride), complete the sketch identifying the transfer of electron(s) (by arrows) to show how positive magnesium (Mg^+) and negative chloride (Cl^-) ions form ionic bonds to create a molecule of $MgCl_2$ (magnesium chloride). [p.20]

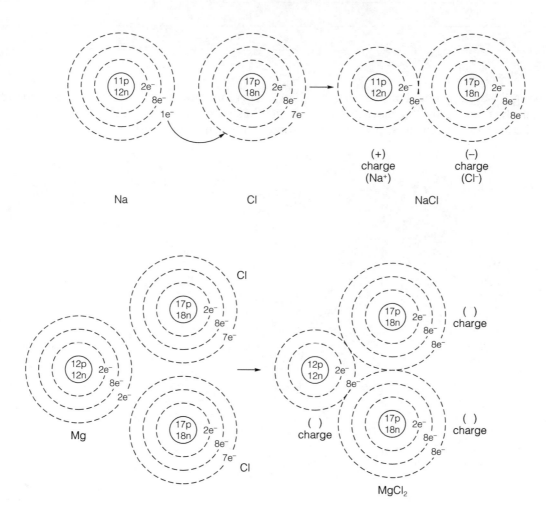

Short Answer

6. Distinguish between a nonpolar covalent bond and a polar covalent bond. Give one example of each. [pp.20–21]

_____

_____

_____

Identification

7. Following the model of hydrogen gas (below), complete the sketch by placing electrons (as dots) in the outer shells to identify the nonpolar covalent bonding that forms oxygen gas; similarly identify polar covalent bonds by completing electron structures to form a water molecule. [pp.20–21]

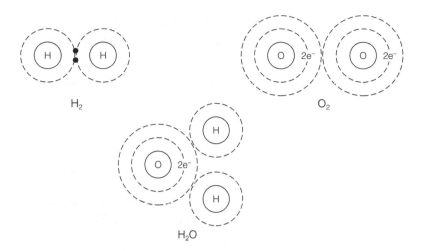

Elimination

For each statement below, cross out the choice that does NOT fit.

8. The octet rule applies to having eight electrons in an atom's outer (shell/orbital/energy level). [p.18]

9. Chemical bonding occurs between the components of a (molecule/mixture/compound). [p.19]

10. In order for atoms to achieve full outer shells, (ionic/covalent/hydrogen) bonding occurs. [pp.20,21]

Multiple Choice

The following statements pertain to the two chemical reactions symbolized below. Circle the choice(s) that correctly completes each sentence. (More than one choice may be circled for each statement.)

$$Na + Cl \longrightarrow NaCl \qquad 2H_2 + O_2 \longrightarrow 2H_2O$$

11. The two sequences above are (chemical equations/physical reactions/molecular processes). [p.19]

12. Na and $2H_2$ are both (products/molecules/reactants). [p.19]

13. NaCl and $2H_2O$ are both (products/molecules/compounds). [p.19]

14. The illustration on the right must include two H_2 and two H_2O in order to be (reversible/properly bonded/balanced). [p.19]

15. The sequence on the left illustrates ionic bonding. In this type of reaction, one or more electrons is (shared/donated/accepted). [p.20]

16. The component atoms of NaCl stay together because they have (the same/opposite/no) charge. [p.20]

17. The sequence on the right illustrates bonding in which electrons are shared, although the charge of the electrons is not distributed equally. The bond that results is (covalent/polar/nonpolar). [pp.20,21]

18. If H_2O were drawn as a structural formula, a line would illustrate each (covalent/ionic/hydrogen) bond. [p.20]

19. The two H_2Os in the second sequence could be attracted to each other by weak (covalent/ionic/hydrogen) bonds. [p.21]

Short Answer

20. Explain how free radicals are produced and why they are a potential threat to human health. [p.22]

_____

_____

_____

21. What is an antioxidant? Name some antioxidants available in our diet. [p.22]

_____

_____

2.6. LIFE DEPENDS ON WATER [p.23]

2.7. ACIDS, BASES, AND BUFFERS: BODY FLUIDS IN FLUX [pp.24–25]

Selected Words: *heat capacity* [p.23], *solvent* [p.23], *acidic* solutions [p.24], *basic* solutions [p.24], *alkaline* fluids [p.24], *coma* [p.25], *tetany* [p.25], *acidosis* [p.25], *alkalosis* [p.25]

Boldfaced, Page-Referenced Terms

[p.23] hydrophilic _____

_____

[p.23] hydrophobic _____

_____

[p.23] solute _____

_____

[p.24] hydrogen ions _____

_____

[p.24] hydroxide ions _____

_____

[p.24] pH scale _____

_____

[p.24] acid _____

_____

[p.24] base _____

_____

[p.25] buffer system _____

_____

[p.25] salts _____

_____

Dichotomous Choice

Circle one of two possible answers given between parentheses in each statement.

1. By weight, your body is about (one-third/two-thirds) water. [p.23]

2. The polarity of water molecules allows them to form (hydrogen/covalent) bonds with one another and with other polar substances. [p.23]

3. Polar molecules are attracted to water and hence are said to be (hydrophobic/hydrophilic). [p.23]

4. The polarity of water repels oil and other nonpolar substances, which are (hydrophobic/hydrophilic). [p.23]

5. Hydrogen bonds between water molecules enable water to absorb a great deal of (cold/heat) energy before it significantly warms or evaporates. [p.23]

6. When the amount of heat energy is raised sufficiently, hydrogen bonds between water molecules break apart and (re-form/do not re-form); water then evaporates. [p.23]

7. Sweating followed by evaporation results in the body (gaining/losing) heat energy. [p.23]

8. Water is a superb (solute/solvent). [p.23]

9. Substances dissolved in water are (solutes/solvents). [p.23]

10. A substance is said to (precipitate/dissolve) as clusters of water molecules form around its individual ions or molecules. [p.23]

Labeling and Analysis

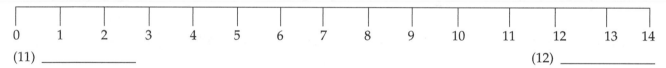

(11) _____   (12) _____

11–12. In the blanks provided above, label the basic (alkaline) and acidic ends of the pH scale above. [p.24]

13. Circle the number on the scale above at which the concentrations of H^+ and OH^- are equal (neutrality). [p.24]

14. A solution with a pH of 8 is ten times more basic (alkaline) than one with a pH of 7. A solution with a pH of 9 is _____ more basic than one with a pH of 7. [p.24]

15. Gastric fluid with a pH of 2 is _____ more acidic than pure water with a pH of 7. [p.24]

16. Blood has a pH of 7.3–7.5. It is slightly _____. [p.24]

$$H_2CO_3 \longrightarrow HCO_3^- + H^+ \qquad\qquad HCO_3^- + H^+ \longrightarrow H_2CO_3$$

17. The equations above represent the reversible reactions of a _____ system. [p.25]

18. _____ is the reactant in the left equation and the product on the right. [p.25] _____ + H^+ are the products on the left and the reactants on the right.

19. The equation on the (left/right) occurs when the pH of blood and tissue fluid begins to become more acidic. This occurs in order to (add/remove) H^+. [p.25]

20. The equation on the (left/right) occurs when the pH of blood and tissue fluid begins to become more basic. This occurs in order to (add/remove) H^+. [p.25]

21. This type of system helps to prevent uncontrolled shifts in pH such as a severe decrease in blood pH called (acidosis/alkalosis), or a severe increase in blood pH called (acidosis/alkalosis). [p.25]

$$HCl \quad + \quad NaOH \quad \longrightarrow \quad NaOH \quad + \quad H_2O$$

(22) _____ (23) _____ (24) _____

22–24. In the blanks provided above, label the salt, acid, and base components of the reaction. [p.25]

25. A (acid/base/salt) is an ionic compound that releases ions other than H^+ and OH^-. [p.25]

Short Answer

26. Explain how milk of magnesia acts as an antacid in the stomach. [p.24]

_____

_____

_____

Complete the Table

27. Fill in the missing information.

Fluid	pH Value	Acid or Base
rainwater [p.24]		
seawater [p.24]		
acid rain [p.24]		

2.8. ORGANIC COMPOUNDS: BUILDING ON CARBON ATOMS [pp.26–27]

Selected Words: "inorganic" [p.26], *dehydration synthesis* [p.27]

Boldfaced, Page-Referenced Terms

[p.26] organic compound _____

_____

[p.26] functional groups _____

_____

[p.27] enzymes _____

_____

[p.27] condensation _____

_____

[p.27] polymer _____

_____

[p.27] monomers _____

_____

[p.27] hydrolysis _____

_____

True/False

If a statement is true, write a "T" in the blank. If it is false, underline the incorrect word and write the correct word in the blank.

1. _____ Organic compounds have hydrogen and often other elements bonded to atoms of carbon by covalent bonds. [p.26]

2. _____ The four classes of biological molecules are carbohydrates, lipids, proteins, and water. [p.26]

3. _____ Carbon forms two covalent bonds (2 pairs of shared electrons). [p.26]

4. _____ Organic molecules can be chains or rings. [p.26]

5. _____ A carbon backbone with only hydrogen atoms attached is a carbohydrate. [p.26]

6. _____ Atoms or clusters of atoms that are covalently bonded to the carbon backbone and influence the chemical behavior of an organic compound are called enzymes. [p.26]

Labeling

Study the following structural formulas of organic compounds. Identify the circled functional groups (sometimes repeated) by entering the correct name in the blanks with matching number; complete the exercise by circling the names in the parentheses following each blank of the compounds in which the functional group might appear.

7. _____ (fats—sugars—amino acids—proteins—phosphate compounds, e.g., ATP) [p.26]

8. _____ (fats—sugars—amino acids—proteins—phosphate compounds, e.g., ATP) [p.26]

9. _____ (fats—sugars—amino acids—proteins—phosphate compounds, e.g., ATP) [p.26]

10. _____ (fats—sugars—amino acids—proteins—phosphate compounds, e.g., ATP) [p.26]

11. _____ (fats—sugars—amino acids—proteins—phosphate compounds, e.g., ATP) [p.26]

12. _____ (fats—sugars—amino acids—proteins—phosphate compounds, e.g., ATP) [p.26]

Short Answer

13. State the general role of enzymes as they relate to organic compounds. [p.27]

_____

_____

Identification

14. The structural formulas for two adjacent amino acids are shown below. By circling an H atom from one amino acid and an −OH group from the other, identify how enzyme action causes formation of a covalent bond and a water molecule (through a condensation reaction). Also, draw an arrow to indicate the covalent bond that formed the bond between the monomers of the dipeptide. [p.27]

amino acid   amino acid   dipeptide

15. a. _____ By what process would the monomers be *separated*? [p.27]

    b. _____ Is an enzyme required for this process? [p.27]

    c. _____ What other molecule is needed? [p.27]

2.9. CARBOHYDRATES: PLENTIFUL AND VARIED [pp.28–29]

Selected Words: "saccharide" [p.28], "fiber" [p.29]

Boldfaced, Page-Referenced Terms

[p.28] carbohydrates _____

_____

[p.28] monosaccharides _____

_____

[p.28] oligosaccharides _____

_____

[p.28] polysaccharides _____

_____

Labeling

1. a. The products of the chemical equation on page 23 include a molecule of H_2O. On the left side of the equation, circle the components of the reactant molecules that are removed and combined to form the water. [p.28]

1. b. On the right-hand side of the equation, draw an arrow to the bond that results from this reaction. [p.28]

1. c. Note that the reaction arrows indicate that this is a reversible reaction. Above the top arrow, write the name of the enzyme-catalyzed reaction occurring (either *hydrolysis* or *condensation*). [p.28]

1. d. Below the lower arrow, write the name of the enzyme-catalyzed reaction occurring (either *hydrolysis* or *condensation*). [p.28]

glucose   glucose   maltose   water
(a monosaccharide)   (a monosaccharide)   (a disaccharide)

Complete the Table

2. In the following table, enter the name of the carbohydrate described by its carbohydrate class and function(s). Select from *glucose, sucrose, lactose, glycogen, cellulose, starch,* and *deoxyribose*.

Carbohydrate	Carbohydrate Class	Function
a. [p.28]	Oligosaccharide (disaccharide)	Most plentiful sugar in nature; table sugar
b. [p.28]	Monosaccharide	Main energy source for body cells; building block of many organic compounds
c. [p.29]	Polysaccharide	Tough, insoluble structural material in plant cell walls
d. [p.28]	Monosaccharide	Five-carbon sugar; occurs in DNA
e. [p.28]	Oligosaccharide (disaccharide)	Sugar present in milk
f. [p.29]	Polysaccharide	Storage form of sugar in animals, including humans
g. [p.29]	Polysaccharide	Storage form of glucose in plants

2.10. LIPIDS: FATS AND THEIR CHEMICAL KIN [pp.30–31]

Selected Words: unsaturated tail [p.30], *saturated* tail [p.30], "vegetable oils" [p.30], *trans-fatty acids* [p.30], "hydrogenated" [p.30]

Boldfaced, Page-Referenced Terms

[p.30] lipid _____

[p.30] fats _____

_____

[p.30] fatty acid _____

_____

[p.30] triglycerides _____

_____

[p.31] phospholipid _____

_____

[p.31] sterols _____

_____

Identification

1. Combine glycerol with three fatty acids (see following illustration) to form a triglyceride by circling the participating atoms that will produce three molecules of H_2O and three covalent bonds; also draw arrows to indicate the covalent bonds in the triglyceride product. [p.30]

glycerol   three fatty acids   triglyceride (a complete fat molecule)

Labeling

2. In the appropriate blanks, label the molecule shown as either *saturated* or *unsaturated*. [p.30]

a. _____

a.

oleic acid

b. _____

b.

stearic acid

Short Answer

3. Define *phospholipid*; describe the structure and biological functions of phospholipids. [p.31]

_____

_____

_____

4. Describe how sterols are structurally different from fats. [p.31]

_____

_____

_____

Matching

5. _____ fatty acids [p.30]

6. _____ triglycerides [p.30]

7. _____ phospholipids [p.31]

8. _____ trans-fatty acids [p.30]

9. _____ sterols [p.31]

A. Partially saturated (hydrogenated)
B. Up to 36 carbons long with unsaturated or saturated tails
C. Vitamin D, bile salts, steroid hormones
D. Main components of cell membranes
E. Body's richest source of energy; e.g., butter, lard, and oils

2.11. PROTEINS: BIOLOGICAL MOLECULES WITH MANY ROLES [pp.32–33]

2.12. A PROTEIN'S FUNCTION DEPENDS ON ITS SHAPE [pp.34–35]

Selected Words: peptide bonds [p.32], *primary* structure [p.32], *secondary* structure [p.34], *tertiary* structure [p.34], *quaternary* structure [p.35], *denatured* [p.35]

Boldfaced, Page-Referenced Terms

[p.32] amino acid _____

_____

[p.32] polypeptide chain _____

_____

[p.35] lipoproteins _____

_____

[p.35] glycoproteins _____

_____

Labeling

For questions 1–3, write the name of the major parts of every amino acid. Choose from *R group*, *carboxyl group*, and *amino group*. [p.32]

1. _____

2. _____

3. _____

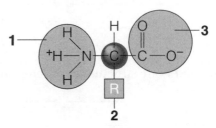

4. _____ Which of these three parts differs from one amino acid to another, and determines the chemical and physical properties of the amino acid?

Identification

5. In the following illustration of four amino acids in cellular solution (ionized state), circle the atoms and ions that form water to allow identification of covalent (peptide) bonds between adjacent amino acids to form a polypeptide. On the completed polypeptide, draw an arrow to indicate each newly formed peptide bond. [p.33]

enzyme action

Matching

6. _____ primary protein structure [p.32]

7. _____ secondary protein structure [p.34]

8. _____ tertiary protein structure [p.34]

9. _____ quaternary protein structure [p.35]

A. A coiled or extended pattern, based on regular hydrogen-bonding interactions

B. Incorporates two or more polypeptide chains, yielding a protein that is globular, fibrous, or both

C. Unique sequence of amino acids in the polypeptide chain of a specific protein

D. Bending and looping of the polypeptide chain, caused by the interaction of R groups

Short Answer

10. Glycoproteins and lipoproteins are both modified proteins. How are they different from each other? [p.35]

_____

_____

_____

11. Describe protein denaturation. [p.35]

_____

_____

_____

2.13. NUCLEOTIDES AND NUCLEIC ACIDS [p.36]

2.14. FOOD PRODUCTION AND A CHEMICAL ARMS RACE [p.37]

Selected Words: herbicides [p.37], insecticides [p.37], fungicides [p.37]

Boldfaced, Page-Referenced Terms

[p.36] nucleotide _____

_____

[p.36] ATP _____

_____

[p.36] coenzyme _____

_____

[p.36] nucleic acids _____

_____

[p.36] DNA _____

_____

[p.36] RNA _____

_____

Labeling

For questions 1–3, write the name of the major parts of a nucleotide. Choose from *five-carbon sugar, phosphate group,* and *nitrogen-containing base.* [p.36]

1. _____

2. _____

3. _____

4. Part 2 differs in the two types of nucleic acid. Beside each nucleic acid listed below, write the name of this part.

     a. DNA _____   b. RNA _____

Identification

5. In the accompanying diagram of a single-stranded nucleic acid molecule, encircle as many *complete* nucleotides as possible. How many complete nucleotides are present? [p.36]

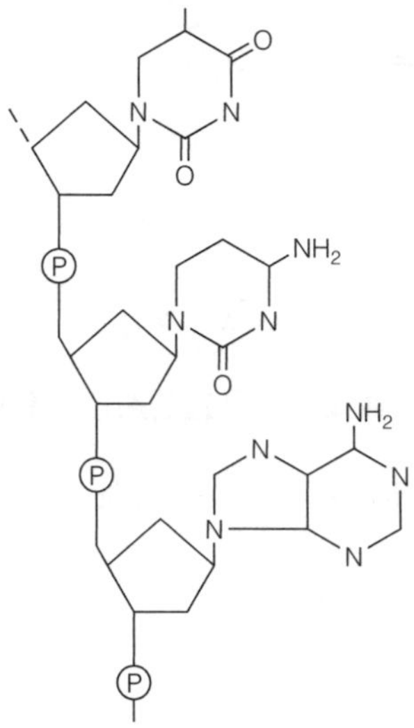

Matching

6. _____ adenosine triphosphate (ATP) [p.36]

7. _____ RNA [p.36]

8. _____ DNA [p.36]

9. _____ NAD$^+$ and FAD [p.36]

A. Single nucleotide strand; plays a key role in processes by which genetic instructions are used to build the body's proteins

B. Nucleotide that provides energy for cellular reactions by the transfer of a phosphate group

C. Coenzymes that have nucleotides as subunits; transport hydrogen ions and their associated electrons from one cell reaction site to another

D. Helical double-nucleotide strand; encodes genetic information with base sequences

Short Answer

10. What is the benefit of chemical agents used to protect crops? What are the drawbacks? [p.37]

_____

_____

_____

Self-Quiz

Are you ready for the exam? Test yourself on key concepts by taking the additional tests linked with your BiologyNow CD-ROM.

Complete the Table

1. Complete the following table by entering the correct name of the major cellular organic compounds suggested in the "types" column (choose from *carbohydrates, proteins, nucleic acids,* and *lipids*).

Cellular Organic Compounds	Types
a. [p.31]	Phospholipids
b. [p.32]	May be fibrous or globular
c. [p.32]	Enzymes, defense, transport, and regulatory molecules
d. [p.36]	Genetic instructions
e. [p.29]	Glycogen, starch, and cellulose
f. [p.30]	Triglycerides
g. [p.35]	Glycoproteins
h. [p.30]	Fats
i. [p.36]	Nucleotides including ATP
j. [p.31]	Cholesterol and steroid hormones
k. [p.35]	Lipoproteins
l. [p.29]	Glucose and sucrose

Multiple Choice

_____ 2. A molecule is _____; a compound is _____. [p.19]
  a. a combination of two or more atoms; a molecule consisting of two or more elements in proportions that never vary
  b. the smallest unit of matter peculiar to a particular element; less stable than its constituent atoms
  c. a combination of two or more atoms; a very large molecule
  d. a carrier of one or more extra neutrons; a substance in which atoms of the same element are present in different proportions

_____ 3. If lithium has an atomic number of 3 and an atomic mass of 7, it has _____ neutrons in its nucleus; if a chlorine atom has an atomic number of 17, it will have _____ shells containing electrons. [pp.16,18]
  a. three; two
  b. four; two
  c. three; three
  d. four; three
  e. seven; three

____ 4. A hydrogen bond is _____; a covalent bond is _____. [pp.20,21]
   a. a sharing of a pair of electrons between a hydrogen nucleus and an oxygen nucleus; a bond in which electrons are gained or lost
   b. a sharing of a pair of electrons between a hydrogen nucleus and either an oxygen or a nitrogen nucleus; a bond that forms ions
   c. a weak attraction between an electronegative atom of a molecule and a neighboring hydrogen atom that is also part of a polar covalent bond; one in which electrons are shared
   d. none of the above

____ 5. The pH scale is a measure of the concentration of _____ in a liquid; a substance that releases H^+ in water is a(n) _____. [p.24]
   a. ionized water; acid
   b. OH^-; base
   c. hydrogen atoms; carbon compound
   d. OH^-; acid
   e. H^+; acid

____ 6. A blood pH of 7.3 is _____; a combination of a weak acid and weak base that helps to counter slight shifts in pH is known as a(n) _____. [pp.24,25]
   a. acid; neutral molecule
   b. basic; salt
   c. neutral; base
   d. acid; acid
   e. basic; buffer

____ 7. Carbon is part of so many different substances because _____; ^{12}C and ^{14}C atoms differ in their numbers of _____. [pp.16,26]
   a. carbon forms two covalent bonds with a variety of other atoms; protons
   b. carbon forms four covalent bonds with a variety of other atoms; neutrons
   c. carbon ionizes easily; electrons
   d. carbon is a polar compound; neutrons

____ 8. Amino, carboxyl, phosphate, and hydroxyl are examples of _____; _____ form the structural elements of bones, muscles, and hair. [pp.26,32]
   a. functional groups; carbohydrates
   b. sugar units; nucleic acids
   c. functional groups; proteins
   d. coenzymes; lipids

____ 9. _____ are compounds used by cells as transportable packets of quick energy, storage forms of energy, and structural materials; fats, oils, and steroids are _____. [pp.28–29, 30–31]
   a. Lipids; carbohydrates
   b. Nucleic acids; proteins
   c. Carbohydrates; lipids
   d. Proteins; nucleic acids

____ 10. Hydrolysis could be correctly described as the _____; glycogen, plant starch, and cellulose are _____. [pp.27,29]
   a. heating of a compound in order to drive off its excess water and concentrate its volume; proteins
   b. breaking of a polymer into its subunits by adding water molecule components to separated monomers; carbohydrates
   c. linking of two or more molecules by the removal of one or more water molecules; lipids
   d. constant removal of hydrogen atoms from the surface of a carbohydrate; nucleic acids

____ 11. Genetic information is encoded in the sequence of the bases of _____; molecules of _____ function in processes using genetic instructions to build the body's proteins. [p.36]
   a. DNA; DNA
   b. DNA; RNA
   c. RNA; DNA
   d. RNA; RNA

Chapter Objectives/Review Questions

This section lists general and detailed chapter objectives that can be used as review questions. You can make maximum use of these items by writing answers on a separate sheet of paper. Fill in answers where blanks are provided. To check for accuracy, compare your answers with information in the chapter or glossary.

1. A(n) _____ is a fundamental form of matter that has mass and takes up space. [p.15]
2. Describe how matter is organized in terms of elements, atoms, molecules, and compounds. [p.15]
3. How does atomic number differ from mass number? [p.16]
4. ^{12}C and ^{14}C represent _____ of carbon. [p.16]
5. _____ are unstable and tend to spontaneously emit subatomic particles or energy in order to achieve stability. [p.16]
6. _____-_____ _____ (PET) provides information about abnormalities in the metabolic functions of specific tissues. [p.17]
7. Explain the relationship between orbitals and shells (energy levels) of an atom. [p.18]
8. Atoms with an unfilled outer shell tend to form chemical _____ with other atoms and so fill their outer shell. [p.18]
9. What is the octet rule? [p.18]
10. Draw the structure of the carbon atom, with the correct spatial arrangement of protons, neutrons, and electrons. [p.19]
11. A(n) _____ bond is an attraction between two oppositely charged ions. [p.20]
12. "A pair of electrons shared between two atoms" defines a(n) _____ bond. [p.20]
13. Describe how hydrogen bonds are formed; cite one example of a large molecule in which the hydrogen bond is important to you. [p.21]
14. Define and list examples of *antioxidants*; what is a free radical? [p.22]
15. Polar molecules are attracted to water and are said to be _____. [p.23]
16. Nonpolar molecules are repelled by water and are known as _____. [p.23]
17. Explain why water is such a good solvent. [p.23]
18. A substance that releases H^+ when it dissolves in water is a(n) _____; any substance that accepts H^+ when it dissolves in water is a(n) _____. [p.24]
19. Describe the structure and use of the pH scale. [p.24]
20. Define *buffer system*; describe how the carbonic acid/bicarbonate buffer system works in the blood. [p.25]
21. Na^+ and K^+ are components of _____, substances that dissociate into ions other than H^+ and OH^- in solution. [p.25]
22. Acids often combine with bases to form _____ and water. [p.25]
23. List the four main types of organic compounds that represent the molecules characteristic of life. [p.26]
24. A(n) _____ is a carbon backbone with only hydrogen atoms attached. [p.26]
25. —OH, —COOH, and —NH$_3$ are examples of _____ groups. [p.26]
26. Define the terms *condensation* and *hydrolysis*, and *polymer* and *monomer*, and explain the first two in terms of their relationship with the second two. [p.27]
27. List the ways that cells use carbohydrates. [p.28]
28. List the three classes of carbohydrates; how is each distinguished from the other two? [pp.28,29]
29. State the major functions of lipids in most organisms. [pp.30–31]
30. Distinguish an unsaturated fatty acid from a saturated fatty acid. [p.30]
31. Butter, lard, and oils are all examples of _____. [p.30]
32. Describe trans-fatty acids, what foods they are found in, and a possible concern about their use. [p.30]
33. The main materials of cell membranes are _____. [p.31]
34. List the major functions of proteins; all proteins are constructed from about 20 different kinds of _____ acids. [p.32]
35. One group of proteins, the _____, make metabolic reactions proceed much faster than they otherwise would. [p.32]
36. Define *peptide bond* and *polypeptide chain*. [p.32]
37. Distinguish between the primary, secondary, tertiary, and quaternary structure of a protein molecule. [pp.32–33]

38. The loss of a protein's three-dimensional shape following disruption of bonds is called _____. [p.35]
39. How are nucleotides related to nucleic acids? Cite the names and functions of two major nucleic acids important to life. [p.36]

Media Menu Review Questions

Questions 1–5 are drawn from the following InfoTrac College Edition article: "Sorting Fat From Fiction." Kellie Fischer. *Prepared Foods,* October 2003.

1. The majority of trans-fatty acids in a modern diet comes from _____ oils.
2. Partial hydrogenation was developed to improve oil stability and change liquid oils into _____.
3. The FDA recently announced that a final rule is pending that will require the amount of _____ fatty acids to be listed on food labels.
4. Reformulating products to have fewer trans-fatty acids will give them a (longer/shorter) shelf life.
5. The challenge for food product developers is to reduce the level of trans-fatty acids without increasing the level of _____ fatty acids.

Questions 6–10 are drawn from the following InfoTrac College Edition article: "The Form Counts: Proteins, Fats and Carbohydrates." Beatrice Trum Hunter. *Consumers' Research Magazine,* August 2001.

6. The human body only possesses enzymes that can metabolize the _____ forms of amino acids.
7. Some amino acids, such as glutamate, behave quite differently when they are in a _____ versus a free form.
8. The *cis* form of fatty acids in butter is converted to the *trans* form in _____.
9. A left-rotating sugar would be a _____ sugar.
10. Raw broccoli is (less/more) effective in raising vitamin C levels in the blood than when the vegetable is briefly cooked.

Integrating and Applying Key Concepts

1. Explain what would happen if water were a nonpolar molecule instead of a polar molecule. Would water be a good solvent for the same kinds of substances? Would the nonpolar molecule's heat capacity likely be higher or lower than that of water? Would its ability to form hydrogen bonds change?
2. Whereas plants store their energy as starch, the long-term energy storage molecule of animals is fat. While starch attracts some water, fat is hydrophobic and thus stores more energy in a smaller space. Why do you think that fat is a better energy storage molecule for animals in spite of the fact that it is more difficult to break down than is starch?
3. Most DNA information tells the cell how to make proteins, while none of this information represents instructions for other types of organic molecules. What type of protein allows reactions to occur that make or break down other organic molecules? Explain why damage to DNA instructions for making a specific enzyme may result in a genetic disease.

3

CELLS AND HOW THEY WORK

Interactive Exercises

Impacts, Issues: When Mitochondria Spin Their Wheels [p.41]

3.1. CELLS: ORGANIZED FOR LIFE [pp.42–43]

3.2. THE PARTS OF A EUKARYOTIC CELL [pp.44–45]

For additional practice, use the interactive vocabulary exercises linked with your BiologyNow CD-ROM.

Selected Words: *Luft's syndrome* [p.41], *Friedreich's ataxia* [p.41]

Boldfaced, Page-Referenced Terms

[p.42] cell theory _____

_____

[p.42] cytosol _____

[p.42] prokaryotic cell _____

_____

[p.42] eukaryotic cells _____

_____

[p.42] organelles _____

_____

[p.43] surface-to-volume ratio _____

_____

[p.43] lipid bilayer _____

_____

[p.44] nucleus _____

_____

[p.44] endoplasmic reticulum (ER) _____

_____

[p.44] Golgi body _____

_____

[p.44] vesicles _____

_____

[p.44] mitochondria _____

_____

[p.44] ribosomes _____

_____

[p.44] cytoskeleton _____

Short Answer

1. State the three parts of the cell theory. [p.42]

_____

_____

_____

_____

Complete the Table

2. To complete the following table, identify the three basic characteristics of any cell. [p.42]

Cell Part	Function
A.	Inherited genetic instructions
B.	Everything between the plasma membrane and the region of DNA; consisting of the cytosol and other components
C.	Thin, outermost covering; allows life-sustaining activities to proceed inside the cell apart from the environment

Short Answer

3. Distinguish prokaryotic cells from eukaryotic cells. [p.42]

_____

_____

_____

4. Explain why most cells are so small. How do cells that are not small compensate for their larger size? [p.43]

_____

_____

_____

5. Describe the "heads-out, tails-in" structure of a cell membrane, including the terms "hydrophobic" and "hydrophilic." [p.43]

_____

_____

_____

6. Why are organelles the solution to the cell's dilemma of having reactions occurring simultaneously that have conflicting purposes? [p.43]

_____

_____

_____

Labeling

7. On the animal cell below, label the indicated parts. [p.44]

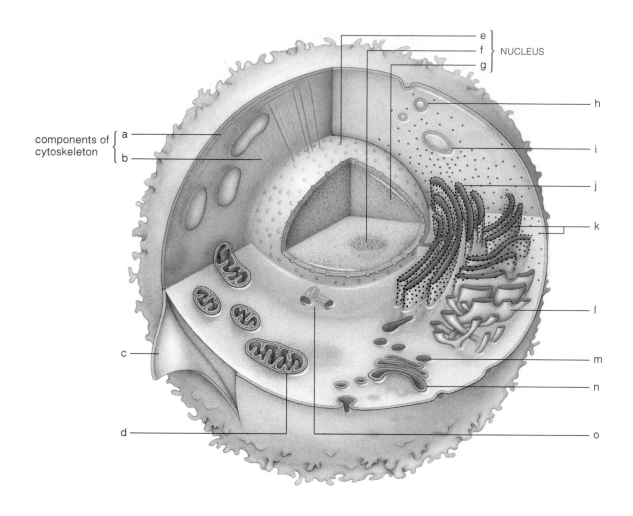

Matching

Match these cell parts to their functions.

8. _____ mitochondria

9. _____ endoplasmic reticulum (ER)

10. _____ Golgi body

11. _____ cytoskeleton

12. _____ vesicles

13. _____ ribosomes

A. Cell support, compartmentalization, and movement
B. Transport and modification of polypeptides; assembly of lipids
C. Transport and storage of substances in cell, breakdown of cell parts
D. Assembly of polypeptide chains
E. ATP production
F. Modification and shipping of proteins and lipids for secretion or use within cell

3.3. THE PLASMA MEMBRANE: A LIPID BILAYER [p.46]

3.4. SUGAR WARS [p.47]

3.5. THE NUCLEUS [pp.48–49]

3.6. THE ENDOMEMBRANE SYSTEM [pp.50–51]

Selected Words: "glycobiology" [p.47], "self" [p.47], "cancer" [p.47], *rough* ER [p.50], *smooth* ER [p.50]

Boldfaced, Page-Referenced Terms

[p.48] nucleus _____

_____

[p.48] nuclear envelope _____

_____

[p.49] nucleolus _____

_____

[p.49] chromatin

_____

[p.49] chromosome _____

_____

[p.50] endomembrane system _____

_____

[p.50] endoplasmic reticulum (ER) _____

_____

[p.50] ribosomes _____

_____

[p.50] Golgi bodies _____

_____

[p.51] vesicles _____

_____

[p.51] lysosome _____

_____

[p.51] peroxisomes _____

_____

Short Answer

1. Why is it necessary for the plasma membrane to be "fluid" rather than rigid? Why is the membrane described as a "mosaic" structure? [p.46]

_____

_____

_____

Dichotomous Choice

Circle one of two possible answers given between parentheses in each statement.

2. The "fluid" property of a cell membrane is created by the (phospholipids/proteins). [p.46]

3. Most cell functions are carried out by (phospholipids/proteins). [p.46]

Complete the Table

4. Study the following illustration and complete the table by entering the name of each membrane structure involved with the function described. Choose from: _lipid bilayer, transport proteins, receptor proteins, recognition proteins,_ and _adhesion proteins._ [p.46]

Membrane Structure   _Function_

a.	Diverse types at the cell surface are like molecular fingerprints
b.	The phospholipid "sandwich" that makes a cell membrane somewhat fluid
c.	Proteins that span the bilayer and allow water-soluble substances to move through their interior
d.	Help cells of the same type stick together and stay positioned within the proper tissue
e.	Bind hormones and other extracellular substances that trigger changes in cell activities; there are different combinations on different cells

Elimination

For each statement below, cross out the choice that does NOT fit.

5. Glycoproteins in the cell membrane are proteins with attached (lipids/sugars). [p.47]

6. Membrane glycoproteins function in ("self" identification/eliciting an immune response/determining blood type/providing energy for transport). [p.47]

7. Among the viruses that could be controlled by a better understanding of carbohydrates are (smallpox/hepatitis B/hepatitis C/Hib meningitis). [p.47]

8. One class of cancer vaccines works by alerting the body that cancer cells are using glycoproteins (as an energy supply/to fool the immune system/to evoke faster cell division). [p.47]

Matching

Match each of the following terms with the appropriate phrase or description.

9. _____ chromosome [p.49]

10. _____ nucleolus [p.49]

11. _____ pore [p.48]

12. _____ chromatin [p.49]

13. _____ nucleus [p.48]

14. _____ nuclear envelope [p.48]

A. Organelle that encloses the DNA of eukaryotic cells
B. A cell's collection of DNA and its associated proteins
C. Consists of two membranes
D. Dense mass within nucleus where subunits of ribosomes are built
E. Passageway across nuclear envelope for movement of ions and small molecules
F. A single DNA molecule with its associated proteins

Matching

Study the following illustration and match each component of the endomembrane system with its description.

15. _____ smooth ER [p.50]

16. _____ rough ER [p.50]

17. _____ Golgi body [p.50]

18. _____ vesicle [p.50]

19. _____ lysosome [p.51]

20. _____ peroxisome [p.51]

A. Type of vesicle containing enzymes that break down fatty acids, amino acids, and alcohol
B. Organelle in which enzymes finish proteins and lipids, sort them, and package them within vesicles
C. Contain enzymes that digest cells, cell parts, and bacteria
D. Tiny membranous sac that moves through the cytoplasm
E. Pipelike structure in which many cells assemble lipids or inactivate drugs and toxins
F. Abundant in cells that secrete finished proteins

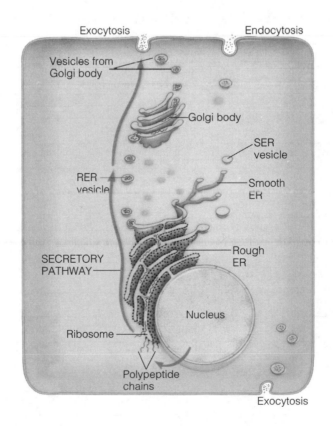

Sequence

21. Assume that the cell diagram on page 39 is of a pancreas cell that secretes digestive enzymes. In the first blank below, list the first cellular structure involved in the production, processing, and secretion of these proteins, followed by the remainder of cell structures or events involved in the correct order. Choose from: *Golgi body, rough ER, exocytosis into small intestine, rough ER vesicle, Golgi body vesicle, ribosome.*

a.   d.

b.   e.

c.   f.

3.7. MITOCHONDRIA: THE CELL'S ENERGY FACTORIES [p.52]

3.8. MICROSCOPES: WINDOWS INTO THE WORLD OF CELLS [p.53]

3.9. THE CYTOSKELETON: CELL SUPPORT AND MOVEMENT [pp.54–55]

Selected Words: *compound light microscope* [p.53], *transmission electron microscope* [p.53], *scanning electron microscope* [p.53], *cell typing* [p.54], *"9 + 2 array"* [p.55], *pseudopods* [p.55]

Boldfaced, Page-Referenced Terms

[p.52] mitochondrion (plural: mitochondria) _____

_____

[p.54] cytoskeleton _____

_____

[p.54] microtubules _____

_____

[p.54] microfilaments _____

_____

[p.54] cell cortex _____

_____

[p.54] intermediate filaments _____

_____

[p.54] motor proteins _____

_____

[p.55] flagella _____

_____

[p.55] cilia _____

_____

[p.55] centrioles _____

_____

[p.55] basal body _____

_____

Fill-in-the-Blanks

The main energy carrier in cells is (1) _____ [p.52]. These molecules form when (2) _____ [p.52] compounds are broken down to (3) _____ _____ [p.52] and (4) _____ [p.52]. Only (5) _____ [p.52] cells contain mitochondria. The kind of ATP-forming reactions that occur in mitochondria cannot be completed without plenty of (6) _____ [p.52] taken in by breathing.

A mitochondrion has a double- (7) _____ [p.52] system. The inner membrane has folds called (8) _____ [p.52]. This membrane forms two separate (9) _____ [p.52] that are critical to the function of the mitochondrion.

Short Answer

10. List three reasons why many biologists believe that mitochondria evolved from ancient bacteria consumed by a larger cell. [p.52]

_____

_____

_____

Matching

11. _____ transmission electron microscope [p.53]

12. _____ compound light microscope [p.53]

13. _____ scanning electron microscope [p.53]

A. lenses enlarge light that passes through a thin specimen
B. uses a magnetic field to channel a stream of electrons
C. produces images with fantastic depth

Short Answer

14. In what way is the cytoskeleton different from your skeleton? [p.54]

_____

_____

_____

15. What are the three main types of fibers in the cytoskeleton? [p.54]

_____

_____

_____

Blanks

Fill in the word that matches the definition.

_____ 16. Proteins such as dynein and myosin that move cell parts [p.54]

_____ 17. Refers to centrioles found at the base cilia or flagella [p.55]

_____ 18. Consist of two strands of actin twisted together; reinforce cell shape [p.54]

_____ 19. Largest cytoskeleton element; made of tubulin [p.54]

_____ 20. Along with cilia, these allow movement with a "9 + 2 array" of microtubules [p.55]

_____ 21. Strengthening protein fibers; used in cell typing [p.54]

_____ 22. System of microtubules, microfilaments, and intermediate filaments within the cytosol [p.54]

3.10. MOVING SUBSTANCES ACROSS MEMBRANES BY DIFFUSION AND OSMOSIS [pp.56–57]

3.11. OTHER WAYS SUBSTANCES CROSS CELL MEMBRANES [pp.58–59]

3.12. REVENGE OF EL TOR [p.59]

Selected Words: "concentration" [p.56], "gradient" [p.56], "dynamic equilibrium" [p.56], _electric gradient_ [p.56], _pressure gradient_ [p.56], _tonicity_ [p.57], _isotonic_ [p.57], _hypotonic_ [p.57], _hypertonic_ [p.57], _osmotic pressure_ [p.57], _hydrostatic pressure_ [p.57], "facilitated diffusion" [p.58], "passive" [p.58], "membrane pumps" [p.58], _Vibrio cholerae_ [p.59], _oral rehydration therapy (ORT)_ [p.59]

Boldfaced, Page-Referenced Terms

[p.56] selective permeability _____

_____

[p.56] concentration gradient _____

_____

[p.56] diffusion _____

_____

[p.57] osmosis _____

_____

[p.58] passive transport _____

_____

[p.58] active transport _____

_____

[p.59] exocytosis _____

_____

[p.59] endocytosis _____

_____

[p.59] phagocytosis _____

_____

Matching

1. _____ osmosis [p.57]
2. _____ pressure gradient [p.56]
3. _____ endocytosis [p.59]
4. _____ selective permeability [p.56]
5. _____ active transport [p.58]
6. _____ diffusion [p.56]
7. _____ electric gradient [p.56]
8. _____ exocytosis [p.59]
9. _____ tonicity [p.57]
10. _____ facilitated diffusion [p.58]
11. _____ phagocytosis [p.59]
12. _____ concentration gradient [p.56]
13. _____ dynamic equilibrium [p.56]
14. _____ passive transport [p.58]

A. Difference in pressure between two adjoining regions
B. Property of allowing some substances but not others to cross a plasma membrane
C. Net movement of like molecules or ions from an area of higher concentration to an area of lower concentration
D. Difference in electric charge across a plasma membrane
E. Diffusion of water across a selectively permeable membrane in response to solute concentration gradients
F. Refers to the relative solute concentrations in two fluids
G. Solutes are made to cross plasma membranes against their own concentration gradients
H. A small section of plasma membrane folds inward and pinches organic matter from outside the cell into a vesicle
I. A vesicle moves to the plasma membrane and fuses with it, releasing its contents to the outside of the cell
J. Difference in the number of molecules or ions of a given substance in two neighboring regions
K. Literally means "coming inside the cell"
L. Movement of solutes that does not require energy from ATP
M. Solutes diffuse across a plasma membrane through a transport protein down their own concentration gradients
N. Refers to the nearly uniform net distribution of molecules throughout two adjoining regions

True/False

If the statement is true, write a "T" in the blank. If the statement is false, underline the incorrect word(s) and write the correct word in the answer blank.

_____ 15. Diffusion occurs faster when the concentration gradient is steep. [p.56]

_____ 16. In a solution with more than one solute, each solute diffuses separately from others, according to its own concentration gradient. [p.56]

_____ 17. In a dynamic equilibrium, the number of molecules moving one way is greater than the number moving the other way. [p.56]

_____ 18. Osmotic pressure refers to water's tendency to pass from a hypotonic solution to a hypertonic one. [p.57]

_____ 19. Osmosis occurs in response to a concentration gradient of solute that results in unequal concentrations of water molecules. [p.57]

_____ 20. If an animal cell were placed in a hypertonic solution, it would swell and perhaps burst. [p.57]

_____ 21. Physiological saline is 0.9 percent NaCl; red blood cells placed in such a solution will not gain or lose water; therefore, one could state that the fluid in red blood cells is hypertonic. [p.57]

_____ 22. Cell membranes display selective permeability. [p.56]

23. _____ Red blood cells shrivel and shrink when placed in a hypotonic solution. [p.57]

24. _____ *Vibrio cholerae* is a fungus that causes disease when it enters human drinking water supplies. [p.59]

25. _____ Cellular products are released to the outside of a cell by phagocytosis. [p.59]

26. _____ Exocytosis occurs when a cell takes in a substance by engulfing it into a vesicle derived from the plasma membrane. [p.59]

Complete the Table

Complete the table by filling in the empty boxes.

Substance	Process	Where Substance Crosses	Active or Passive
small nonpolar molecules like CO_2 or O_2	(27)	(28)	passive [p.56]
(29)	osmosis	(30)	passive [p.57]
solutes with charge	(31)	transport proteins	passive [p.58]
solutes with charge	(32)	(33)	active [p.58]

3.13. METABOLISM: DOING CELLULAR WORK [pp.60–61]

Selected Words: metabolism [p.60], "phosphorylation" [p.60], *intermediate* [p.60], *end products* [p.60], *energy carriers* [p.60], "gastric juice" [p.61]

Boldfaced, Page-Referenced Terms

[p.60] ATP/ADP cycle _____

_____

[p.60] metabolic pathways _____

_____

[p.60] anabolism _____

_____

[p.60] catabolism _____

_____

[p.60] substrate _____

_____

[p.61] enzymes _____

_____

[p.61] active site _____

_____

[p.61] coenzymes _____

_____

[p.61] NAD$^+$ _____

_____

[p.61] FAD _____

_____

Matching

1. _____ active site [p.61]
2. _____ catabolism [p.60]
3. _____ energy carrier [p.60]
4. _____ metabolic pathway [p.60]
5. _____ anabolism [p.60]
6. _____ substrate [p.60]
7. _____ intermediate [p.60]
8. _____ phosphorylation [p.60]
9. _____ enzyme [p.61]

A. Catalytic molecule that speeds up the rate of a specific chemical reaction
B. Transfer of a phosphate group to a molecule
C. A surface crevice on an enzyme where it interacts with a substrate
D. Pathway where large molecules are broken down into products that have less energy
E. ATP and several other compounds that activate other substances
F. Any substance that forms between the start and end of a pathway
G. A series of reactions occurring in orderly steps that are catalyzed by enzymes
H. Any substance that enters a reaction; also called a reactant or a precursor
I. Pathway in which small molecules are put together into more complex molecules

Short Answer

10. List the four features that all enzymes share. [p.61]

_____

_____

_____

_____

Matching

Match the items on the following sketch with the correct description. [p.57]

11. _____
12. _____
13. _____
14. _____

A. Substrate molecules
B. Active site
C. Enzyme
D. End product

Circle one of two possible answers given between parentheses in each statement.

15. The reaction diagrammed above is (anabolic/catabolic). [p.60]

16. In the human body, enzymes function best within a pH range of (7.35–7.4/6.35–8.4). [p.61]

17. The pathways that build complex carbohydrates, proteins, and other large molecules are all (anabolic/catabolic). [p.60]

18. Nearly all enzymes are (carbohydrates/proteins). [p.61]

19. NAD$^+$ and FAD are examples of (transport proteins/coenzymes). [p.61]

3.14. HOW CELLS MAKE ATP [pp.62–63]

3.15. SUMMARY OF AEROBIC RESPIRATION [p.64]

3.16. ALTERNATIVE ENERGY SOURCES IN THE BODY [p.65]

Selected Words: "aerobic" [p.62], "phosphorylations" [p.62], *adipose* tissue [p.65]

Boldfaced, Page-Referenced Terms

[p.62] aerobic respiration _____

_____

[p.62] glycolysis _____

_____

[p.62] pyruvate _____

_____

[p.62] PGAL _____

_____

[p.62] Krebs cycle _____

_____

[p.63] electron transport systems _____

_____

Dichotomous Choice

1. Cells make ATP by breaking (ionic/covalent) bonds in carbohydrates especially, but also proteins and lipids. [p.62]

2. During breakdown reactions, energy associated with (calories/electrons) drives ATP formation. [p.62]

3. Cells of the human body typically form ATP by way of (anaerobic/aerobic) respiration, which requires oxygen. [p.62]

4. Aerobic pathways start with a set of reactions known as (electron transport/glycolysis). [p.62]

5. Glycolysis occurs in the (mitochondria/cytoplasm). [p.62]

6. In glycolysis, glucose is partially broken down into two molecules of (pyruvate/ATP). [p.62]

7. For every glucose molecule entering glycolysis, (two/four) three-carbon pyruvate molecules are produced. [p.62]

8. The first steps of glycolysis are (energy releasing/energy requiring). [p.62]

9. The first steps of glycolysis proceed only when two ATP molecules each transfer a phosphate group to (pyruvate/glucose), donating energy to it. [p.62]

10. The glucose molecule is then split into two molecules of (PGAL/pyruvate), which is then converted to an intermediate. [p.62]

11. During glycolysis, intermediates donate phosphate groups to ADP, resulting in (four/36) ATP molecules. [p.62]

12. The *net* yield of ATP from glycolysis is (two/four) ATP molecules. [p.62]

13. NAD^+ picks up the electrons and hydrogen ions released from (end products/intermediates) and becomes NADH, which can give up H^+ and electrons to become NAD^+ again. [p.62]

14. The oxygen needed to make ATP is used in the (cytoplasm/mitochondria). [p.62]

15. Following glycolysis, the two pyruvates enter the (inner/outer) compartment of a mitochondrion. [p.62]

16. In the preparatory steps, a carbon is removed from each pyruvate, leaving a fragment that is combined with (coenzyme A/carbon dioxide), resulting in acetyl CoA. [p.62]

17. The two-carbon fragment is transferred to (pyruvate/oxaloacetate), which starts the Krebs cycle. [p.62]

18. The carbons in the two pyruvates are lost in the Krebs cycle as ($NADH/CO_2$). [p.62]

19. Reactions during the Krebs cycle produce (two/36) molecules of ATP directly. [p.62]

20. Reactions prior to and during the Krebs cycle produce a large number of (phosphate/coenzyme) molecules such as NADH and $FADH_2$. [p.62]

21. $FADH_2$ and NADH molecules from the first and second stages of aerobic respiration are used in the (Krebs cycle/electron transport system). [p.63]

22. Electron transport systems and neighboring transport proteins are embedded in the (outer/inner) membrane that divides the mitochondrion into two compartments. [p.63]

23. The transfer of electrons in the electron transport systems results in the movement of (water/H^+) from the inner to the outer compartment of a mitochondrion. [p.63]

24. An H^+ gradient forms in the outer compartment of the mitochondrion after the pumping of H^+ from the inner compartment; H^+ then flows back into the inner compartment, where (ADP/ATP) is formed. [p.63]

25. At the end of the electron transport systems, oxygen withdraws electrons and then combines with H^+, resulting in the formation of (carbon dioxide/water). [p.63]

26. For each molecule of glucose entering the aerobic pathway, including glycolysis, (32/36) total molecules of ATP are formed. [p.64]

Complete the Table

Complete the table by filling in the empty boxes. [p.63]

Stage	Location in Cell	Net ATP	NADH/FADH$_2$
(27)	cytoplasm	(28)	(29)
Krebs cycle	(30)	(31)	(32)
(33)	(34)	(35)	0

Sequencing

36. In the blanks beside the events listed, write a number to indicate the order in which they occur, with "1" occurring first. [p.65]

_____ Excess glucose units are combined into glycogen.

_____ The pancreas releases insulin.

_____ Glucose from digested food enters the bloodstream.

_____ The pancreas releases glucagons.

_____ Cells take in glucose and convert it to glucose-6-phosphate.

_____ Liver cells break down glycogen to increase the glucose level in the blood.

Short Answer

37. What percentage of the human body's total energy is stored as fat? In what tissue is most of the body's fat stored? [p.65]

_____

_____

_____

38. When are triglycerides used as an energy source? [p.65]

_____

_____

_____

39. Why does a fatty acid yield more ATP than a glucose molecule? [p.65]

_____

_____

_____

40. What percentage of the human body's total energy is stored as protein? In what form is the ammonia produced by protein breakdown excreted? [p.65]

_____

_____

_____

Self-Quiz

Are you ready for the exam? Test yourself on key concepts by taking the additional tests linked with your BiologyNow CD-ROM.

_____ 1. _____ are organelles of digestion within a cell; they bud from _____ membranes. [p.51]
   a. Peroxisomes; ER
   b. Mitochondria; cytoskeleton
   c. Lysosomes; Golgi
   d. Centrioles; basal body

_____ 2. The _____ is free of ribosomes, curves through the cytoplasm, and is the main site of lipid synthesis. [p.50]
   a. lysosome
   b. Golgi body
   c. smooth ER
   d. rough ER

_____ 3. Which of the following is NOT one of the three fundamental features of cells? [p.42]
   a. cell wall
   b. plasma membrane
   c. cytoplasm
   d. DNA

_____ 4. Mitochondria convert energy obtained from _____ to forms that the cell can use, principally ATP. [p.52]
   a. water
   b. glucose and other organic molecules
   c. $NADPH_2$
   d. carbon dioxide

_____ 5. In a lipid bilayer, phospholipid tails point inward and form a(n) _____ layer that excludes water. [p.43]
   a. acidic
   b. hydrophilic
   c. basic
   d. hydrophobic

_____ 6. A protist adapted to life in a freshwater pond is collected in a bottle and transferred to a saltwater bay. Which of the following is likely to happen? [p.57]
   a. The cell will burst.
   b. Salts will flow out of the protist cell.
   c. The cell will shrink.
   d. Enzymes will flow out of the protist cell.

_____ 7. Calcium pumps move calcium ions against their concentration gradient, keeping a much lower concentration inside muscle cells than outside. This is an example of _____. [p.58]
   a. passive transport
   b. endocytosis
   c. exocytosis
   d. active transport

_____ 8. O_2 and CO_2 and other small, nonpolar molecules move across the plasma membrane by _____. [p.56]
   a. facilitated diffusion
   b. endocytosis
   c. diffusion
   d. active transport

_____ 9. Ions and small, water-soluble molecules cross the plasma membrane down their concentration gradients by _____, also called "facilitated diffusion." [p.58]
   a. passive transport
   b. endocytosis
   c. simple diffusion
   d. active transport

_____ 10. Phagocytosis is a special type of _____ used by immune system cells to engulf harmful viruses and bacteria. [p.59]
   a. exocytosis
   b. passive transport
   c. endocytosis
   d. hydrolysis

_____ 11. Aerobic respiration would quickly halt if the process ran out of _____, which serves as the hydrogen and electron acceptor. [pp.62–63]
   a. coenzyme A
   b. oxygen
   c. NAD^+
   d. H_2O

___ 12. The final electron acceptor at the end of aerobic respiration is _____. [p.63]
  a. NADH
  b. carbon dioxide (CO_2)
  c. ATP
  d. oxygen

___ 13. When glucose is used as an energy source, the largest amount of ATP is generated by the _____ portion of the respiratory process. [p.63]
  a. glycolysis
  b. acetyl-CoA
  c. Krebs cycle
  d. electron transport systems

___ 14. Which of these is NOT necessary for aerobic respiration to occur? [pp.62–63]
  a. glucose
  b. oxygen
  c. mitochondrion
  d. carbon dioxide

Chapter Objectives/Review Questions

This section lists general and detailed chapter objectives that can be used as review questions. You can make maximum use of these items by writing answers on a separate sheet of paper. Fill in answers where blanks are provided. To check for accuracy, compare your answers with information in the chapter or glossary.

1. List the three generalizations that together constitute the cell theory. [p.42]
2. Name and describe the three basic parts of cells. [p.42]
3. Describe and contrast the distinguishing features of prokaryotic and eukaryotic cells. [p.42]
4. _____ physically separate incompatible cellular chemical reactions in time and cytoplasmic space. [p.44]
5. Phospholipid molecules have _____ heads and _____ tails. [p.43]
6. Describe the structure of a cell membrane. [p.46]
7. Give the functions of the following basic eukaryotic organelles and structures: nucleus, nucleolus, nuclear envelope, chromosomes, chromatin, endoplasmic reticulum (rough and smooth), Golgi bodies, lysosomes, peroxisomes, vesicles, mitochondria, ribosomes, cytoskeleton, microtubules, microfilaments, intermediate filaments, flagellum, cilium, centrioles, and basal body. [pp.48–55]
8. The outermost part of the nucleus, the nuclear _____, has two lipid bilayers. [p.48]
9. Assembly of protein and RNA subunits of ribosomes occurs in the _____. [p.49]
10. A DNA molecule and its associated proteins is a(n) _____. [p.49]
11. Explain how the ER, Golgi bodies, and certain vesicles function as the endomembrane skeleton. [pp.50–51]
12. _____ are enzyme sacs that break down fatty acids and amino acids. [p.51]
13. _____ are the ATP-producing powerhouses of eukaryotic cells. [p.52]
14. The _____ gives cells their shape and internal organization. [p.54]
15. When molecules move down a concentration gradient, they move from a region where they are _____ concentrated to a region where they are _____ concentrated. [p.56]
16. Define *electric gradient* and *pressure gradient*. [p.56]
17. Compare and contrast the processes of osmosis and diffusion. [pp.56–57]
18. Two fluids are said to be _____ when solute concentrations are equal on both sides of a cell membrane; when solute concentrations are unequal, one fluid is _____ (fewer solutes) and the other is _____ (more solutes). [p.57]
19. Describe the mechanisms involved in active and passive transport (facilitated diffusion) across the cell membrane. [p.58]
20. Explain the role of vesicles during exocytosis and endocytosis. [p.59]
21. Dynamic cellular chemical activity is called _____. [p.60]

22. Distinguish between biosynthetic metabolic pathways (anabolism) and degradative metabolic pathways (catabolism). [p.60]
23. What is a metabolic pathway? What is necessary for each step in the pathway to occur? [p.60]
24. Define and state the role of each of the following participants in a metabolic pathway: *substrate, intermediate, enzyme, cofactor, coenzyme, energy carrier,* and *end products.* [pp.60–61]
25. _____ are molecules that an enzyme can chemically recognize, bind, and modify. [p.61]
26. Substrates interact with enzymes on the _____ sites of enzymes. [p.61]
27. The main energy carrier in cells is _____. [p.62]
28. What is the role of NAD^+ in ATP production? [p.62]
29. Where in a eukaryotic cell do glycolysis, the Krebs cycle, and electron transport occur? [p.63]
30. Describe how ATP is produced with a pumping mechanism across the inner membrane of the mitochondrion. [p.63]
31. List (in order) and cite the highlights of the three major stages of aerobic respiration. [pp.62–64]
32. Describe how alternative energy sources such as fats and proteins may enter the energy-releasing pathways. [p.65]

Media Menu Review Questions

Questions 1–3 are drawn from the following InfoTrac College Edition article: "Celle Fantastyk." *Natural History*, May 2000.

1. Just as a single molecule of H_2O doesn't make water, a single cell from the brain of a mouse contains no
   _____.
2. Memory and skills may have to do with the end of the _____ of one nerve cell making a connection with another nerve cell.
3. A single cell is said to occupy a spot between the microcosmic world of genes and proteins, and the visible world of the _____.

Questions 4–10 are drawn from the following InfoTrac College Edition article: "Mitochondria: Cellular Energy Co." *The Scientist,* June 2002.

4. Where fossil fuels power the world, _____ power the cell.
5. High-energy _____ are transported along the electron transport chain, which acts like a bucket brigade for electrons.
6. Because of their high energy needs, what type of cell is particularly susceptible to damage from low levels of ATP?
7. Mitochondria cannot work in an environment high in the element _____.
8. When mutations impair mitochondrial function, disorders like Parkinson's and _____ disease may result.
9. The first step in drug discovery for conditions caused by malfunctioning mitochondria is to identify all of the mitochondrial _____.
10. The hope of creatine research is that it will be found to increase the amount of _____ in a cell so it would be better able to regulate calcium levels.

Integrating and Applying Key Concepts

1. If there were no such thing as active transport, how would the lives of organisms be affected?
2. Exactly how does being deprived of oxygen, such as being suffocated with a pillow, kill a person?

4

TISSUES, ORGAN SYSTEMS, AND HOMEOSTASIS

Interactive Exercises

Impacts, Issues: Open or Close the Stem Cell Factories? [p.69]

4.1. EPITHELIUM: THE BODY'S COVERING AND LININGS [pp.70–71]

4.2. CONNECTIVE TISSUE: BINDING, SUPPORT, AND OTHER ROLES [pp.72–73]

4.3. MUSCLE TISSUE: MOVEMENT [p.74]

4.4. NERVOUS TISSUE: COMMUNICATION [p.75]

4.5. THE HUMAN BODY SHOP: ENGINEERING NEW TISSUES AND ORGANS [pp.76–77]

 For additional practice, use the interactive vocabulary exercises linked with your BiologyNow CD-ROM.

Selected Words: simple epithelium [p.70], stratified epithelium [p.70], *pseudostratified* epithelium [p.70], *squamous epithelium* [p.70], *cuboidal epithelium* [p.70], *columnar epithelium* [p.70], "ground substance" [p.72], *hyaline cartilage* [p.73], *elastic cartilage* [p.73], *fibrocartilage* [p.73], *striated* [p.74], "a muscle" [p.74], "involuntary" [p.74], "harvest" [p.77], "organ donor" [p.77]

Boldfaced, Page-Referenced Terms

[p.69] stem cells _____

_____

[p.69] tissue _____

_____

[p.69] organ _____

_____

[p.69] organ system _____

_____

[p.69] internal environment _____

_____

[p.70] epithelium _____

_____

[p.70] microvilli _____

_____

[p.70] basement membrane _____

_____

[p.70] gland _____

_____

[p.70] exocrine glands _____

_____

[p.70] endocrine glands _____

_____

[p.72] connective tissue _____

_____

[p.72] loose connective tissue _____

_____

[p.72] dense, irregular connective tissue _____

_____

[p.72] dense, regular connective tissue _____

_____

[p.72] cartilage _____

_____

[p.73] bone tissue _____

_____

[p.73] adipose tissue _____

_____

[p.73] blood _____

_____

[p.74] muscle tissue _____

_____

[p.74] skeletal muscle _____

_____

[p.74] smooth muscle _____

_____

[p.74] cardiac muscle _____

_____

[p.75] nervous tissue _____

_____

[p.75] neurons _____

_____

[p.75] nerve _____

_____

[p.75] neuroglia _____

_____

True/False

Place a "T" or "F" in the blank beside each statement.

_____ 1. Stem cells can divide indefinitely in a laboratory setting. [p.69]

_____ 2. Stem cells harvested from adults have produced more tissue types than those from embryos. [p.69]

_____ 3. Stem cells are the first to form when a fertilized egg starts dividing. [p.69]

_____ 4. Stem cells may be the answer to diseases like Parkinson's and to paralysis. [p.69]

_____ 5. Stem cell research in the United States is much more advanced than in the rest of the world. [p.69]

Sequence

Arrange the following levels of organization in nature in the correct hierarchic order. Find the letter of the simplest level and write it next to the number 6; write the letter of the most complex level next to the number 9; and so forth.

6. _____   A. organ [p.69]

7. _____   B. cell [p.69]

8. _____   C. tissue [p.69]

9. _____   D. organ system [p.69]

Short Answer

10. Define homeostasis [p.69]

_____

_____

_____

Fill-in-the-Blanks

(11) _____ [p.70] tissue has a free surface, which faces either the outside environment or a body fluid. Epithelial cells are arranged closely together in one or more (12) _____ [p.70]. Cells at the surface may have cilia or protrusions called (13) _____ [p.70], which absorb substances into the cells. The other surface adheres to a(n) (14) _____ [p.70] membrane, a noncellular layer packed with proteins and polysaccharides. There are two basic types of epithelial tissue that differ by the number of cell layers present: (15) _____ [p.70] epithelium with one cell layer and (16) _____ [p.70] epithelium with more than one. (17) _____ [p.70] epithelium is simple epithelium that appears to be more than one layer due to the staggering of nuclei in adjacent cells.

Complete the Table

18. Complete the following table to associate the major types of epithelium (use terms for layer number and shapes, such as stratified cuboidal) with sites where they may be found in the human body. [p.70]

Epithelium Type	Typical Location in the Human Body
a.	Male urethra, ducts of salivary glands
b.	Glands and their ducts, surface of ovaries, pigmented epithelium of eye
c.	Ducts of sweat glands
d.	Throat, nasal passages, sinuses, trachea, male genital ducts
e.	Linings of blood vessels, lung alveoli (sites of gas exchange)
f.	Skin (keratinized), mouth, throat, esophagus, vagina (nonkeratinized)
g.	Stomach, intestines, uterus

Fill-in-the-Blanks

A gland makes and (19) _____ [p.70] products such as saliva or mucus. (20) _____ [p.70] glands secrete substances onto an epithelial surface through ducts or tubes. Examples of exocrine gland secretions include (21) _____ [p.70], (22) _____ [p.70], (23) _____ [p.70], (24) _____ [p.70], (25) _____ [p.70], and (26) _____ [p.70]. (27) _____ [p.70] glands do not secrete substances through tubes or ducts. Their products, (28) _____ [p.70], are secreted directly into the extracellular fluid bathing the glands. Examples include (29) _____ [p.71], (30) _____ [p.71], and (31) _____ [p.71] glands.

Fill-in-the-Blanks

The most abundant and widely distributed body tissue is (32) _____ [p.72] tissue. The two general types of connective tissue are (33) _____ [p.72] and (34) _____ [p.72]. In all connective tissue except blood, the cells produce a (35) _____ _____ [p.72] of polysaccharides. The cells also secrete fibrous proteins called (36) _____ [p.72] and (37) _____ [p.72]. The ground substance and protein fibers form a (38) _____ [p.72] around the cells that gives each connective tissue its properties.

Choice

For questions 39–44, choose from the following:

a. dense, regular connective tissue   b. loose connective tissue   c. dense, irregular connective tissue

39. _____ Often supports epithelia [p.72]

40. _____ Consists of fibroblasts and fibers of mostly collagen oriented at different angles [p.72]

41. _____ Has rows of fibroblasts between parallel bundled fibers; resistant to tearing [p.72]

42. _____ Includes loosely arranged fibers, fibroblasts, and infection-fighting white blood cells [p.72]

43. _____ Connective tissue that composes elastic ligaments that attach bone to bone, and collagen-containing tendons, which attach muscle to bone [p.72]

44. _____ Connective tissue in skin that forms protective capsules around organs [p.72]

Short Answer

45. Describe the texture of cartilage and name the cells that produce the cartilage matrix. [p.72]

_____

_____

_____

46. Why is cartilage so slow to heal when injured? [p.73]

_____

_____

_____

Choice

For questions 47–53, choose from the following:

a. hyaline cartilage   b. elastic cartilage   c. fibrocartilage

47. _____ The most common type of cartilage [p.73]

48. _____ Forms cartilage "cushions" in joints such as the knee and in the disks that separate vertebrae of the spinal column [p.73]

49. _____ Serves as the forerunner of bone in the developing skeleton of an embryo [p.73]

50. _____ Provides a friction-reducing cover at the ends of freely movable mature bones where they articulate in joints [p.73]

51. _____ Occurs in places where a flexible yet rigid structure is required; has collagen and elastic fibers [p.73]

52. _____ Packed with collagen fibers arranged in thick bundles to withstand tremendous pressure [p.73]

53. _____ Type of cartilage found in the ear and the epiglottis [p.73]

Fill-in-the-Blanks

Bone tissue has (54) _____ [p.73] fibers and ground substance strengthened by

(55) _____ [p.73] salts. Different bones support and (56) _____ [p.73] softer tissues

and organs. Limb bones bear weight and interact with skeletal (57) _____ [p.73] to move body

parts. Bones also store calcium and some produce (58) _____ [p.73] cells.

    Adipose tissue is specialized for (59) _____ [p.73] storage. Most adipose tissue is located just

beneath the (60) _____ [p.73].

    Blood is classified as (61) _____ [p.73] tissue because it is derived mainly from this tissue

type. Its biological role is (62) _____ [p.73]. The fluid matrix of blood is blood (63) _____

[p.73]. It contains red and white blood cells and (64) _____ [p.73], along with dissolved

substances.

Short Answer

65. Describe what occurs at the cellular level that allows muscle tissue to maintain and change the positions of body parts. [p.74]

_____

_____

_____

Choice

For questions 66–72, choose from the following:

a. skeletal   b. smooth   c. cardiac

66. _____ Found in the walls of blood vessels, the stomach, and the intestines [p.74]

67. _____ Found only in the heart [p.74]

68. _____ Found in muscles attached to bones [p.74]

69. _____ Bundled groups of fascicles [p.74]

70. _____ Contractile cells tapered at both ends [p.74]

71. _____ Communication junctions allow contraction as a unit [p.74]

72. _____ Typically striated [p.74]

Choice

For questions 73–76, choose from the following:

a. neurons   b. neuroglia   c. nerve

73. _____ Some receive and integrate sensory input, while others relay signals to muscles and glands to carry out responses [p.75]

74. _____ A cluster of processes from several neurons that conduct messages between the central nervous system (CNS) and muscles and glands [p.75]

75. _____ Accessory cells located around neurons; provide physical support and insulation [p.75]

76. _____ Possess cell bodies and two types of processes called dendrites and axons [p.75]

Short Answer

77. Explain how cartilage and bone tissues are grown in a desired shape. [p.76]

_____

_____

_____

78. Why are stem cells from early embryos valued more in research than are adult stem cells? [p.77]

_____

_____

_____

4.6. CELL JUNCTIONS: HOLDING TISSUES TOGETHER [p.78]

4.7. MEMBRANES: THIN, SHEETLIKE COVERS [p.79]

Selected Words: epithelial membranes [p.79], connective tissue membranes [p.79], mucous membranes [p.79], serous membranes [p.79], cutaneous membranes [p.79], synovial membranes [p.79]

Boldfaced, Page-Referenced Terms

[p.78] tight junctions _____

_____

[p.78] adhering junctions _____

_____

[p.78] gap junctions _____

_____

Labeling-Matching

Most epithelial cells and cells of other tissues adhere strongly to one another by means of specialized attachment sites. Identify each example of cell junctions by entering the correct name in the blank below each sketch. Complete the exercise by matching and entering the letter of the correct description in the parentheses following each label.

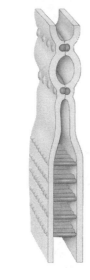

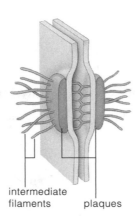

intermediate
filaments   plaques

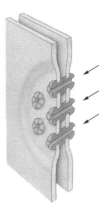

1. _____ junction   2. _____ junction   3. _____ junction
   (___) [p.78]   (___) [p.78]   (___) [p.78]

A. Spot-weld junctions holding together the plasma membranes of two adjacent cells; important in tissues subject to stretching, such as epithelium of the skin and stomach
B. Protein strands forming seals that help stop substances from leaking across a tissue; important in control of what enters the body
C. Channels directly linking the cytoplasm of adjacent cells; allow rapid communication by the direct transfer of ions and small molecules from cell to cell

Complete the Table

4. Complete the following table by filling in the type of membrane to which each phrase refers. [p.79]

Membrane Type	Description
a.	Occur in paired sheets separated by a thin film of fluid; help anchor and lubricate organs
b.	Secrete fluid to lubricate the ends of moving bones; no epithelial cells present
c.	Hardy, dry membrane; part of the integumentary system
d.	Pink, moist membranes; most have glands that secrete substances such as mucus

Matching

Match the types of membranes with the locations in which they may be found.

5. _____ lining the sheaths of tendons [p.79]

6. _____ found where absorption and glandular secretion occur, such as lining the stomach [p.79]

7. _____ lining capsulelike cavities such as around the knee joint [p.79]

8. _____ covering the outside of the body [p.79]

9. _____ lining the digestive, respiratory, urinary, and reproductive tracts [p.79]

10. _____ enclosing organs such as the heart and lungs [p.79]

A. serous
B. cutaneous
C. synovial
D. mucous

4.8. ORGAN SYSTEMS: BUILT FROM TISSUES [p.80]

4.9. THE SKIN—EXAMPLE OF AN ORGAN SYSTEM [pp.81–82]

4.10. SUN, SKIN, AND THE OZONE LAYER [p.83]

Selected Words: ectoderm [p.80], mesoderm [p.80], endoderm [p.80], *Langerhans cells* [p.82], *Granstein cells* [p.83], "stretch marks" [p.83], *sebaceous glands* [p.83], *acne* [p.83], "split ends" [p.83], *herpes simplex* [p.83]

Boldfaced, Page-Referenced Terms

[p.80] cranial cavity _____

_____

[p.80] spinal cavity _____

_____

[p.80] thoracic cavity _____

_____

[p.80] abdominal cavity _____

_____

[p.80] pelvic cavity _____

_____

[p.82] integument _____

_____

[p.82] epidermis _____

_____

[p.82] dermis _____

_____

[p.82] keratinocytes _____

_____

[p.82] melanocytes _____

_____

[p.83] hair _____

_____

Fill-in-the-Blanks

The (1) _____ [p.80] cavity and the (2) _____ [p.80] cavity house the central nervous system—the brain and spinal cord. The heart and lungs are located in the (3) _____ [p.80] cavity. A muscular diaphragm separates the thoracic cavity from the (4) _____ [p.80] cavity, which holds the stomach, liver, most of the intestine, and other organs. Reproductive organs, bladder, and rectum are located within the (5) _____ [p.80] cavity.

The (6) _____ [p.80] plane divides the body into right and left halves. The (7) _____ [p.80] plane divides the body into (8) _____ [p.80] (front) and posterior (back) parts. The (9) _____ [p.80] plane divides it into superior (upper) and (10) _____ [p.80] (lower) parts.

Complete the Table

11. In the left column of the table below, fill in the three types of primary tissues that form in an embryo. In the right column, list the adult tissues derived from them. [p.80]

Primary Tissue	Adult Tissue Types
a.	
b.	
c.	

Matching

Match the most appropriate function with the letter for each system shown on the accompanying illustrations.

12. _____ Female: production of eggs; provision of a protected nutritive environment for development. Male: production and transfer of sperm to the female. Both systems have hormonal influences on other organs. [p.81]

13. _____ Ingestion of food and water; breakdown and absorption of food molecules; elimination of food residues from the body [p.81]

14. _____ Movement of body and its internal parts; maintenance of posture; generation of heat [p.81]

15. _____ Detection of external and internal stimuli; control and coordination of responses to stimuli; integration of activities of all organ systems [p.81]

16. _____ Protection from injury and dehydration; some defense against pathogens; body temperature control; excretion of some wastes; reception of external stimuli [p.81]

17. _____ Delivery of oxygen to tissue fluids; removal of carbon dioxide wastes produced by cells; regulation of pH of body fluids [p.81]

18. _____ Support and protect body parts; sites for muscle attachment, red blood cell production, and calcium and phosphorus storage [p.81]

19. _____ Hormonal control of body functioning; works with nervous system in integrating short-term and long-term activities [p.81]

20. _____ Maintenance of volume and composition of fluids in the internal environment; excretion of excess fluid and blood-borne wastes [p.81]

21. _____ Rapid internal transport of many materials to and from cells; helps stabilize internal temperature and pH [p.81]

22. _____ Return of some tissue fluid to blood; roles in immunity (defense against infection and tissue damage) [p.81]

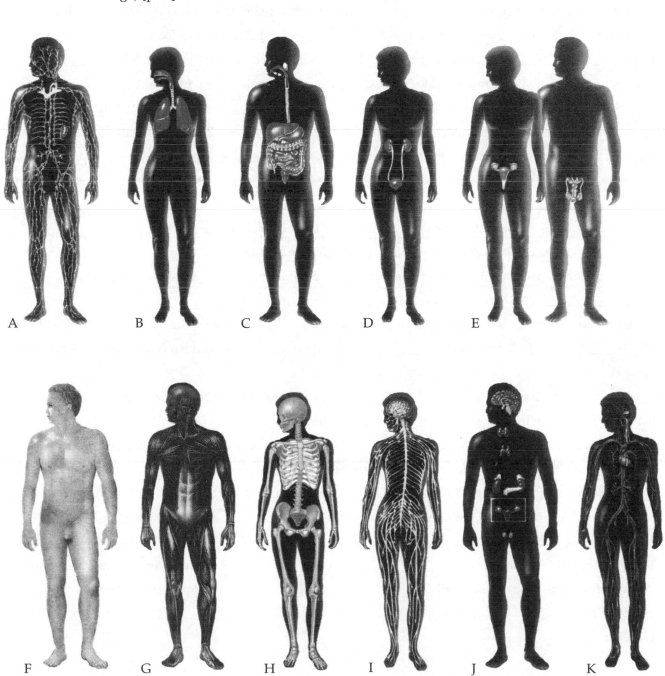

A   B   C   D   E

F   G   H   I   J   K

Fill-in-the-Blanks

Skin is a part of the organ system called the (23) _____ [p.82]. Skin holds its shape, blocks harmful rays from the (24) _____ [p.82], kills many (25) _____ [p.82], holds in (26) _____ [p.82], and fixes small cuts and burns. It also helps control the internal body (27) _____ [p.82]. Signals from the skin's (28) _____ [p.82] receptors help the brain assess what's going on in the outside world. Skin makes cholecalciferol, a precursor of vitamin (29) _____ [p.82] required for calcium absorption in the intestine.

Skin has two distinct regions—an outer (30) _____ [p.82] and an underlying (31) _____ [p.82] made mainly of dense (32) _____ [p.82] tissue containing elastin and fibrin. Beneath this cutaneous layer is a subcutaneous layer, the (33) _____ [p.82]. This is a loose connective tissue that (34) _____ [p.82] the skin and yet allows it some freedom of movement. (35) _____ [p.82] stored in the hypodermis insulates the body and cushions some body parts.

Epidermis consists of layers; it is a (36) _____ _____ _____ [p.82]. Most cells of the epidermis are keratinocytes that make (37) _____ [p.82], a tough, water-insoluble protein. The outermost layer of skin is the tough, waterproof (38) _____ _____ [p.82].

In the deepest epidermal layer, cells called (39) _____ [p.82] produce the brown-black pigment called melanin. This is transferred to keratinocytes, forming a shield against harmful (40) _____ [p.82] radiation. All humans have the same number of melanocytes, but (41) _____ _____ [p.82] varies due to differences in distribution and activity.

The epidermis also contains two types of cells involved in immunity. They are the phagocytic (42) _____ [p.82] cells and the (43) _____ [p.83] cells.

Dense connective tissue makes up most of the (44) _____ [p.83]. The dermis is packed with (45) _____ [p.83] and (46) _____ [p.83] vessels, as well as the receptor ends of sensory nerves.

Sweat glands, oil glands, or (47) _____ [p.83] glands, and the keratinized structures called hair reside mostly in the dermis, even though they are derived from epidermis.

Tanning results from increased (48) _____ [p.83] levels in the epidermis and provides some protection from UV radiation, but after many years, tanning causes (49) _____ [p.83] fibers in the dermis to clump together. The skin loses resiliency and begins to look leathery.

Excessive exposure to UV radiation also suppresses the (50) _____ [p.83] system. UV radiation from sunlight or from the lamps of tanning salons can activate (51) _____ [p.83] in the skin. These bits of DNA can trigger (52) _____ [p.83]. With loss of the ozone layer in the stratosphere, the rate of skin cancers now is rapidly (53) _____ [p.83]. Experts estimate that even if all ozone-depleting substances were banned tomorrow, it would take about (54) _____ [p.83] years for the planet to recover.

Labeling

Label the numbered parts of the accompanying illustration. [p.82]

55. _____ _____

56. _____ _____

57. _____

58. _____

59. _____ _____

60. _____ _____

61. _____ _____

62. _____ _____

63. _____ _____

64. _____

65. _____

66. _____

4.11. HOMEOSTASIS: THE BODY IN BALANCE [pp.84–85]

Selected Words: "internal environment" [p.84], *interstitial* [p.84], "set points" [p.84]

Boldfaced, Page-Referenced Terms

[p.84] extracellular fluid _____

_____

[p.84] homeostasis _____

_____

[p.84] sensory receptors _____

_____

[p.84] stimulus _____

_____

[p.84] integrator _____

_____

[p.84] effectors _____

_____

[p.84] negative feedback _____

_____

[p.85] positive feedback _____

_____

Fill-in-the-Blanks

(1) _____ [p.84] fluid is the fluid outside of cells. Much of this fluid is (2) _____ [p.84], meaning it occupies spaces between cells and tissues. Substances move between the interstitial fluid and the (3) _____ [p.84] that it bathes as the cells draw nutrients from it and dump metabolic wastes into it.

Drastic changes in its composition can have drastic effects on cell activities. The number and type of (4) _____ [p.84] are especially crucial. (5) _____ [p.84] means "staying the same." Homeostatic mechanisms operate to maintain (6) _____ [p.84] in the volume and composition of extracellular fluid.

Sensory (7) _____ [p.84] are cells or cell parts that can detect a(n) (8) _____ [p.84], a specific change in the environment. Your brain is a(n) (9) _____ [p.84], a control point where different bits of information are pulled together in selecting a response. It can send signals to muscles and glands (or both). Muscles and glands are (10) _____ [p.84]—they carry out the response. When information from receptors shows that conditions have deviated from (11) "_____ _____" [p.84], the brain functions to bring conditions back within proper operating range.

Mechanisms for (12) _____ [p.84] help keep physical and chemical aspects of the body within tolerable ranges. In (13) _____ [p.84] feedback, an activity alters a condition in the internal environment, and this triggers a response that reverses the altered condition. An example is a furnace with a thermostat or the body temperature of humans being maintained within a normal range. In (14) _____ [p.85] feedback, a chain of events is set in motion that intensifies a change from an original condition; childbirth is an example.

Labeling-Matching

Identify the numbered items on the diagram of components necessary for negative feedback. Enter your choices in the numbered blanks. Choose from *effector, receptor, stimulus, integrator,* and *response.* Complete the exercise by matching the lettered examples to each component. [p.80]

15. _____ (___)

16. _____ (___)

17. _____ (___)

18. _____ (___)

19. _____ (___)

15
(input)
↓

| 16 | → | 17 | → | 18 | → | **19** (output) |

Response leads to change. Change is "fed back" to receptor. In *negative* feedback, the response of the system cancels or counteracts original stimulus.

A. Brain
B. Excessively salty food
C. Muscles related to the function of the jaw, esophagus, and stomach
D. Regurgitate food
E. Taste buds

Choice

For questions 20–25, choose from the following:

        a. negative feedback mechanisms   b. positive feedback mechanisms

20. _____ Set in motion a chain of events that intensify a change from an original condition [p.81]

21. _____ A furnace with a thermostat [p.84]

22. _____ After a limited time, the intensification reverses the change [p.85]

23. _____ Childbirth [p.85]

24. _____ Maintaining body temperature in a normal range [p.84]

25. _____ An activity alters a condition in the internal environment, and this triggers a response that reverses the altered condition [p.84]

Self-Quiz

Are you ready for the exam? Test yourself on key concepts by taking the additional tests linked with your BiologyNow CD-ROM.

Labeling-Matching

Identify each of the following illustrations by labeling it with one of the following: *connective, epithelial, muscle,* or *nervous.* Complete the exercise by matching and entering the capital letter from the first group of choices in the first set of parentheses and a matching lowercase letter from the second group of choices in the set of second parentheses following each label.

1. _____ (____) (____) [p.72]

2. _____ (____) (____) [pp.70–71]

3. _____ (____) (____) [p.74]

4. _____ (____) (____) [p.74]

5. _____ (____) (____) [p.72]

6. _____ (____) (____) [pp.72–73]

7. _____ (____) (____) [pp.72–73]

8. _____ (____) (____) [pp.70–71]

9. _____ (____) (____) [p.73]

10. _____ (____) (____) [p.75]

11. _____ (____) (____) [p.74]

12. _____ (____) (____) [pp.70–71]

13. _____ (____) (____) [p.73]

A. Adipose
B. Bone
C. Cardiac
D. Dense, regular
E. Loose
F. Simple columnar
G. Simple cuboidal
H. Simple squamous
I. Smooth
J. Skeletal
K. Blood
L. Cartilage
M. Neurons

a. Pumps circulatory blood; involuntary and striated
b. Single layer of cubelike cells that function in secretion and absorption
c. Communication by means of electrochemical signals
d. Fibers loosely arranged; cells include fibroblasts; functions in elasticity and diffusion
e. Contract to propel substances along internal pathways; involuntary
f. Transport of nutrients and waste products to and from cells
g. Attaches muscle to bone and bone to bone
h. Large, densely clustered cells specialized for fat storage; insulation and padding
i. Chondroblasts found inside solid matrix; hyaline, elastic, and fibrocartilage types
j. A single layer of flattened cells; found in blood vessel walls and lung air sacs; functions in diffusion
k. Provides support, protection, and attachment sites for skeletal muscles
l. Contraction for voluntary movements; striated and voluntary
m. Single layer of tall cells that may be ciliated; found in parts of the gut, uterus, and respiratory linings; function in secretion and absorption

1.

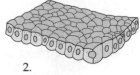

2.

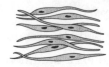

3.

4.

5.

6.

7.

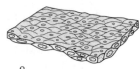

8.

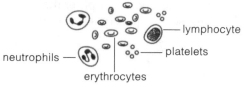

neutrophils — erythrocytes — lymphocyte — platelets

9.

dendrite

axon

10.

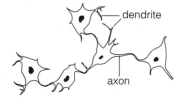

11.

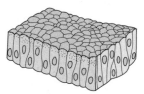

12.

13.

Multiple Choice

___ 14. Which of the following is
NOT a connective tissue?
[pp.72–73]
a. bone
b. adipose
c. cartilage
d. skeletal muscle

___ 15. All connective tissues contain cells
separated by _____. [p.72]
a. muscle and nerves
b. fluid and proteins
c. fibers and ground substance
d. blood

___ 16. Blood is considered a(n) _____
tissue. [p.73]
a. epithelial
b. muscular
c. connective
d. none of these

___ 17. Which group is arranged correctly from
smallest structure to largest? [p.74]
a. muscle cells, muscle bundle,
muscle
b. muscle cells, muscle, muscle
bundle
c. muscle bundle, muscle cells,
muscle
d. none of the above

___ 18. Involuntary muscle consisting of tapered
cells is _____. [p.74]
a. cardiac
b. skeletal
c. striated
d. smooth

___ 19. An organized group of cells and intercellular substances that take part in one or
more tasks is a(n) _____. [p.69]
a. organ
b. organ system
c. tissue
d. cuticle

___ 20. A tissue whose cells are striated and
bound by specialized junctions that allow
many cells to contract as a single unit is
_____ tissue. [p.74]
a. smooth muscle
b. dense fibrous connective
c. supportive connective
d. cardiac muscle

___ 21. Most cells of the epidermis produce
the protein _____; the dermis
contains the fibrous proteins and
_____. [p.82]
a. melanin; keratin and collagen
b. keratin; collagen and elastin
c. melanin; keratin and hemoglobin
d. keratin; hemoglobin and melanin

___ 22. Which of the following is NOT associated
with a negative feedback mechanism?
[pp.84–85]
a. maintains tolerable ranges of physical
and chemical aspects of the body
b. is similar to the function of a thermostat
and furnace
c. a chain of events is set in motion that
intensifies a change from an original
condition
d. mechanisms that keep the body from
overheating

Chapter Objectives/Review Questions

This section lists general and detailed chapter objectives that can be used as review questions. You can
make maximum use of these items by writing answers on a separate sheet of paper. To check for accuracy,
compare your answers with the information given in the chapter or glossary.

1. Cells are the basic units of life; in a multicellular animal, similar interacting cells and their intercellular
substances are grouped into a(n) _____. [p.69]
2. State the structural relationships among cells, tissues, organs, and organ systems. [p.69]
3. _____ always has a free surface, which faces a body cavity or the outside environment. [p.70]
4. _____ glands have no ducts; the thyroid gland is an example. [pp.70–71]

5. Describe the general characteristics of connective tissue and explain how these enable connective tissue to carry out its various tasks. [p.72]
6. Describe the structural differences and functions of loose connective tissue and dense, irregular connective tissue. [p.72]
7. _____ connective tissue is the type found in ligaments and tendons. [p.72]
8. The most common type of cartilage in the human body is _____ cartilage. [p.73]
9. Cartilage in the human ear is _____ cartilage. [p.73]
10. Cartilage that cushions the knee and other joints is _____. [p.73]
11. The specialized function of cells in adipose tissue is for _____ storage. [p.73]
12. Distinguish among skeletal, cardiac, and smooth muscle in terms of location, structure, and function. [p.74]
13. Muscle tissues contain cells that _____ when they receive outside stimulation. [p.74]
14. Neurons can transmit nerve impulses and thus serve as lines of _____. [p.75]
15. Neuron branch processes called _____ pick up incoming chemical messages; others called _____ conduct outgoing messages. [p.75]
16. A(n) _____ is a cluster of processes from several neurons. [p.75]
17. State the general functions of tight junctions, adhering junctions, and gap junctions. [p.78]
18. Various types of thin, sheetlike _____ cover or line organs and may be of the mucous or serous type. [p.79]
19. List the cavities of the human body and generally name the organs found in each. [p.80]
20. List each of the eleven principal organ systems in humans and match each to its main task. [p.81]
21. Explain how keratin in the epidermis protects the rest of your body. [p.82]
22. Describe the two-layered structure of human skin and identify the structures located in each layer. [pp.82–83]
23. Describe the ways by which extracellular fluid helps cells survive. [p.84]
24. Describe the relationships among sensory receptors, integrators, and effectors in a negative feedback system. [pp.84–85]
25. Describe a positive feedback system and give an example. [p.85]

Media Menu Review Questions

Questions 1–5 are drawn from the following InfoTrac College Edition article: "The Human Body Shop." Doug Garr. *Technology Review,* April 2001.

1. Designing and growing living replacement body parts is called _____.
2. Within _____ years, replacement bladders and ventricles will quite possibly be available.
3. _____ needed to supply laboratory-grown organs can be grown artificially using computer chip etching techniques.
4. The artificially grown organ that is almost to the stage of human trials is the _____.
5. A three-dimensional printer may be useful in building complex organs like livers because it builds complex structure one _____ at a time.

Question 6 is drawn from the following InfoTrac College Edition article: "Spring in Your Step? The Forces in Cartilage." *Science News,* July 6, 2002.

6. The ability of molecules to resist compression is being studied in order to understand how diseases affect the resilience of _____.

Questions 7–10 are drawn from the following InfoTrac College Edition article: "Researchers Create Functioning Heart Cells in the Lab." *Blood Weekly*, August 16, 2001.

7. To understand human heart disorders, the stem cells of _____ are being mutated to simulate human problems.

8. Stem cell therapy may be useful in repairing and preventing reclogging of blood vessels subjected to balloon _____.

9. Stem cells could be designed to be sensitive to a certain _____ so that they would die on command if something went wrong.

10. The number of stem cells a person has decreases about _____ percent for each decade of life.

Integrating and Applying Key Concepts

The condition known as osteoporosis primarily affects postmenopausal women. It involves a reduction in the deposition of calcium into the body's bones. Explain how this affects the functions of bones.

5

THE SKELETAL SYSTEM

Interactive Exercises

Impacts, Issues: Hold the Hype [p.89]

5.1. BONE—MINERALIZED CONNECTIVE TISSUE [pp.90–91]

5.2. THE SKELETON: THE BODY'S BONY FRAMEWORK [pp.92–93]

For additional practice, use the interactive vocabulary exercises linked with your BiologyNow CD-ROM.

Selected Words: osteoblasts [p.90], lacunae [p.90], *epiphyseal plate* [p.90], osteoclasts [p.91], *osteoporosis* [p.91], "irregular" bone [p.92]

Boldfaced, Page-Referenced Terms

[p.89] skeletal system _____

_____

[p.90] osteocytes _____

_____

[p.90] compact bone _____

_____

[p.90] osteon _____

_____

[p.90] spongy bone _____

_____

[p.91] bone remodeling _____

_____

[p.92] bone marrow _____

_____

[p.92] axial skeleton _____

_____

[p.92] appendicular skeleton _____

_____

[p.92] ligaments _____

_____

[p.92] tendons _____

_____

Short Answer

1. Name the two kinds of bone tissue and explain how they differ. [p.90]

_____

_____

_____

Sequence

2. Indicate the order of the events that occur in compact bone formation by writing a "1" beside the first event, a "2" by the second, and so on.

_____ Osteoblasts mature into osteocytes within a developing osteon, and mineralization proceeds.

_____ A cartilage "model" of a bone develops in an embryo.

_____ Growth continues in long bones until the epiphyseal plates are replaced by bone.

_____ Osteoblasts deposit a matrix of collagen along with other components.

Matching

Write the letter of the matching definition by the terms below.

3. _____ marrow cavity [p.90]

4. _____ osteoblast [p.90]

5. _____ phosphate [p.91]

6. _____ human growth hormone [p.90]

7. _____ spongy bone [p.90]

8. _____ compact bone [p.90]

9. _____ osteoclast [p.91]

10. _____ osteoporosis [p.91]

11. _____ osteocyte [p.90]

12. _____ lacunae [p.90]

13. _____ collagen fibers [p.90]

14. _____ bone remodeling [p.91]

15. _____ osteon [p.90]

16. _____ periosteum [p.90]

17. _____ epiphysis [p.90]

18. _____ epiphyseal plate [p.90]

19. _____ mechanical stress [p.91]

A. constant and dynamic depositing and withdrawing of minerals in bone
B. flaring end of a long bone
C. delicate-looking bone inside long bones
D. cell that secretes collagen in developing bone
E. mature bone cell
F. mineral found in bone along with calcium
G. forms the shaft and ends of a long bone cavity inside a long bone
H. membrane that covers a long bone
I. triggers production of denser, stronger bones
J. progressive deterioration of bone, especially in women past menopause
K. cartilage plate that separates the epiphysis from the bone shaft to allow growth
L. layered compact bone with canal in center (Haversian system)
M. connective tissue fibers in compact bone that withstand mechanical stress
N. triggers the development of epiphyseal plates and maintains them
O. spaces in developing bone where osteoblasts are found
P. secretes enzymes that break down the organic matrix of bone
Q. space within a bone where bone marrow is located

Labeling

Identify each indicated part of the accompanying illustrations. [pp.90,92]

20. _____ _____

21. _____ _____

22. _____ _____

23. _____ _____

24. _____ _____ _____

25. _____

26. _____ _____

27. _____

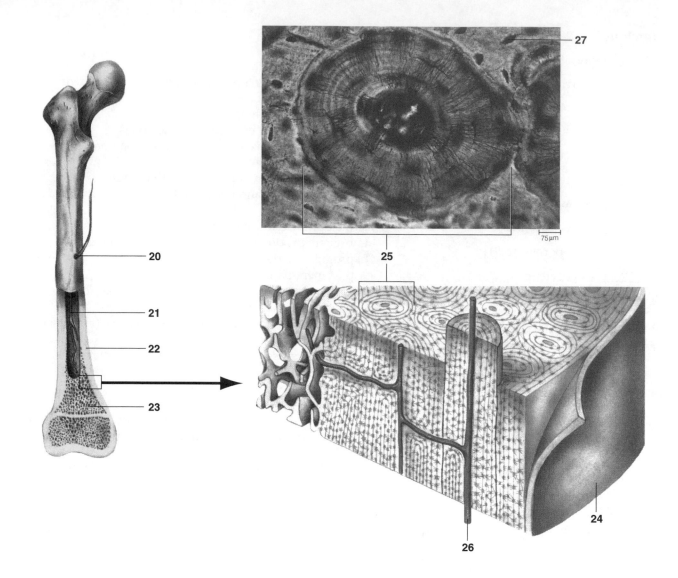

Dichotomous Choice

Circle one of the two possible answers given between parentheses in each statement.

28. The long bones of an adult contain mainly (red marrow/yellow marrow). [p.92]

29. In an adult, most blood cells are formed in (long/irregular) bone. [p.92]

30. Bones are connected at joints by (tendons/ligaments). [p.92]

31. Muscles are held to bone or to each other by (tendons/ligaments). [p.92]

Short Answer

32. List the five functions of bone. [p.92]

_____

_____

_____

Identification

Write the scientific name of the numbered bones on this page in the blanks provided. [p.93]

33. _____ _____

34. _____

35. _____

36. _____

37. _____

38. _____

39. _____

40. _____

41. _____

42. _____

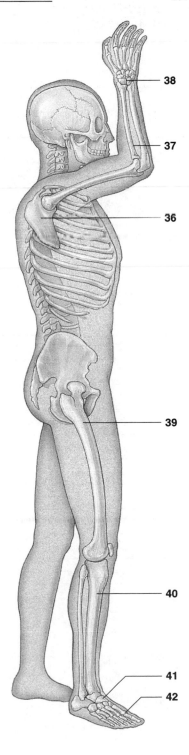

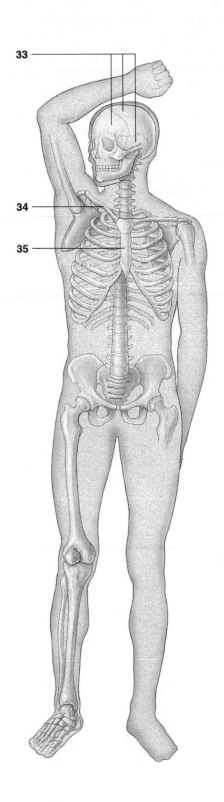

5.3. THE AXIAL SKELETON [pp.94–95]

5.4. THE APPENDICULAR SKELETON [pp.96–97]

Selected Words: *frontal bone* [p.94], *sinusitis* [p.94], *temporal bones* [p.94], *sphenoid bone* [p.94], *ethmoid bone* [p.94], *parietal bones* [p.94], *occipital bone* [p.94], *foramen magnum* [p.94], *maxillary bones* [p.94], *zygomatic bones* [p.94], *lacrimal bone* [p.94], *palatine bones* [p.94], *vomer bone* [p.94], *cervical* vertebrae [p.95], *thoracic* vertebrae [p.95], *lumbar* vertebrae [p.95], sacrum [p.95], coccyx [p.95], *herniate* [p.95], *carpal* bones [p.96], *carpal tunnel syndrome* [p.96], *metacarpals* [p.96], *phalanges* [p.96], *coxal bones* [p.97], *pubic arch* [p.97], *iliac* [p.97], *pelvis* [p.97], *tibia* [p.97], *fibula* [p.97], patella [p.97], *tarsal* [p.97], *metatarsals* [p.97]

Boldfaced, Page-Referenced Terms

[p.94] brain case _____

_____

[p.94] sinuses _____

_____

[p.94] mandible _____

_____

[p.95] intervertebral disks _____

_____

[p.95] ribs _____

_____

[p.95] sternum _____

_____

[p.96] pectoral girdle _____

_____

[p.96] scapula _____

_____

[p.96] clavicle _____

_____

[p.96] humerus _____

_____

[p.96] radius _____

_____

[p.96] ulna _____

_____

[p.97] pelvic girdle _____

_____

[p.97] femur _____

_____

Identification

1. _____ bone [p.94]
2. _____ bone [p.94]
3. _____ bone [p.94]
4. _____ bone [p.94]
5. _____ [p.94]
6. _____ [p.94]
7. _____ bone [p.94]
8. _____ bone [p.94]
9. _____ _____ [p.94]
10. _____ vertebrae [p.95]
11. _____ vertebrae [p.95]
12. _____ vertebrae [p.95]
13. _____ (vertebrae) [p.95]
14. _____ (vertebrae) [p.95]

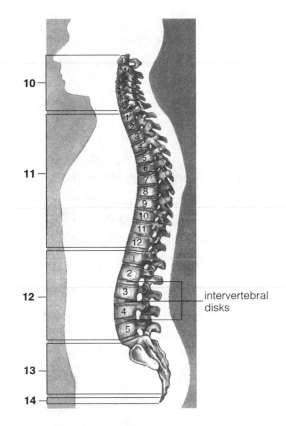

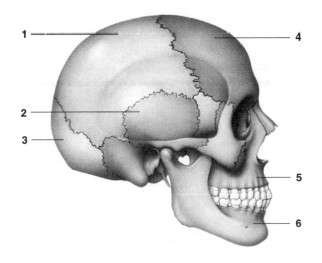

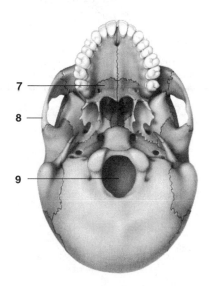

Dichotomous Choice

Circle one of the two possible answers given between parentheses in each statement.

15. The (appendicular/axial) skeleton includes the skull, vertebral column, ribs, and sternum. [p.94]

16. The frontal bone contains air spaces called (foramina/sinuses) lined with mucous membrane. [p.94]

17. (Occipital/Temporal) bones surround the ear canals. [p.94]

18. The inner eye socket is formed by the (sphenoid/parietal) bone. [p.94]

19. The ethmoid bone helps support the (jaw/nose). [p.94]

20. The parietal bones form a large part of the skull, as does the (zygomatic/occipital) bone. [p.94]

21. The lower jaw is the (maxilla/mandible). [p.94]

22. The (zygomatic/temporal) bones form the "cheekbones" and outer eye sockets. [p.94]

23. Tear ducts pass between the maxillary bones and the (palatine/lacrimal) bones. [p.94]

24. The hard palate, or "roof" of the mouth is formed by extensions of the palatine bones, together with the (parietal/maxillary) bones. [p.94]

25. The nasal cavity is divided into two sections by the thin nasal septum, formed partially by the (mandible/vomer) bone. [p.94]

26. Humans have seven (cervical/thoracic) vertebrae. [p.95]

27. The thoracic vertebrae are located in the (neck/chest). [p.95]

28. The (lumbar/sacral) vertebrae are in the lower back. [p.95]

29. The "tailbone" is properly called the (sacrum/coccyx). [p.95]

30. About one-fourth of the length of the human vertebral column consists of tough, compressible (intervertebral disks/lumbar vertebrae). [p.95]

31. A "slipped" disk is said to be (herniated/ruptured). [p.95]

32. The "rib cage" consists of twelve pairs of ribs and the paddle-shaped (sternum/scapula). [p.95]

33. The humerus is a part of the (pelvic/pectoral) girdle. [p.96]

34. The humerus joins the radius and ulna, which join eight small wristbones known as (carpals/phalanges). [p.96]

35. "Hipbones" are the upper iliac regions of the (pubic/coxal) bones. [p.97]

36. The legs include the femur, tibia, and (shin/fibula). [p.97]

37. The ankle is made up of (tarsal/metatarsal) bones. [p.97]

38. The foot contains five long bones called (phalanges/metatarsals). [p.97]

5.5. JOINTS—CONNECTIONS BETWEEN BONES [pp.98–99]

5.6. SKELETAL DISEASES, DISORDERS, AND INJURIES [pp.100–101]

5.7. REPLACING JOINTS [p.101]

Selected Words: joints [p.98], synovial fluid [p.98], sutures [p.98], osteoarthritis [p.100], rheumatoid arthritis [p.100], tendinitis [p.100], carpal tunnel syndrome [p.100], strain [p.100], sprain [p.100], prosthesis [p.101]

Boldfaced, Page-Referenced Terms

[p.98] synovial joint _____

_____

[p.98] cartilaginous joint _____

_____

[p.98] fibrous joint _____

_____

[p.99] flexion _____

_____

[p.99] extension _____

_____

[p.99] circumduction _____

_____

[p.99] rotation _____

_____

[p.99] abduction _____

_____

[p.99] adduction _____

_____

[p.99] supination _____

_____

[p.99] pronation _____

_____

Fill-in-the-Blanks

(1) _____ [p.98] joints are freely movable and are lubricated by (2) _____ [p.98] fluid secreted into a capsule made of dense connective tissue that surrounds the bones of the joint. Synovial joints include (3) _____ [p.98] joints such as the knee and elbow, as well as (4) _____-_____-_____ [p.98] joints such as the hip. (5) _____ [p.98] joints such as

between vertebrae allow only slight movement. An adult's (6) _____ [p.98] joints, such as those holding teeth in their sockets, don't allow movement.

Identification

7. _____ [p.99]

8. _____ [p.99]

9. _____ [p.99]

10. _____ [p.99]

11. _____ [p.99]

12. _____ [p.99]

13. _____ [p.99]

14. _____ [p.99]

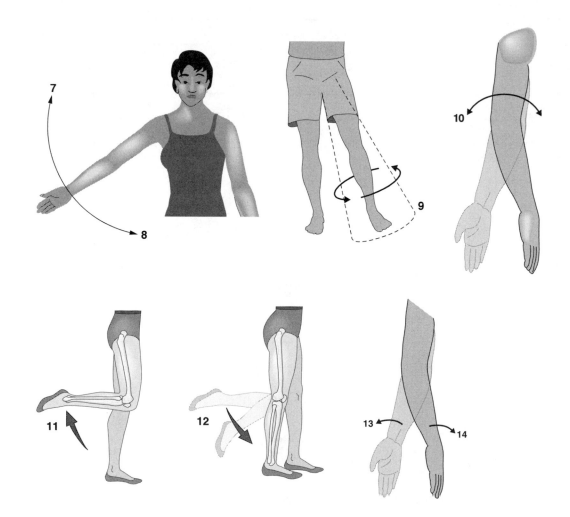

Fill-in-the-Blanks

As a person ages, the (15) _____ [p.100] covering the ends of bones at freely movable joints may wear away, resulting in (16) _____ [p.100]. When a person's immune system attacks tissues in a joint and causes the synovial membrane to thicken and cartilage to erode away, the degenerative condition (17) _____ _____ [p.100] results. (18) _____ [p.100] develops when tendons and synovial membranes around joints become inflamed due to repeated movements. One of the most common repetitive motion injuries is (19) _____ _____ _____ [p.100]. Twisting a joint too far causes a (20) _____ [p.100]. Tearing a ligament or tendon leads to a (21) _____ [p.100]. When a joint is (22) _____ [p.100], the two bones will no longer be in contact. A (23) _____ [p.101] fracture is just a crack in a bone. In a (24) _____ [p.101] fracture, the bone is completely separated into two pieces. A (25) _____ [p.101] fracture involves broken ends of bone puncturing the skin. When a joint such as a hip or knee is seriously damaged, it can be replaced with an artificial joint called a(n) (26) _____ [p.101].

Self-Quiz

Are you ready for the exam? Test yourself on key concepts by taking the additional tests linked with your BiologyNow CD-ROM.

Choice

For questions 1–10, choose from the following answers:

a. carpals
b. metatarsals
c. femur
d. humerus

e. ulna
f. tarsals
g. radius
h. fibula

i. phalanges (used twice)
j. metacarpals
k. tibia

The upper arm bone (1. _____ [p.96]) is connected to the lower arm bones (2. _____ and _____ [p.96]). The lower arm bones are connected to the wristbones (3. _____ [p.96]). The wristbones are connected to the hand bones (4. _____ [p.96]). The hand bones are connected to the finger bones (5. _____ [p.96]).

The upper leg bone (6. _____ [p.97]) is connected to the lower leg bones (7. _____ and _____ [p.97]). The lower leg bones are connected to the anklebones (8. _____ [p.97]). The anklebones are connected to the foot bones (9. _____ [p.97]). The foot bones are connected to the toe bones (10. _____ [p.97]).

Choice

For questions 11–15, choose from the following answers:

a. ligaments
b. osteoblasts
c. osteoclasts
d. bone marrow
e. tendons
f. osteon

11. _____ Secrete bone-dissolving enzymes [p.91]

12. _____ Major site of blood cell formation [p.92]

13. _____ Secrete matrix that will become bone [p.90]

14. _____ Compact bone surrounding central canal [p.90]

15. _____ Attach bone to bone [p.92]

For questions 16–19, choose from the following answers:

a. synovial   b. cartilaginous   c. fibrous

16. _____ Nonmoving joints (sutures) between skull bones [p.98]

17. _____ Hips and shoulders [p.98]

18. _____ Knees and elbows [p.98]

19. _____ Connections of some ribs to the breastbone [p.98]

Chapter Objectives/Review Questions

This section lists general and detailed chapter objectives that can be used as review questions. You can make maximum use of these items by writing answers on a separate sheet of paper. To check for accuracy, compare your answers with information given in the chapter or glossary.

1. Explain the various roles of osteoblasts, osteoclasts, cartilage models, long bones, and epiphyseal plates in the development of human bones. [pp.90–91]
2. Identify the human bones by name and location, including the bones of the skull, the rib cage, the vertebral column, the pectoral girdle and upper appendages, and the pelvic girdle and lower appendages. [p.93] Which of these are parts of the axial skeleton? [pp.94–95] Which are parts of the appendicular skeleton? [pp.96–97]
3. Give an example of each of these synovial joint movements: flexion, extension, circumduction, rotation, abduction, adduction, supination, pronation. [p.99]
4. Explain the difference between tendinitis and a sprain. [p.100]

Media Menu Review Questions

Questions 1–4 are drawn from the following InfoTrac College Edition article: "Use Them or Lose Them: Exercise to Keep Bones Strong." *Environmental Nutrition*, March 2002.

1. For bones to stay strong, demands must be placed on them with both _____ and _____ exercise.
2. In _____ exercise, bones and muscles work against gravity.
3. _____ exercise usually involves weight lifting.
4. Ideally, you should aim to exercise at least _____ minutes a day.

Questions 5–10 are drawn from the following InfoTrac College Edition article: "The Age of Arthritis: We're Headed for an Epidemic of Joint Disease. What You Can Do to Protect Yourself." *Time*, December 9, 2002.

5. _____ is a degenerative disorder in which the cartilage that acts as a cushion in the joints breaks down.
6. Sometime between ages 40 and 55, the activity of cartilage-building cells called _____ starts slowing down.
7. Bjorn Olsen's Harvard research group has identified at least three _____ variations that make people more susceptible to arthritis.
8. When the body's immune system becomes involved in arthritis, the _____ process causes swelling and more damage.
9. The problem with replacing cartilage in joints affected by arthritis is that the new cartilage is subject to the same _____ forces as the original.
10. In arthritis, no matter what the complications are, they will be magnified a hundred times without _____.

Integrating and Applying Key Concepts

Explain why a fracture at an epiphyseal plate in a child might be more serious than a fracture in the center of a long bone.

6

THE MUSCULAR SYSTEM

Interactive Exercises

Impacts, Issues: Pumping Up Muscles [p.105]

6.1. THE STRUCTURE AND FUNCTIONING OF SKELETAL MUSCLES [pp.106–107]

6.2. HOW MUSCLES CONTRACT [pp.108–109]

For additional practice, use the interactive vocabulary exercises linked with your BiologyNow CD-ROM.

Selected Words: "muscle fiber" [p.106], *reciprocal innervation* [pp.106–107]

Boldfaced, Page-Referenced Terms

[p.105] muscular system _____

_____

[p.106] skeletal muscle _____

_____

[p.106] myofibrils _____

_____

[p.106] tendon _____

_____

[p.106] origin _____

_____

[p.106] insertion _____

_____

[p.108] sarcomere _____

_____

[p.108] actin _____

_____

[p.108] myosin _____

_____

[p.108] sliding-filament model _____

_____

Short Answer

1. Of the three muscle types, only skeletal muscle is included in the "muscular system." Why is this type of muscle called "skeletal"? What are its main jobs in the body? [p.106]

_____

_____

Dichotomous Choice

Circle one of two possible answers given between parentheses in each statement.

2. The end of the muscle that is attached to a bone that remains fairly motionless during a movement is the (origin/insertion). [p.106]

3. The end that is attached to the bone that moves the most is the (origin/insertion). [p.106]

4. When a skeletal muscle contracts, it (pushes/pulls) on the bones to which it is attached. [p.106]

Fill-in-the-Blanks

Muscle cells, also called muscle (5) _____ [p.106], are bundled together into groups by

(6) _____ [p.106] tissue. A (7) _____ [p.106] is a cord of dense connective tissue that

extends from a group of bundled muscle cells and attaches the muscle to a (8) _____ [p.106].

Many muscles are arranged as (9) _____ [p.106] or groups. While some muscles work syner-

gistically, others work antagonistically or in (10) _____ [p.106]. For example, if the biceps muscle

contracts, the triceps will (11) _____ [p.106] and the arm will (12) _____ [p.106]. If the

triceps muscle contracts, the biceps will (13) _____ [p.106] and the arm will (14) _____

[p.106].

Identification

Name each numbered muscle on the accompanying illustration. In the parentheses provided, write the letter of its specific movement. [p.107]

15. _____ (   )

16. _____ (   )

17. _____ _____ (   )

18. _____ _____ (   )

19. _____ (   )

20. _____ _____ (   )

21. _____ _____ (   )

22. _____ (   )

23. _____ _____ (   )

24. _____ _____ (   )

A. Bends lower leg at the knee when walking, extends the foot when jumping

B. Flexes the foot toward the shin

C. Raises the arm

D. Depresses the thoracic cavity, compresses the abdomen, bends the backbone

E. Bends the thigh at the hip, bends lower leg at the knee, rotates thigh in an outward direction

F. Draws the arm forward and in toward the body

G. Extends and rotates thigh outward when walking, running, and climbing

H. Lifts the shoulder blade, braces the shoulder, draws the head back

I. Flexes the thigh at the hips, extends the leg at the knee

J. Draws thigh backward, bends the knee

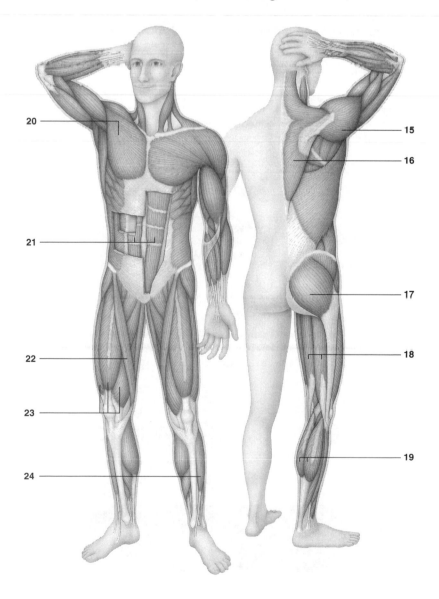

Fill-in-the-Blanks

A skeletal muscle contains bundles of muscle cells that run (25) _____ [p.108] within the muscle. Within a muscle cell, each myofibril is divided into (26) _____ [p.108]. Because this pattern is repeated in all myofibrils, a skeletal muscle cell looks (27) _____ [p.108]. The dark lines of each myofibril are (28) _____ [p.108] bands. They mark the ends of each (29) _____ [p.108], the basic unit of contraction. The thin filaments within a sarcomere are twisted strands of (30) _____ [p.108], a contractile protein. These are attached at one end to a Z band.

The thick filaments are bundled molecules of the protein (31) _____ [p.108] with heads projecting outward. According to the (32) _____ _____ [p.108] model, a contraction occurs when (33) _____ [p.108] heads connect to a nearby (34) _____ [p.108] filament and pull the filament toward the middle of the sarcomere. Because the actin is attached to a Z band, this action shrinks the width of the (35) _____ [p.108]. Energy for the contraction is provided by (36) _____ [p.108].

Labeling

Label the numbered parts of the accompanying illustrations. [pp.106,108]

37. _____ _____   40. _____

38. _____ _____ _____   41. _____

39. _____ _____   42. _____ _____

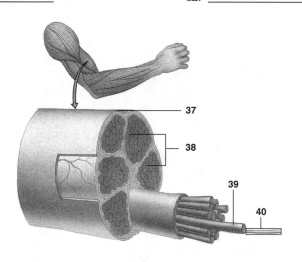

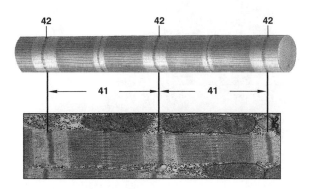

6.3. ENERGY FOR MUSCLE CELLS [p.109]

6.4. HOW THE NERVOUS SYSTEM CONTROLS MUSCLE CONTRACTION [pp.110–111]

6.5. PROPERTIES OF WHOLE MUSCLES [pp.112–113]

6.6. MUSCLE MATTERS [pp.114–115]

Selected Words: axons [p.111], *synapse* [p.111], *neurotransmitter* [p.111], *isometrically* [p.112], *isotonically* [p.112], *lengthening* contraction [p.112], *tetanus* [p.113], "slow" or "red" muscle [p.112], "fast" or "white" muscle [p.112], *atrophy* [p.114], *aerobic exercise* [p.114], *strength training* [p.114]

Boldfaced, Page-Referenced Terms

[p.109] creatine phosphate _____

_____

[p.109] muscle fatigue _____

_____

[p.109] oxygen debt _____

_____

[p.110] T tubules _____

_____

[p.110] sarcoplasmic reticulum _____

_____

[p.111] neuromuscular junctions _____

_____

[p.112] muscle tension _____

_____

[p.112] motor unit _____

_____

[p.112] muscle twitch _____

_____

[p.112] temporal summation _____

_____

[p.113] tetany _____

_____

[p.113] all-or-none principle _____

_____

[p.113] muscle tone _____

_____

[p.114] exercise _____

_____

Short Answer

1. Explain the function of creatine phosphate in muscle cells. [p.109]

   _____

   _____

   _____

2. During prolonged, moderate exercise, what does the muscle cell use as a source of energy during the first five to ten minutes? What source then becomes available that lasts for the next half hour? What source of energy is used after that? [p.109]

   _____

   _____

   _____

3. During hard exercise, why must muscle cells depend only on the small amount of ATP made by glycolysis? [p.109]

   _____

   _____

4. What is muscle fatigue? What probably causes it? [p.109]

   _____

   _____

   _____

Fill-in-the-Blanks

The nervous system delivers orders to muscles by way of (5) _____ [p.110] neurons. Neural

signals spread across the plasma membrane of a muscle cell, reaching tubelike extensions called

(6) _____ _____ [p.110]. These are entrances into the (7) _____ _____

[p.110], which is modified endoplasmic reticulum. An incoming nerve impulse triggers the release of

(8) _____ [p.110] ions from the SR. These ions diffuse into myofibrils and bind to

(9) _____ [p.110] on the surface of (10) _____ [p.110] filaments and expose binding sites.

This allows (11) _____ _____-_____ [p.110] to attach to the exposed sites and

cause a contraction. The muscle cell relaxes when the nerve impulse ends and (12) _____ [p.110]

is actively transported back into the SR. The actin binding site is covered again so that (13) _____

[p.110] can no longer form cross-bridges.

A(n) (14) _____ _____ [p.111] is a place where motor nerve endings called (15) _____ [p.111] abut a muscle cell membrane. The slight gap between the endings and the muscle cell membrane is called a(n) (16) _____ [p.111]. The nervous message is sent across this gap by way of a (17) _____ [p.111] called ACh (acetylcholine). When ACh binds to receptors on the muscle cell membrane, it sets in effect the events that cause the muscle cell to (18) _____ [p.111].

Labeling

Identify the numbered parts of the accompanying illustration. [p.110]

19. _____ _____

20. _____ _____

21. _____ _____

22. _____ _____

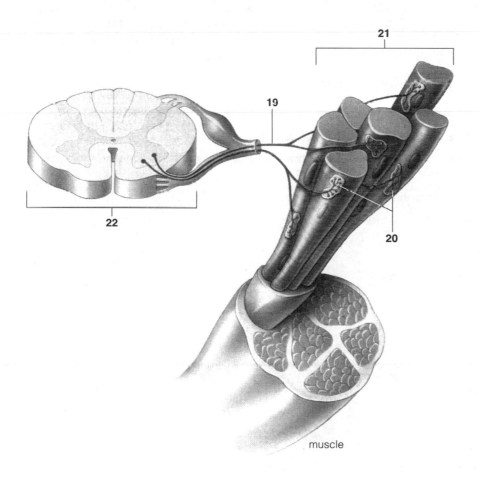

muscle

Matching

Match each of the following terms with the appropriate phrase or description.

23. _____ motor unit [p.112]

24. _____ all-or-none principle [p.113]

25. _____ temporal summation [p.112]

26. _____ "red" or "slow" muscle [p.113]

27. _____ isometric contraction [p.112]

28. _____ isotonic lengthening [p.112]

29. _____ isotonic contraction [p.112]

30. _____ "white" or "fast" muscle [p.113]

31. _____ tetany [p.113]

32. _____ muscle tension [p.112]

33. _____ muscle tone [p.113]

34. _____ tetanus [p.113]

35. _____ muscle twitch [p.112]

A. steady, low-level contracted state that helps stabilize joints and maintain muscle health
B. develops tension but does not shorten
C. the muscle cells controlled by a given motor neuron
D. muscles remain contracted due to a bacterial toxin
E. effect of new contraction is added to contraction already under way
F. capable of fast, powerful contraction but cannot sustain contraction for long
G. mechanical force that a contracting muscle exerts on an object such as a bone
H. a muscle cell contracts fully or not at all
I. muscle shortens and moves a load
J. normal operation of muscles near or at maximum temporal summation
K. increase, peak, and declining of tension
L. muscle lengthens during contraction
M. contain lots of myoglobin; capable of sustained contraction

Dichotomous Choice

Circle one of two possible answers given between parentheses in each statement.

36. Atrophy refers to a muscle (getting stronger/wasting away). [p.114]

37. Physical activity done at a rate at which the muscles are well supplied with oxygen is (aerobic/strength-training) exercise. [p.114]

38. Intense, short-duration exercise is (aerobic/strength-training) exercise. [p.114]

39. Starting at about age 30, muscle (circulation/tension) begins to decrease. [p.114]

Short Answer

40. List three benefits of aerobic exercise. [p.114]

_____

_____

_____

41. Strength training produces muscles that are large and strong. What do muscles produced in this way lack? [p.114]

_____

_____

42. Because their structure is similar to that of testosterone, anabolic steroids have undesirable effects. List some of these in both males and females. [pp.114–115]

_____

_____

_____

Self-Quiz

Choice

For questions 1–5, choose from the following answers:

a. isometric contraction(s)
b. cross-bridge formation
c. the sliding-filament model
d. reciprocal innervation
e. isotonic contraction(s)

1. _____ Helps coordinate the contraction and relaxation of antagonistic pairs of muscles [pp.107–108]

2. _____ Muscle shortens and moves a load (e.g., walking downstairs) [p.112]

3. _____ Muscle develops tension; does not shorten during contraction (e.g., holding a glass of lemonade in one position) [p.112]

4. _____ Process assisted by calcium ions and ATP [p.108]

5. _____ Explains how myosin filaments move to the center of a sarcomere and back [p.108]

For questions 6–9, choose from the following answers:

a. actin
b. myofibril
c. myosin
d. sarcomere
e. sarcoplasmic reticulum

6. _____ Each consists of many repetitive units of muscle contraction; many found in a muscle cell [p.108]

7. _____ Basic unit of muscle contraction [p.108]

8. _____ Thin filaments within a sarcomere, attached to Z bands [p.108]

9. _____ Stores calcium ions and releases them in response to neural signals [p.110]

Chapter Objectives/Review Questions

This section lists general and detailed chapter objectives that can be used as review questions. You can make maximum use of these items by writing answers on a separate sheet of paper. To check for accuracy, compare your answers with information given in the chapter or glossary.

10. Define *reciprocal innervation*, and explain how it helps coordinate motor elements. [pp.107–108]
11. Refer to Figure 6.3 of your main text and name (a) a muscle used in sit-ups, (b) another used in dorsally flexing the foot, and (c) another used in flexing the elbow joint. [p.107]
12. Describe the fine structure of a muscle cell, using the terms *myofibril, sarcomere, actin,* and *myosin.* [p.108]
13. Explain what occurs in skeletal muscle cells when there is not enough oxygen to allow ATP production by aerobic respiration. When does muscle fatigue occur and why? [p.109]
14. List, in sequence, the biochemical and mechanical events that occur following an incoming neural signal to contract, and explain how the cell relaxes. [p.110]
15. Distinguish a twitch contraction from tetany, using the term *temporal summation* [pp.112–113]
16. Name two natural substances that are used by athletes to increase muscle mass and strength, and explain how each works. [p.115]

Media Menu Review Questions

Questions 1–6 are drawn from the following InfoTrac College Edition article: "Creatine Use by Athletes Overshadow Potential to Avoid Muscular Disorder." *Environmental Nutrition,* October 1999.

1. Creatine is an _____ made by the body and stored predominantly in skeletal _____.
2. Creatine serves as a reservoir to replenish adenosine triphosphate (_____).
3. Creatine supplements have been shown to enhance performance in activities requiring _____ bursts of _____ energy (but not in endurance sports).
4. Creatine may be beneficial in treating muscular disorders such as amyotrophic lateral sclerosis (_____'s disease) and muscular _____.
5. Because there is little data on potential damage from long-term use, you should consult your _____ before using creatine, and drink plenty of fluids to avoid "_____."
6. Taking more creatine than what is recommended only results in the excess being _____ while raising the risk of _____.

Integrating and Applying Key Concepts

Explain how an individual's genetically determined amounts of "fast" versus "slow" muscle cells may influence success at sprinting versus long-distance athletic events.

7

DIGESTION AND NUTRITION

Interactive Exercises

Impacts, Issues: Hips and Hunger [p.119]

7.1. THE DIGESTIVE SYSTEM: AN OVERVIEW [pp.120–121]

7.2. CHEWING AND SWALLOWING: FOOD PROCESSING BEGINS [pp.122–123]

7.3. THE STOMACH: FOOD STORAGE, DIGESTION, AND MORE [p.124]

For additional practice, use the interactive vocabulary exercises linked with your BiologyNow CD-ROM.

Selected Words: *lumen* [p.120], *mucosa* [p.121], *submucosa* [p.121], *smooth muscle* [p.121], *serosa* [p.121], *sphincters* [p.121], "primary teeth" [p.122], *caries* [p.122], *gingivitis* [p.122], *periodontal membrane* [p.122], *parotid gland* [p.122], *submandibular glands* [p.122], *sublingual glands* [p.122], *mucins* [p.122], *trachea* [p.122], swallowing reflex [p.123], *epiglottis* [p.123], *intrinsic factor* [p.124], "heartburn" [p.124], *gastrin* [p.124], "gastric mucosal barrier" [p.124], *Helicobacter pylori* [p.124], *peptic ulcer* [p.124], *rugae* [p.124]

Boldfaced, Page-Referenced Terms

[p.120] digestive system _____

_____

[p.120] gastrointestinal (GI) tract _____

_____

[p.121] mechanical processing _____

_____

[p.121] secretion _____

_____

[p.121] digestion _____

_____

[p.121] absorption _____

_____

[p.121] elimination _____

_____

[p.122] oral cavity _____

_____

[p.122] salivary glands _____

_____

[p.122] salivary amylase _____

_____

[p.122] bolus _____

_____

[p.122] palate _____

_____

[p.122] pharynx _____

_____

[p.122] esophagus _____

_____

[p.123] stomach _____

_____

[p.123] pepsins _____

[p.123] gastric fluid _____

[p.123] chyme _____

Labeling

Identify the numbered parts of the accompanying illustration. [p.120]

1. _____ _____
2. _____ _____
3. _____
4. _____ _____
5. _____
6. _____ _____
7. _____
8. _____
9. _____
10. _____
11. _____

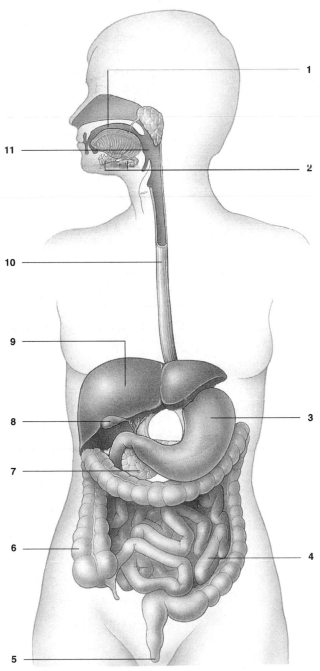

Complete the Table

12. Complete the following table by naming the digestive system components described. [p.120]

Organ	Main Functions
a.	Food is moistened and chewed; polysaccharide digestion starts
b.	Secrete saliva containing digestive enzymes, buffers, and mucus
c.	Passageway to both the tubular part of the digestive system and to the respiratory system; moves food forward by contracting sequentially
d.	Muscular tube, moistened by saliva, that moves food from the pharynx to the stomach
e.	Stores food; kills many microorganisms; starts protein digestion
f.	Digests and absorbs most nutrients
g.	Secretes enzymes that break down all major food molecules; produces buffers against hydrochloric acid from stomach
h.	Secretes bile for fat emulsification; roles in metabolism of carbohydrates, fats, and proteins
i.	Stores and concentrates bile produced by liver
j.	Concentrates and stores undigested matter by absorbing ions and water
k.	Distension stimulates expulsion of feces
l.	Terminal opening for expelling feces

Short Answer

13. List the five tasks that the various parts of the human digestive system accomplish by working together. [p.121]

_____

_____

_____

14. Name each of the four basic layers of the digestive tube and tell what each contributes to help digest food. [p.121]

_____

_____

_____

_____

Matching

15. _____ bolus [p.122]

16. _____ caries [p.122]

17. _____ esophagus [p.122]

18. _____ gingivitis [p.122]

19. _____ oral cavity [p.122]

20. _____ palate [p.122]

21. _____ periodontal disease [p.122]

22. _____ pharynx [p.122]

23. _____ salivary amylase [p.122]

24. _____ parotid, submandibular, and sublingual [p.122]

A. Destruction of the bone around the teeth that loosens the teeth
B. A muscular tube leading from the pharynx to the stomach
C. A softened, lubricated ball of food formed in the mouth
D. A term synonymous with *throat*
E. Tooth decay caused by bacteria living on food residues in the mouth
F. Three pairs of saliva-producing glands in the vicinity of the ears, lower jaw, and under the tongue
G. An enzyme that breaks down starch
H. A term synonymous with *mouth*
I. An inflammation of the gums caused by bacterial infection
J. Hardened roof of the mouth providing a hard surface against which the tongue can press food as it mixes with saliva

Dichotomous Choice

Circle one of two possible answers given between parentheses in each statement about the sequence of events called *swallowing*.

25. Swallowing begins when voluntary movements push a bolus into the (esophagus/pharynx), stimulating sensory receptors in its wall. [p.123]

26. The sensory receptors trigger (voluntary/involuntary) muscle contractions that prevent food from entering the nose and the trachea. [p.123]

27. The vocal cords stretch across the entrance to the larynx, and a flaplike valve, the (uvula/epiglottis) closes the opening to the respiratory tract while food moves into the esophagus. [p.123]

28. The (Heinrich/Heimlich) maneuver is an emergency procedure that can save a person choking on food. [p.123]

Fill-in-the-Blanks

The (29) _____ [p.124] is a muscular, expandable sac that stores food, helps break it down, and controls its passage into the (30) _____ _____ [p.124]. (31) _____ [p.124] fluid contains hydrochloric acid (HCl), mucus, and pepsinogens (precursors of digestive enzymes known as (32) _____ [p.124]). Gland cells in the stomach lining also secrete (33) _____ [p.124] factor, a protein necessary for vitamin B_{12} absorption in the small intestine. Combined with stomach contractions, the acidity in the stomach helps convert swallowed boluses into a thick mixture called (34) _____ [p.124]. (35) _____ [p.124] digestion begins in the stomach. The high acidity due to HCl secretion (36) _____ [p.124] proteins and exposes their peptide bonds. The acidity also converts inactive pepsinogens to active (37) _____ [p.124], which break the peptide bonds. The secretion of HCl and pepsinogen is stimulated by the hormone (38) _____ [p.124] that is secreted from gland cells. The (39) "_____ _____ _____" [p.124] of mucus and bicarbonate prevents the HCl and pepsin from breaking down the inner surface of the stomach lining. When the barrier is disrupted, a particular (40) _____ [p.124], *Helicobacter pylori*, may infect the stomach wall and cause the inner surface to break down, thus allowing hydrogen ions and pepsins to diffuse into the lining. The resulting open sore is a(n) (41) _____ _____ [p.124]. In the stomach, waves of contraction and relaxation called (42) _____ [p.124] mix the chyme and build up force as they approach the (43) _____ [p.124] sphincter. Strong contractions of the stomach wall (44) _____ [p.124] the sphincter. This squeezes most of the chyme back so that only a small amount moves into the (45) _____ _____ [p.124] with each contraction. From two to six hours later, the stomach is empty and its walls crumple into folds called (46) _____ [p.124]. Only (47) _____ [p.124] and a few other substances begin to be absorbed across the stomach wall.

7.4. THE SMALL INTESTINE: A HUGE SURFACE AREA FOR DIGESTION AND ABSORPTION [p.125]

7.5. HOW ARE NUTRIENTS DIGESTED AND ABSORBED? [pp.126–127]

7.6. THE MULTIPURPOSE LIVER [p.128]

7.7. THE LARGE INTESTINE [p.129]

7.8. DIGESTION CONTROLS AND DISRUPTIONS [p.130]

Selected Words: "brush border" [p.125], "pancreatic juice" [p.126], *micelle* [p.127], *gallstones* [p.128], *cecum* [p.129], *Escherichia coli* [p.129], "coliform bacteria" [p.129], *constipation* [p.129], *appendicitis* [p.129], *peritonitis* [p.129], *bulk* [p.129], *insoluble fiber* [p.129], *soluble* fiber [p.129], *diverticulitis* [p.125], *diarrhea* [p.129], *gastrin* [p.130], *somatostatin* [p.130], *secretin* [p.130], *cholecystokinin* [p.130], *glucose insulinotropic peptide* (GIP) [p.130], *cystic fibrosis* [p.130], *Crohn's disease* [p.130]

Boldfaced, Page-Referenced Terms

[p.125] villus (plural: villi) _____

_____

[p.125] microvillus _____

_____

[p.126] pancreas _____

_____

[p.126] liver _____

_____

[p.126] gallbladder _____

_____

[p.126] segmentation _____

_____

[p.127] lacteals _____

_____

[p.128] hepatic portal vein _____

_____

[p.129] large intestine _____

_____

[p.129] colon _____

_____

[p.129] rectum _____

_____

[p.129] anal canal _____

_____

[p.129] anus _____

_____

[p.129] appendix _____

_____

[p.129] malabsorption disorder _____

_____

Fill-in-the-Blanks

The key to the small intestine's ability to absorb nutrients is the structure of its (1) _____ [p.125]. The mucosa of the small intestine is densely folded, tremendously increasing the (2) _____ _____ [p.125] available for absorbing nutrients. The folds have smaller projections, about a millimeter long each, called (3) _____ [p.125]. The epithelial cells covering each villus usually have threadlike projections of the plasma membrane called (4) _____ [p.125].

The small intestine has three sections, the (5) _____ [p.126], the jejunum, and the ileum. Digestion in the small intestine depends on secretions from three accessory organs: the (6) _____ [p.126], the (7) _____ [p.126], and the (8) _____ [p.126]. A part of "pancreatic juice," the enzymes trypsin and chymotrypsin, digest the polypeptide chains of proteins into (9) _____ [p.126] fragments. These fragments are further degraded to (10) _____ _____ [p.126] by the enzymes carboxypeptidase and aminopeptidase. The pancreas also secretes the buffer (11) _____ [p.126], which helps neutralize HCl arriving from the stomach. This maintains a favorable environment in which pancreatic (12) _____ [p.126] can function. Besides enzymes, fat digestion requires (13) _____ [p.126] secreted by the liver and stored in the (14) _____ [p.126]. Bile salts speed up fat digestion by a process known as (15) _____ [p.126].

(16) _____ [p.126] moves nutrients, water, salts, and vitamins into the internal environment. The action of smooth muscle arranged in rings, referred to as (17) _____ [p.126], facilitates absorption. (18) _____ [p.127] proteins in the plasma membrane of brush border cells actively

move monosaccharides and amino acids across the intestinal lining. (19) _____ [p.127] salts assist the absorption of fatty acids and monoglycerides following lipid digestion by clumping together with cholesterol and other substances to form tiny droplets known as (20) _____ [p.127]. Nutrient molecules diffuse from the droplets into (21) _____ [p.127] cells. There, fatty acids and monoglycerides recombine into (22) _____ [p.127]. Then triglycerides combine with proteins into particles that leave cells by (23) _____ [p.127] and enter tissue fluid. Once glucose and amino acids are absorbed, they enter (24) _____ [p.127] vessels directly. The triglycerides enter lymph vessels called (25) _____ [p.127] that drain into blood vessels.

Dichotomous Choice

Circle one of two possible answers given between parentheses in each statement.

26. Nutrient-laden blood in intestinal villi flows to the (hepatic portal vein/hepatic vein). [p.128]

27. In the liver, excess glucose is taken up before a (hepatic portal vein/hepatic vein) returns the blood to the general circulation. [p.128]

28. The liver converts and stores much of the glucose to (glycogen/glucagon). [p.128]

29. The digestive role of the liver is to secrete (glycogen/bile). [p.128]

30. When there is no food moving through the GI tract, a sphincter closes off the main bile duct and bile backs up into the (small intestine/gallbladder) for temporary storage. [p.128]

31. Liver cells use (glycogen/cholesterol) to synthesize bile salts. [p.128]

32. Excess cholesterol that precipitates in the gallbladder forms (gallstones/urea). [p.128]

33. Toxic ammonia (NH_3) produced when cells break down amino acids is carried by the circulatory system to the liver, where it is converted to less toxic (glucose/urea) that is excreted in the urine. [p.128]

Matching

Choose the most appropriate description for each term.

34. _____ anal canal [p.129]

35. _____ appendicitis [p.129]

36. _____ appendix [p.129]

37. _____ bulk [p.129]

38. _____ cecum [p.129]

39. _____ constipation [p.129]

40. _____ diarrhea [p.129]

41. _____ *Escherichia coli* [p.129]

42. _____ feces [p.129]

43. _____ insoluble fiber [p.129]

44. _____ colon (or large intestine) [p.129]

A. Feces become dry and hard, and defecation becomes difficult
B. The volume of fiber and other undigested food material that cannot be decreased by absorption in the colon
C. Connects with the sigmoid colon; feces here move toward the outside of the body through the anal canal
D. A serious internal infection resulting from an infected appendix that bursts, releasing bacteria into the abdominal cavity
E. Plant carbohydrates that swell or dissolve in water
F. Receives material not absorbed in the small intestine
G. An irritated mucosal lining that secretes more water and salts than the large intestine can absorb
H. A slender projection from the cecum with no known digestive function
I. Feces within this structure are moving directly toward the outside of the body

45. _____ peritonitis [p.129]

46. _____ rectum [p.129]

47. _____ soluble fiber [p.129]

J. Normally inhabit intestines for nourishment but also furnish useful fatty acids and vitamins, such as vitamin K

K. Cellulose and other plant compounds that do not easily dissolve in water and cannot be digested by humans

L. A serious inflammation due to fecal obstruction of blood flow to the appendix

M. A mixture of undigested and unabsorbed food material, water, and bacteria

N. A blind pouch that is the beginning of the large intestine

Matching

Choose the most appropriate description for each term.

48. _____ cholecystokinin (CCK) [p.130]

49. _____ cystic fibrosis [p.130]

50. _____ Crohn's disease [p.130]

51. _____ gastrin [p.130]

52. _____ glucose insulinotropic peptide [p.130]

53. _____ malabsorption disorder [p.130]

54. _____ nervous system, local nerve network, and endocrine system [p.130]

55. _____ secretin [p.130]

56. _____ somatostatin [p.130]

A. Hormone that stimulates the pancreas to secrete bicarbonate

B. Gastrointestinal hormone released in response to the presence of glucose and fat in the small intestine; stimulates insulin release

C. Gastrointestinal hormone that inhibits acid secretion

D. Sufferers do not produce pancreatic enzymes necessary for normal digestion and absorption of fats and other nutrients

E. Disorder that causes severe damage to the intestinal lining

F. The digestive controls that act before food is absorbed

G. Gastrointestinal hormone released in response to fat in the small intestine; enhances the actions of secretin and stimulates gallbladder contractions

H. Hormone that stimulates the secretion of acid into the stomach in the presence of peptides and amino acids

I. Anything that interferes with the uptake of nutrients across the lining of the small intestine, such as lactose intolerance

7.9. NUTRIENT PROCESSING AFTER AND BETWEEN MEALS [p.131]

7.10. THE BODY'S NUTRITIONAL REQUIREMENTS [pp.132–133]

7.11. DIET ALTERNATIVES [p.134]

7.12. MALNUTRITION AND UNDERNUTRITION [p.135]

7.13. VITAMINS AND MINERALS [pp.136–137]

7.14. CALORIES COUNT: FOOD ENERGY AND BODY WEIGHT [pp.138–139]

Selected Words: complete proteins [p.132], *incomplete* proteins [p.132], "food group" approach [p.132], *Mediterranean diet* [p.132], "white foods" [p.134], *glycemic index* [p.134], *malnutrition* [p.135], *undernutrition* [p.135], starvation [p.135], kwashiorkor [p.135], marasmus [p.135], xerophthalmia [p.135], "fat epidemic" [p.138], *body mass index* (BMI) [p.138], *set point* [p.139], *anorexia nervosa* [p.139], *bulimia* [p.139]

Boldfaced, Page-Referenced Terms

[p.132] essential fatty acids _____

_____

[p.132] essential amino acids _____

_____

[p.132] food pyramid _____

_____

[p.136] vitamins _____

_____

[p.136] minerals _____

_____

[p.138] obesity _____

_____

[p.138] basal metabolic rate (BMR)_____

_____

[p.138] kilocalories (kcal) _____

_____

Dichotomous Choice

Circle one of two possible answers given between parentheses in each statement.

1. Living cells (store/break apart) most of their carbohydrates, lipids, and proteins, and use the products as energy sources or building blocks. [p.131]

2. Carbohydrates that are not required by the cell are transformed into (glycogen/fats) that are stored in adipose tissue. [p.131]

3. Cells first use (protein/glucose) as an energy source before tapping into stores of fat. [p.131]

Fill-in-the-Blanks

Starch and, to a lesser extent, glycogen should be the main (4) _____ [p.132] in the human diet.

Complex carbohydrates are preferable to simple sugars because they are high in (5) _____ [p.132],

and contain (6) _____ [p.132] and minerals. Lipids, including fats, are essential in the diet because

they are a necessary component of cell (7) _____ [p.132], serve as (8) _____ [p.132]

reserves, (9) _____ [p.132] many organs, provide (10) _____ [p.132] beneath the skin, and store fat-soluble (11) _____ [p.132]. Essential fatty acids are those that cannot be produced by the (12) _____ [p.132], and therefore must come from the diet. Butter and other (13) _____ [p.132] fats raise the level of cholesterol in the blood, which can result in serious problems with circulation. Of the twenty common amino acids in proteins, (14) _____ [p.132] are essential amino acids. (15) _____ [p.132] proteins have ratios of amino acids that match human nutritional needs. Plant proteins are (16) _____ [p.132] because they lack one or more of the essential amino acids.

Short Answer

17. What is the basic premise of a low-fat, high-carbohydrate diet? [p.132]

_____

_____

18. What is the main benefit associated with the Mediterranean diet? [p.132]

_____

_____

19. On the food pyramid illustrating a low-fat, high-carb diet, what kind of foods should you have the most servings of each day? [p.133]

_____

_____

20. What size is one serving of poultry? [p.133]

_____

_____

21. How many servings of vegetables a day should an adult male have each day on this diet? [p.133]

_____

22. Does an adult male or a teenager require more servings of milk, cheese, or yogurt? [p.133]

_____

23. By contrast, by observing the food pyramid of the Mediterranean diet, how often should red meat be eaten? [p.133]

_____

24. How must vegetarians and vegans supplement their diet to insure health? [p.134]

_____

_____

Fill-in-the-Blanks

Nutritionists and endocrinologists often support (25) _____-_____ [p.134] diets. Refined carbohydrates have a high (26) _____ [p.134] index. This means that upon absorption, they cause a surge in blood levels of (27) _____ [p.134]. This causes (28) _____ [p.134] to be taken up quickly. Any that is not needed is stored as (29) _____ [p.134]. A drop in blood sugar causes us to feel (30) _____ [p.134], which sets us up for more eating, and the cycle repeats itself. Fat is stored mainly as (31) _____ [p.134], which can lead to heart disease and type 2 (32) _____ [p.134]. Low-carb diets limit (33) _____ [p.134] surges, so a person does not get hungry as often, and begins to use up stored (34) _____ [p.134]. The risk of a low-carb diet is that too much protein and too few complex carbohydrates can cause (35) _____ [p.134] damage.

Matching

Choose the most appropriate description for each term.

36. _____ kwashiorkor [p.134]

37. _____ iron deficiency [p.135]

38. _____ malnutrition [p.135]

39. _____ marasmus [p.135]

40. _____ beriberi [p.135]

41. _____ rickets [p.135]

42. _____ undernutrition or starvation [p.134]

43. _____ xerophthalmia [p.135]

44. _____ scurvy [p.135]

A. Child consumes near-normal amounts of calories but has chronic protein deficiency; swollen abdomen, sickly, many infections
B. Body-wasting disease in very young children; both protein and food calories extremely low; skin and hair dry; growth and mental development retarded
C. Vitamin C deficiency
D. A state in which body functions or development suffer due to inadequate or unbalanced food intake
E. Lack of vitamin A is responsible for this form of preventable blindness
F. Vitamin B$_1$ deficiency
G. The individual lacks food and thus does not obtain sufficient kilocalories or nutrients to sustain proper growth, body functioning, and development
H. Causes weakness, weight loss, impaired immunity, and other symptoms in adults; especially devastating to infants and children
I. Vitamin D deficiency

Dichotomous Choice

Circle one of two possible answers given between parentheses in each statement.

45. Organic substances essential for growth and survival are (minerals/vitamins). [p.136]

46. Inorganic substances that are essential in the diet are (minerals/vitamins). [p.136]

47. There are at least 13 (minerals/vitamins) that our cells require. [p.136]

48. A balanced diet (can/cannot) provide all the needed minerals and vitamins. [p.136]

Matching

Choose the most appropriate description for each term.

49. _____ percent body fat that defines obesity [p.138]

50. _____ percent of U.S. population that is overweight [p.138]

51. _____ BMI of this or higher is a health risk [p.138]

52. _____ basal metabolic rate (BMR) [p.138]

A. 27
B. 60
C. 20 for men, 25 for women
D. usually higher in male

Complete the Table

53. Complete the following table by determining how many kilocalories the people described should take in daily, given the stated exercise level, to maintain their weight. Consult page 138 of the text.

Height	Age	Sex	Level of Physical Activity	Present Weight	Number of Kilocalories
5'6"	25	Female	Moderately active	140	
5'10"	37	Male	Very active	140	
5'8"	59	Female	Not very active	140	

Blanks

54. _____ Which of the individuals from the chart above is overweight? [p.138]

55. _____ Which is underweight? [p.138]

56. _____ Which is at the ideal weight? [p.138]

57. _____ How many hours would the 25-year-old female have to jog to lose one pound? [p.139]

Short Answer

58. What factors could cause a person's set point to result in obesity? [p.139]

59. What are the effects of the hormones ghrelin, leptin, and PYY3-36 on appetite? [pp.119,139]

60. Distinguish between the eating disorders anorexia nervosa and bulimia. [p.139]

Self-Quiz

Are you ready for the exam? Test yourself on key concepts by taking the additional tests linked with your BiologyNow CD-ROM.

Multiple Choice

____ 1. The process that moves nutrients into the blood or lymph is _____. [p.121]
   a. ingestion
   b. absorption
   c. assimilation
   d. digestion
   e. none of the above

____ 2. The enzymatic digestion of proteins begins in the _____. [p.122]
   a. mouth
   b. stomach
   c. liver
   d. pancreas
   e. small intestine

_____ 3. The enzymatic digestion of starches begins in the _____. [p.122]
   a. mouth
   b. stomach
   c. liver
   d. pancreas
   e. small intestine

_____ 4. A bolus moves from the pharynx to the _____. [p.123]
   a. trachea
   b. larynx
   c. glottis
   d. oral cavity
   e. esophagus

_____ 5. Digestion of fats requires bile and _____. [p.126]
   a. lecithin
   b. cholesterol
   c. pancreatic enzymes
   d. pigments
   e. *Escherichia coli*

_____ 6. Water moves through the membranes of the small intestine by _____. [p.127]
   a. peristalsis
   b. osmosis
   c. diffusion
   d. active transport
   e. bulk flow

_____ 7. Which one of the following does NOT apply to the large intestine? [p.129]
   a. It contains large populations of bacteria.
   b. It is divided into the duodenum, jejunum, and ileum.
   c. It concentrates undigested matter.
   d. It stores undigested matter.
   e. It absorbs water.

_____ 8. Of the following, _____ has (have) the highest amounts of all eight essential amino acids. [p.134]
   a. sunflower seeds
   b. cream cheese
   c. eggs
   d. black-eyed peas
   e. mushrooms

_____ 9. Males tend to burn more kilocalories than females because men _____. [p.139]
   a. are more active
   b. are taller
   c. weigh more
   d. have more muscle
   e. have a higher BMR

_____ 10. _____ is an eating disorder in which an individual purposely starves and overexercises. [p.139]
   a. Beriberi
   b. Marasmus
   c. Bulimia
   d. Anorexia nervosa
   e. Xerophthalmia

Matching

11. _____ gallbladder [p.120]

12. _____ large intestine [p.120]

13. _____ liver [p.120]

14. _____ oral cavity [p.120]

15. _____ pancreas [p.120]

16. _____ small intestine [p.120]

17. _____ stomach [p.120]

A. Secretes bile
B. Secretes an enzyme for each major food category; secretes insulin
C. Where most digestion and absorption occurs
D. Where water and salts are absorbed, where indigestible food is concentrated and stored
E. Stores bile
F. Holds food temporarily; secretes ghrelin
G. Where salivary amylase works

Chapter Objectives/Review Questions

This section lists general and detailed chapter objectives that can be used as review questions. You can make maximum use of these items by writing answers on a separate sheet of paper. Fill in answers where blanks are provided. To check for accuracy, compare your answers with information given in the chapter or glossary.

1. List all specialized regions (in order) of the human gastrointestinal tract through which food actually passes. Then list the accessory structures that contribute one or more substances to the digestive process. [pp.118–125]
2. Define and distinguish among *mechanical processing and motility*, *secretion*, *digestion*, *absorption*, and *elimination*. [p.117]
3. Briefly describe the four-layered wall of the gastrointestinal tract. [p.117]
4. What can cause "heartburn" and stomach ulcers? [p.120]
5. Describe the role of the stomach's acidity in protein digestion. [p.120]
6. Describe how the digestion and absorption of fats differ from the digestion and absorption of carbohydrates and proteins. [pp.121–123]
7. List the enzyme(s) that act in (a) the oral cavity, (b) the stomach, and (c) the small intestine. Then tell where each enzyme was originally produced. [p.121]
8. Describe the cross-sectional structure of the small intestine, and explain how its structure is related to its function. [pp.121–122]
9. List the molecules that leave the digestive system and enter the circulatory system during the process of absorption. [p.123]
10. Tell which foods undergo digestion in each of the following parts of the human digestive system, and state what kinds of simple biological molecules they are broken into: oral cavity, stomach, small intestine, large intestine. [pp.118–125]
11. Summarize the processes that occur in the colon. [p.125]
12. Explain how, during digestion, food is mechanically broken down. Then explain how it is chemically broken down. [pp.118–123]
13. Explain how the human body manages to meet the energy needs of various body parts even though the person may be feasting sometimes and fasting at other times. [p.127]
14. Compare the contributions of carbohydrates, proteins, and fats to human nutrition with the contributions of vitamins and minerals. [pp.128–131]
15. Distinguish vitamins from minerals. [p.130]
16. Name four minerals that are important in human nutrition and state the specific role of each. [p.131]
17. Explain how a diet low in carbohydrates, but high in proteins and fats, can lead to weight loss. [p.134]

Media Menu Review Questions

Questions 1–4 are drawn from the following InfoTrac College Edition article: "A Hungry Hormone." Josh Fischman. *U.S. News & World Report*, June 3, 2002.

1. One of the strongest appetite stimulants known is called _____ and is produced in the gut.
2. Ghrelin may link the stomach to the _____ chemically, letting the stomach say "feed me" when food is scarce.
3. Evidence for the role of ghrelin in hunger comes from much-reduced appetites in patients who had _____ bypass surgery.
4. David Cummings, researcher, says, "Look, hunger is important in our evolution. We have many backup mechanisms to keep us from _____, and we may need to deal with all of them."

Questions 5–15 are drawn from the following InfoTrac College Edition article: "Losing Weight: More than Counting Calories." Linda Bren. *FDA Consumer*, January/February 2002.

5. More than _____ percent of U.S. adults are overweight or obese.
6. According to the Centers for Disease Control, excess weight and physical _____ account for more than 300,000 deaths each year in the U.S., second only to deaths related to _____.
7. Whereas overweight refers to an excess of body _____, obesity refers specifically to an excess of body _____.
8. Multiplying your weight in pounds by 700, then dividing by your height in inches, then dividing by your height in inches a second time, gives you a figure called BMI, which stands for _____.
9. In addition to a high BMI (25 or over), having excess _____ body fat is a health risk.
10. Successful dieters report four common behaviors: a low-fat, high-carbohydrate diet, regular weight monitoring, physical activity, and eating _____.
11. The National Heart, Lung, and Blood Institute (NHLBI) recommends a weight loss of _____ pounds per week.
12. Weight-conscious consumers should aim for a daily fat intake of no more than _____ percent of total calories, and intake of saturated fat of less than _____ percent.
13. On a food label, required by the FDA and USDA, there are facts about nutrients by weight, as well as by _____ of Daily Value.
14. All adults should get at least _____ minutes of moderate physical activity on most, and preferably all, days of the week.
15. Prescription weight-loss drugs are approved only for people with a BMI of _____ and above.

Questions 16–18 are drawn from the following InfoTrac College Edition article: "American Kidney Fund Warns about Impact of High-Protein Diets on Kidney Health." *Obesity, Fitness & Wellness Week*, June 29, 2002.

16. High _____ diets place a significant strain on the kidneys.
17. A study of five fit endurance runners showed that increasing protein intake led to a progression toward _____, and a greater strain on the _____ due to an excessive amount of protein.
18. Increased protein leads to a buildup of _____ in the blood. The excess ends up at the kidney in the form of _____. The resulting increase in _____ can cause dehydration, further straining the kidneys.

Integrating and Applying Key Concepts

Based on the current world distribution of basic foods versus human populations, suggest ways to prepare for an almost doubling of the world population within the next three decades. Include mention of how nutrition, not just food quantity, is relevant.

8

BLOOD

Interactive Exercises

Impacts, Issues: Chemical Queries [p.143]

8.1. BLOOD: PLASMA, BLOOD CELLS, AND PLATELETS [pp.144–145]

8.2. HOW BLOOD TRANSPORTS OXYGEN [p.146]

8.3. LIFE CYCLE OF RED BLOOD CELLS [p.147]

For additional practice, use the interactive vocabulary exercises linked with your BiologyNow CD-ROM.

Selected Words: edema [p.144], hemoglobin [p.146]

Boldfaced, Page-Referenced Terms

[p.144] blood _____

_____

[p.144] plasma _____

_____

[p.144] red blood cells _____

_____

[p.144] stem cell _____

_____

[p.145] white blood cells _____

_____

[p.145] granulocytes _____

_____

[p.145] agranulocytes _____

_____

[p.145] platelets _____

_____

[p.146] oxyhemoglobin _____

_____

[p.146] cell count _____

_____

Complete the Table

1. Complete the following table, which describes the components of blood. [p.144]

Components	Relative Amounts	Functions
Plasma Portion (50%–60% of total volume):		
Water	a. _____	Solvent
b. _____ (albumin, globulins, fibrinogen, HDLs, LDLs, VLDLs, etc.)	7%–8%	Defense, clotting, lipid transport, roles in extracellular fluid volume, etc.
Ions, sugars, lipids, amino acids, hormones, vitamins, dissolved gases	1%–2%	Roles in extracellular fluid volume, pH, etc.
Cellular Portion (40%–50% of total volume, *per microliter*):		
White blood cells		
c. _____	3,000–6,750	Phagocytosis; inflammation
d. _____	1,000–2,700	Immunity
Monocytes (macrophages)	150–720	e. _____
Eosinophils	100–360	Inflammation
Basophils	25–90	Inflammation
f. _____ _____ cells	4,800,000–5,400,000	O_2, CO_2 transport
g. _____	250,000–300,000	Role in clotting

Fill-in-the-Blanks

For the average-sized female adult, blood volume is generally about (2) _____ [p.144] liters, which is 6 to 8 percent of body weight. About 55 percent of whole blood is (3) _____ [p.144], which is mostly (4) _____ [p.144] and serves as a transport medium for blood cells, platelets, and other substances. Two-thirds of all plasma proteins are molecules of (5) _____ [p.144], which plays an important role in the movement of (6) _____ [p.144] between the bloodstream and interstitial fluid. Other plasma (7) _____ [p.144] include hormones, immune system proteins, and proteins involved in blood clotting. Ions in the plasma help maintain the (8) _____ [p.144] and (9) _____ of extracellular fluid [p.144].

(10) _____ _____ [p.144] cells, or erythrocytes, make up about 45 percent of whole blood. Each is a biconcave disk carrying the iron-containing protein (11) _____ [p.144]. These cells transport oxygen used in aerobic respiration as well as some carbon dioxide wastes. Red blood cells are derived from unspecialized (12) _____ _____ [p.144] located in the bone marrow.

Leukocytes, or (13) _____ _____ [p.145] cells, make up a tiny fraction of whole blood, but have vital functions in day-to-day housekeeping and defense. They scavenge dead or worn-out cells, as well as any (14) _____ [p.145] material such as specific bacteria, viruses, or other disease agents. They circulate in the blood, but most squeeze out of blood vessels and do their work after they enter (15) _____ [p.145]. Leukocytes arise from stem cells located in the (16) _____ _____ [p.145]. There are five types of white blood cells divided into two major classes. The (17) _____ [p.145] contain various granules in the cytoplasm. This group includes (18) _____ [p.145], (19) _____ [p.145], and (20) _____ [p.145]. Leukocytes called (21) _____ [p.145] have no visible granules in the cytoplasm. The first of two types are called (22) _____ [p.145], which differentiate into macrophages that engulf invaders and cellular debris. The second type, (23) _____ [p.145] (B cells and T cells), carry out specific immune responses.

Some stem cells in bone marrow develop into "giant" cells that shed fragments of cytoplasm that become enclosed in a bit of plasma membrane. The fragments are called (24) _____ [p.145]. They release substances that initiate blood (25) _____ [p.145].

Labeling-Matching

Identify the numbered cell types in the following illustration. Complete the exercise by matching and entering the letter of the appropriate function in the parentheses following the given cell types. (The functions of the last two structures are not included.) [p.145]

26. _____ (____)
27. _____ (____)
28. _____ (____)
29. _____ (____)
30. _____ (____)
31. _____
32. _____

A. Found in infected, inflamed, or damaged tissues
B. Engulf invading microbes
C. Involved in clotting
D. Specific immune responses
E. O_2, CO_2 transport

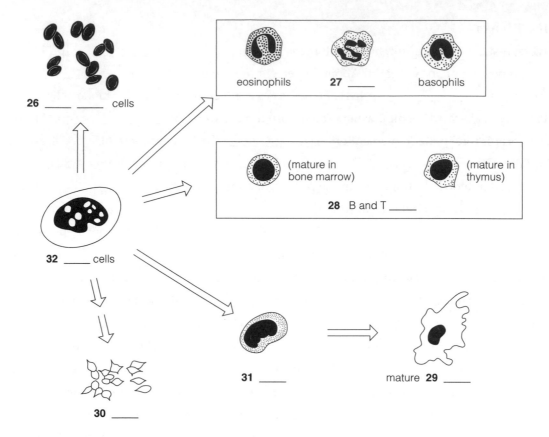

26 ____ ____ cells

eosinophils   **27** ____   basophils

(mature in bone marrow)   (mature in thymus)

28 B and T ____

32 ____ cells

31 ____   mature **29** ____

30 ____

Interpreting Diagrams

33. Study each diagram below in order to answer the questions that follow. [p.146]

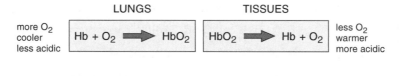

a. What is "Hb" in the diagram? _____

b. What is HbO_2? _____

c. Under what conditions (oxygen, temperature, pH) does oxygen tend to bind to hemoglobin?

_____   _____   _____

d. Where in the body do these conditions exist? _____

e. Under what conditions (oxygen, temperature, pH) does hemoglobin tend to give up oxygen?

_____   _____   _____

f. Where in the body do these conditions exist? _____

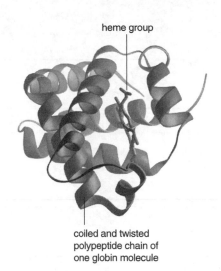

heme group

coiled and twisted
polypeptide chain of
one globin molecule

g. What is the molecule diagrammed above? _____

h. In what type of blood cells are molecules like this found? _____

i. How many polypeptide chains make up the globin component of this molecule? _____

j. What inorganic element is found at the center of each heme group? _____

k. What is the function of this substance? _____

l. What is this molecule called when it binds oxygen? _____

Matching

34. _____ erythropoietin [p.147]

35. _____ bilirubin [p.147]

36. _____ red marrow [p.147]

37. _____ 120 days [p.147]

38. _____ stem cell [p.147]

39. _____ cell count [p.147]

40. _____ "blood doping" [p.147]

41. _____ spleen [p.147]

A. Found in skull, vertebrae, and sternum
B. Average life span of red blood cell
C. Cells in marrow that give rise to red blood cells
D. Average per microliter is 5.4 million in adult males and 4.8 million in adult females
E. Where macrophages remove and recycle old or damaged red blood cells
F. Hormone made in kidneys that signals production of red blood cells
G. Orangish remnant of heme group removed from bloodstream by liver
H. Injection of stored blood of an athlete

8.4. BLOOD TYPES—GENETICALLY DIFFERENT RED BLOOD CELLS [pp.148–149]

8.5. HEMOSTASIS AND BLOOD CLOTTING [p.150]

8.6. BLOOD DISORDERS [p.151]

Selected Words: "self" [p.148], "nonself" [p.148], "universal donors" [p.148], "universal recipients" [p.148], "Rh factor" [p.149], *hemolytic disease of the newborn* [p.149], *cross-matching* [p.149], "intrinsic" clotting mechanism [p.150], "extrinsic" clotting mechanism [p.150], *thrombus* [p.150], *thrombosis* [p.150], *embolus* [p.150], *embolism* [p.150], *iron-deficiency anemia* [p.151], *megaloplastic anemia* [p.151], *aplastic anemia* [p.151], *hemolytic anemias* [p.151]

Boldfaced, Page-Referenced Terms

[p.148] agglutination _____

_____

[p.149] Rh blood typing _____

_____

[p.150] hemostasis _____

_____

[p.151] anemias _____

_____

[p.151] infectious mononucleosis _____

_____

[p.151] leukemias _____

_____

Matching

Choose the most appropriate description for each term.

1. _____ antibodies [p.148]
2. _____ antigen [p.148]
3. _____ ABO typing [p.148]
4. _____ Rh factor [p.149]
5. _____ hemolytic disease of the newborn [p.149]
6. _____ cross-matching [p.149]
7. _____ agglutination [p.148]
8. _____ Type AB [p.148]
9. _____ Type O [p.148]
10. _____ Types A and B [p.148]

A. A person whose blood type is positive carries this marker
B. Blood analysis based on presence or absence of glycoproteins A and B
C. Theoretical "universal recipient"
D. Theoretical "universal donor"
E. Blood types with one marker from ABO
F. May occur in offspring produced by an Rh⁻ woman and an Rh⁺ man
G. Method used to insure compatibility of blood marker forms
H. Clumping when antibodies attack foreign cells
I. Immune system proteins that recognize and organize an attack on most foreign entities
J. Molecule with "nonself" protein markers that prompt a defensive attack by antibodies

Sequence

The following lettered stages of the "intrinsic" clotting mechanism occur in a specific sequence. Place the letter of the first event in number 11, and continue until the last event is placed in blank number 16.

11. _____ [p.150]
12. _____ [p.150]
13. _____ [p.150]
14. _____ [p.150]
15. _____ [p.150]
16. _____ [p.150]

A. Fibrinogen proteins stick together, forming an insoluble net of fibrin.
B. Blood vessel ruptures.
C. Platelets aggregate and form a temporary plug.
D. An activated blood protein triggers reactions that form thrombin.
E. Smooth muscle in a damaged blood vessel contracts, constricting the vessel and slowing blood flow.
F. Blood cells and platelets become entangled in the fibrin net, forming a clot.

Short Answer

17. How does the "extrinsic" clotting mechanism differ from the sequence above? [p.150]

_____

_____

_____

18. How are a thrombus and an embolus the same? How are they different? Which is more dangerous? [p.150]

_____

_____

_____

Matching

19. _____ iron-deficiency anemia [p.151]

20. _____ aplastic anemia [p.151]

21. _____ infectious mononucleosis [p.151]

22. _____ hemolytic anemia [p.151]

23. _____ megaloplastic anemia [p.151]

24. _____ leukemia [p.151]

A. Runaway multiplication of abnormal white blood cells
B. Premature destruction of red blood cells
C. Hemoglobin cannot be formed due to low iron supply
D. Destruction of red bone marrow by radiation or toxins
E. Deficiency of folic acid or vitamin B_{12}
F. Overproduction of agranulocytes, caused by Epstein-Barr virus

Self-Quiz

Are you ready for the exam? Test yourself on key concepts by taking the additional tests linked with your BiologyNow CD-ROM.

____ 1. Most of the oxygen in human blood is transported by _____. [p.146]
a. plasma
b. serum
c. platelets
d. hemoglobin
e. leukocytes

____ 2. The two kinds of _____ operate against specific invaders in an immune response. [p.145]
a. basophils
b. eosinophils
c. monocytes
d. neutrophils
e. lymphocytes

____ 3. Oxygen becomes bound to _____ in hemoglobin molecules. [p.146]
a. iron in heme groups
b. polypeptides
c. plasma
d. globin
e. carbon dioxide molecules

____ 4. The process that stops bleeding is called _____. [p.150]
a. thrombosis
b. phagocytosis
c. inflammation
d. agglutination
e. hemostasis

_____ 5. Oxyhemoglobin is hemoglobin combined with _____. [p.146]
a. carbon dioxide
b. red blood cells
c. plasma
d. calcium
e. oxygen

_____ 6. Megakaryocytes shed millions of _____ into the blood that last about a week. [p.145]
a. platelets
b. red blood cells
c. lymphocytes
d. macrophages
e. neutrophils

_____ 7. Too little albumin can cause swelling called _____. [p.144]
a. thrombosis
b. edema
c. embolism
d. hemostasis
e. granuloma

_____ 8. What combination can lead to hemolytic disease of the newborn? [p.149]
a. Rh^- mother, Rh^- fetus
b. Rh^+ mother, Rh^+ fetus
c. Rh^+ mother, Rh^- fetus
d. Rh^- mother, Rh^+ fetus
e. any of these

_____ 9. Red blood cells and white blood cells develop from _____ cells in the bone marrow [pp.144–145]
a. stem
b. hemolytic
c. megakaryocyte
d. oxyhemoglobin
e. albumin

_____ 10. About how many liters of blood does an adult woman have? [p.144]
a. 1–2
b. 4–5
c. 10–12
d. 15–20
e. 24–45

Chapter Objectives/Review Questions

This section lists general and detailed chapter objectives that can be used as review questions. You can make maximum use of these items by writing answers on a separate sheet of paper. Fill in answers where blanks are provided. To check for accuracy, compare your answers with information given in the chapter or glossary.

1. Name and describe the composition of the two major components of human blood, using percentages of volume. [p.144]
2. Name the most common plasma protein and describe its functions. [p.144]
3. State where erythrocytes, leukocytes, and platelets are produced. [pp.144–145]
4. Contrast the two main types of leukocytes in terms of cell structure. Then, name and state the general functions of the three types of granulocytes and the two types of agranulocytes. [p.145]
5. A hormone produced by the kidneys, _____, stimulates certain stem cells to produce red blood cells. [p.147]
6. Describe what happens to red blood cells when they are old or damaged. [p.147]
7. Describe how blood is typed for the ABO blood group and for the Rh factor. [pp.148–149]
8. List in sequence the chemical events that occur in the formation of a blood clot. [p.150]
9. When red blood cells contain a less-than-normal amount of hemoglobin, it is termed _____. [p.151]

Media Menu Review Questions

Questions 1–3 are drawn from the following InfoTrac College Edition article: "Blood Work: Scientists Seek to Identify All the Proteins in Plasma." John Travis. *Science News*, March 15, 2003.

1. The clear fluid in blood, consisting of water with salts, hormones, enzymes, antibodies, and other proteins, is called _____.
2. In December 2002, a team at Pacific Northwest National Laboratory reported that it had identified almost _____ proteins in a sample of human plasma.
3. Plasma stripped of its clotting factors is called _____.

Questions 4–8 are drawn from the following InfoTrac College Edition article: "Living with Leukemia: Scientists Are Finding Better Ways to Treat Leukemia, and the Chances of Recovery Keep Improving." Carol Lewis. *FDA Consumer*, March/April 2002.

4. The overall five-year survival rate for people with leukemia has _____ over the past 40 years.
5. Leukemia is characterized by the uncontrolled growth of developing _____ cells, resulting in production of excess white blood cells.
6. In leukemia, the increase in white blood cells means not enough _____ blood cells and not enough _____.
7. The main treatment for nearly all types of leukemia is _____, which is the use of drugs to kill cancer cells.
8. Biological therapy (or immunotherapy) uses the body's _____ system to fight cancer or to lessen the side effects of some cancer treatments.

Questions 9–10 are drawn from the following InfoTrac College Edition article: "Platelets in Blood May Guide Immune Response." John Travis. *Science News*, July 26, 2003.

9. Michael Yeaman and his colleagues have shown that platelets can release proteins that rapidly _____ bacteria and some other microbes.
10. Platelets may participate in the immune response by releasing _____ chemicals that beckon macrophages and neutrophils.

Integrating and Applying Key Concepts

For each blood cell marker, there is a gene that codes for it. A baby is type AB⁺. The mother is type B⁻. Which blood types would be possible for the father? Which would not? What other means could be used to determine the identity of the father?

9

CIRCULATION—THE HEART AND BLOOD VESSELS

Interactive Exercises

Impacts, Issues: The Breath of Life [p.153]

9.1. THE CARDIOVASCULAR SYSTEM—MOVING BLOOD THROUGH THE BODY [pp.154–155]

9.2. THE HEART: A DOUBLE PUMP [pp.156–157]

9.3. THE TWO CIRCUITS OF BLOOD FLOW [pp.158–159]

9.4. HEART-SAVING DRUGS [p.159]

9.5. HOW DOES CARDIAC MUSCLE CONTRACT? [p.160]

For additional practice, use the interactive vocabulary exercises linked with your BiologyNow CD-ROM.

Selected Words: "cardiovascular" [p.154], *capillary beds* [p.154], *tricuspid valve* [p.156], *bicuspid valve* or *mitral valve* [p.156], "coronary circulation" [p.157], "heartbeat" [p.157], "lub-dup" [p.157], *renal arteries* [p.159], *superior vena cava* [p.159], *inferior vena cava* [p.159], "true capillaries" [p.159], "thoroughfare channels" [p.159], *hepatic portal vein* [p.159], *hepatic vein* [p.159], *hepatic artery* [p.159], *statins* [p.159], "bad" vs. "good" cholesterol [p.159], *intercalated discs* [p.160], "pacemaker" cells [p.160], *Purkinje fibers* [p.160]

Boldfaced, Page-Referenced Terms

[p.154] arteries _____

_____

[p.154] arterioles _____

_____

[p.154] capillaries _____

_____

[p.154] venules _____

_____

[p.154] veins _____

_____

[p.156] myocardium _____

_____

[p.156] septum _____

_____

[p.156] atrium (plural: atria) _____

_____

[p.156] ventricle _____

_____

[p.156] atrioventricular (AV) valve _____

_____

[p.156] semilunar valve _____

_____

[p.157] coronary arteries _____

_____

[p.157] aorta _____

_____

[p.157] systole _____

_____

[p.157] diastole _____

_____

[p.157] cardiac cycle _____

_____

[p.158] pulmonary circuit _____

_____

[p.159] systemic circuit _____

_____

[p.159] capillary beds _____

_____

[p.160] cardiac conduction system _____

_____

[p.160] sinoatrial (SA) node _____

_____

[p.160] atrioventricular (AV) node _____

_____

[p.160] cardiac pacemaker _____

_____

Fill-in-the-Blanks

The (1) _____ [p.154] system, also called the circulatory system, is built to circulate blood to every living cell in the body. The two main elements in the system are the (2) _____ [p.154] and (3) _____ _____ [p.154], which are tubes of different diameters. The heart pumps blood into large-diameter (4) _____ [p.154]. From there the blood flows into smaller (5) _____ [p.154], which branch into even narrower (6) _____ [p.154]. Blood flows from capillaries into small (7) _____ [p.154], then into large-diameter (8) _____ [p.154] that return blood to the heart. Blood flows (9) _____ [p.154] through arteries, but in capillary beds, it must flow (10) _____ [p.154].

Blood is called the "river of life" because it brings cells essentials such as (11) _____ [p.155], (12) _____ [p.155] from food, and secretions. It also takes away the (13) _____ [p.155] produced by metabolism. (14) _____ [p.155] would be impossible were it not for our circulating blood.

As the heart's pumping keeps pressure on blood flowing through the cardiovascular system, some water and proteins are forced out of the vast network of capillaries and become part of the

(15) _____ [p.155] fluid. A network called the (16) _____ [p.155] system picks up excess interstitial fluid and reclaimable solutes, and returns them to the cardiovascular system.

Labeling

In the following illustration, color in red all vessels that carry oxygen-rich blood (including all parts indicated by "aorta" or "artery," except for the pulmonary arteries that carry oxygen-poor blood from the heart to the lungs). Then fill in all the blanks on the right side of the diagram. Next, color in blue all vessels that carry oxygen-poor blood (including all parts indicated by "vena cava" or "vein," except for the pulmonary veins that return oxygen-rich blood to the heart). Then fill in all the blanks on the left side of the diagram. [p.154]

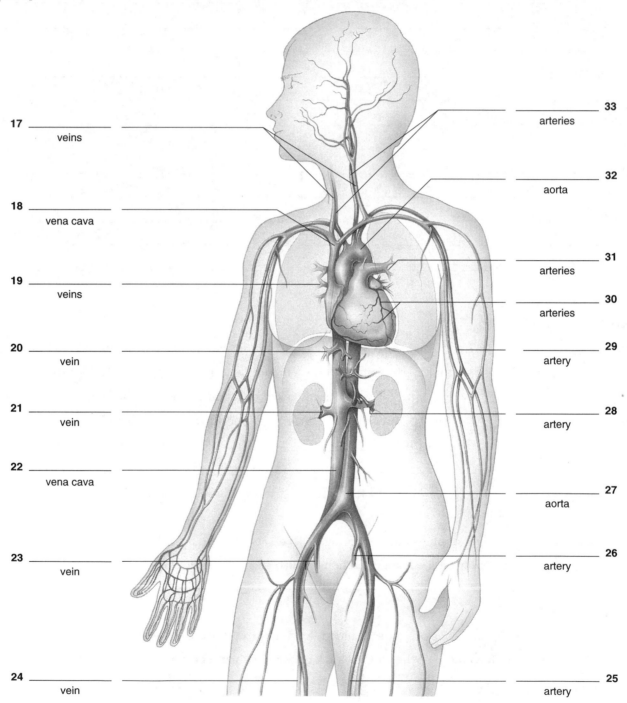

17 _____ _____
        veins

18 _____ _____
        vena cava

19 _____ _____
        veins

20 _____ _____
        vein

21 _____ _____
        vein

22 _____ _____
        vena cava

23 _____ _____
        vein

24 _____ _____
        vein

_____ 33
        arteries

_____ 32
        aorta

_____ 31
        arteries

_____ 30
        arteries

_____ 29
        artery

_____ 28
        artery

_____ 27
        aorta

_____ 26
        artery

_____ 25
        artery

Matching

34. _____ tricuspid valve [p.156]
35. _____ endocardium [p.156]
36. _____ atrioventricular (AV) valve [p.156]
37. _____ aorta [p.157]
38. _____ bicuspid (mitral) valve [p.156]
39. _____ myocardium [p.156]
40. _____ chordae tendineae [p.156]
41. _____ pericardium [p.156]
42. _____ semilunar valve [p.156]
43. _____ septum [p.156]
44. _____ ventricles [p.156]
45. _____ coronary arteries [p.157]
46. _____ atria [p.156]
47. _____ endothelium [p.156]

A. Upper chambers of the heart
B. Tough, fibrous sac surrounding the heart
C. Thick wall dividing the heart into two halves
D. Three-flap AV valve; in heart's right half
E. Major artery carrying oxygenated blood away from the heart
F. Cardiac muscle tissue of the heart
G. Either of two one-way structures found between atrium and ventricle
H. Collagen-reinforced strands that connect AV valve flaps to muscles in ventricle wall
I. One-way structure found between ventricle and arteries leading away from it
J. Two-flap AV valve; in heart's left half
K. Smooth lining of heart's chambers; composed of connective tissue and epithelial cells
L. Lower chambers of heart
M. Two of these service the cardiac muscle cells
N. Epithelial cell layer of the endocardium

Labeling

Identify each indicated part of the accompanying illustration. [p.156]

48. _____ _____ _____
49. _____ _____ _____
50. _____ _____ _____
51. _____
52. _____ _____ _____
53. _____
54. _____ _____
55. _____

48 —
55 —
trunk of pulmonary arteries
49 —
left semilunar valve
right pulmonary veins (from lungs)
right atrium
left pulmonary veins
left atrium
50 —
left AV valve
right ventricle
54 —
51 —
52 —
septum (partition between heart's two halves)
apex
53

Sequencing

56. Show the correct sequence of the path of blood flow through the heart by numbering the events listed in the order that they occur. The first event, "1," is given. [p.157]

___1___ Relaxed atria fill with blood.

_____ Ventricles contract and AV valves close ("lub").

_____ Atrioventricular (AV) valves open.

_____ Blood enters aorta and pulmonary artery.

_____ Ventricles relax and SL valves close ("dup").

_____ Blood flows into ventricles.

_____ Semilunar (SL) valves open.

_____ Atria contract.

Sequencing

Show the correct sequence of the path of blood flow through the pulmonary and systemic circuits by placing the letter of the part of the heart that receives blood from body tissues in the blank beside number 57, then continuing the sequence until you are back at the starting point.

57. _____
58. _____
59. _____
60. _____
61. _____
62. _____
63. _____
64. _____
65. _____
66. _____
67. _____

A. Lungs
B. Aorta
C. Main pulmonary artery
D. Right atrium (use twice)
E. Left atrium
F. Right ventricle
G. Left ventricle
H. Pulmonary veins
I. Torso (systemic circulation)
J. Right and left pulmonary arteries

Dichotomous Choice

Circle the correct answer of the two choices given between parentheses in each statement.

68. Events 57–63 are parts of the (systemic/pulmonary) circuit. [p.158]

69. Events 63–67 are parts of the (systemic/pulmonary) circuit. [p.159]

Fill-in-the-Blanks

As the aorta descends into the torso, major (70) _____ [p.159] branch off it, funneling blood to

(71) _____ [p.159] and tissues. In both the pulmonary and systemic circuits, blood travels

through (72) _____ [p.159], (73) _____ [p.159], (74) _____ [p.159], and

(75) _____ [p.159], and finally returns to the heart in (76) _____ [p.159]. Blood from the

head, arms, and chest arrives through the (77) _____ _____ _____ [p.159], and the (78) _____ _____ _____ [p.159] collects blood from the lower body. The actual exchange of substances between blood and tissues occurs in (79) _____ _____ [p.159]. "Thoroughfare channels" connect (80) _____ [p.159] and (81) _____ [p.159]. Blood flow into "true capillaries" is controlled by collars of smooth muscle cells called precapillary (82) _____ [p.159]. When the CO_2 level rises above a set point, the sphincter (83) _____ [p.159] so that blood flows through the capillary. When the CO_2 level falls, the sphincter (84) _____ [p.159].

After a meal, blood passing through capillary beds in the GI tract detours through the (85) _____ _____ _____ [p.159] to the liver. The liver removes (86) _____ [p.159] and processes absorbed substances. Blood leaves the liver's capillary bed through a (87) _____ [p.159] vein. The liver receives (88) _____ [p.159] blood via the hepatic artery.

Part of the processing that occurs in the liver synthesizes cholesterol. In the 1980s, drugs called (89) _____ [p.159] were shown to greatly reduce the amount of "bad" or (90) _____ [p.159] cholesterol in the blood. Other experiments showed that these drugs also raise the blood level of "good" cholesterol, called (91) _____ [p.159], and lower the levels of (92) _____ [p.159]. The effects of these drugs translate into dramatically reduced risks of (93) _____ _____ [p.159] and (94) _____ _____ [p.159], both of which can be caused when blood flow is blocked by the buildup of fatty, cholesterol-rich plaques in blood vessels.

Dichotomous Choice

Circle the correct answer of the two choices given between parentheses in each statement.

95. A signal to contract spreads rapidly through the heart because of communication junctions between abutting cells called (Purkinje fibers/intercalated discs). [p.160]

96. The (cardiac conduction/intercalated disc) system produces the electrical impulses that stimulate contraction of the heart. [p.160]

97. Contraction follows a wave of excitation that begins at a mass of cells in the upper wall of the right atrium called the (atrioventricular/sinoatrial) node. [p.160]

98. After causing both atria to contract, the wave of excitation reaches the (atrioventricular/sinoatrial) node in the septum between the two atria, where it slows down momentarily to give the atria time to finish contracting. [p.160]

99. The ventricles then contract when the wave continues through conducting bundles that extend from the AV node and make contact with the muscle cells of each ventricle by means of conducting cells called (Purkinje fibers/sinoatrial nodes). [p.160]

100. An artificial pacemaker is implanted when the cardiac pacemaker or (AV/SA) node malfunctions. [p.160]

101. The (cardiovascular/nervous) system can adjust the rate and strength of cardiac muscle contraction. [p.160]

9.6. BLOOD PRESSURE [p.161]

9.7. THE STRUCTURE AND FUNCTIONS OF BLOOD VESSELS [pp.162–163]

9.8. HEART-HEALTHY EXERCISE [p.164]

9.9. EXCHANGES AT CAPILLARIES [pp.164–165]

9.10. CARDIOVASCULAR DISORDERS [pp.166–167]

9.11. THE MULTIPURPOSE LYMPHATIC SYSTEM [pp.168–169]

Selected Words: *systolic pressure* [p.161], *diastolic pressure* [p.161], *hypertension* [pp.161,166], *hypotension* [p.161], *varicose vein* [p.163], *stroke volume* [p.164], "*bulk flow*" [p.165], *ultrafiltration* [p.165], *reabsorption* [p.165], *atherosclerosis* [p.166], *heart attacks* [p.166], *strokes* [p.166], "*heart attack gene*" [p.166], *arteriosclerosis* [p.166], *angina pectoris* [p.166], *coronary bypass surgery* [p.166], *laser angioplasty* [p.166], *balloon angioplasty* [p.166], *low-density lipoproteins* [p.167], *high-density lipoproteins* [p.167], *bradycardia* [p.167], *tachycardia* [p.167], *ventricular fibrillation* [p.167], "*pulp*" [p.169]

Boldfaced, Page-Referenced Terms

[p.161] blood pressure _____

_____

[p.162] pulse _____

_____

[p.163] vasodilation _____

_____

[p.163] vasoconstriction _____

_____

[p.163] baroreceptor reflex _____

_____

[p.163] carotid arteries _____

_____

[p.166] heart failure _____

_____

[p.166] atherosclerotic plaque _____

_____

[p.167] LDLs _____

_____

[p.167] HDLs _____

_____

[p.167] electrocardiogram _____

_____

[p.167] arrhythmias _____

_____

[p.168] lymphatic system _____

_____

[p.168] lymph _____

_____

[p.169] lymph vascular system _____

_____

[p.169] lymph nodes _____

_____

[p.169] spleen _____

_____

[p.169] thymus _____

_____

Fill-in-the-Blanks

(1) _____ _____ [p.161] is the fluid pressure that blood exerts against vessel walls. Blood pressure is at its highest in the (2) _____ [p.161], then it drops along the (3) _____ [p.161] circuit. Consider an adult with a blood pressure of 120/80; 120 is the (4) _____ [p.161] pressure, which is the peak of pressure in the (5) _____ [p.161] when the heart's left ventricle pushes blood into it. The number 80 is the (6) _____ [p.161] pressure, which is the lowest blood pressure in the (7) _____ [p.161] when the heart is relaxed. Elevated blood pressure, or (8) _____ [p.161], is associated with atherosclerosis and kidney disease. Low blood pressure is called (9) _____ [p.161], and is usually not a cause for worry.

Choice

For questions 10–23, choose from the following:

a. arteries   b. arterioles   c. capillaries   d. venules   e. veins

10. _____ Those near the body surface provide a "pulse." [p.162]

11. _____ These merge into venules. [p.163]

12. _____ Their walls have rings of smooth muscle over a single layer of elastic fibers. [p.162]

13. _____ These serve as diffusion zones for exchanges between blood and interstitial fluid. [p.162]

14. _____ Their bulging walls keep blood flowing on through the system. [p.162]

15. _____ These offer more resistance to blood flow than other vessels do. [p.162]

16. _____ These present less total resistance to flow than do the arterioles leading into them; the total drop in blood pressure is more gradual in this region. [p.162]

17. _____ These serve as blood volume reservoirs. [p.163]

18. _____ Red blood cells must squeeze through them single file. [p.162]

19. _____ These have valves that prevent backflow. [p.163]

20. _____ These branch into arterioles. [p.162]

21. _____ Weak valves in these lead to pooled blood and a varicose condition. [p.163]

22. _____ Contractions of smooth muscle in their thin walls occur when blood must circulate faster. [p.163]

23. _____ These merge into veins. [p.163]

Labeling

Identify each vessel and the parts indicated by asterisks in the accompanying illustrations. [p.162]

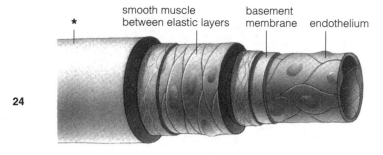

smooth muscle between elastic layers   basement membrane   endothelium

*

24

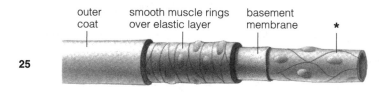

outer coat   smooth muscle rings over elastic layer   basement membrane   *

25

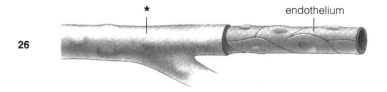

*   endothelium

26

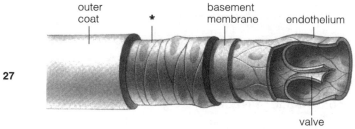

outer coat   *   basement membrane   endothelium

valve

27

24. _____ ; *_____ _____

25. _____ ; *_____

26. _____ ; *_____ _____

27. _____ ; *_____ _____ with _____ _____

Matching

28. _____ vasodilation [p.163]

29. _____ vasoconstriction [p.163]

30. _____ baroreceptors [p.163]

31. _____ medulla oblongata [p.163]

32. _____ stroke volume [p.164]

A. Pressure receptors in the carotid arteries, the aorta, and elsewhere that monitor changes in mean arterial pressure

B. Part of the brain that coordinates the rate and strength of heartbeats with changes in the diameter of arterioles and veins

C. A decrease in blood vessel diameter brought about by brain centers when an abnormal decrease in blood pressure is detected

D. The amount of blood pumped with a heart contraction; increases with physical conditioning

E. Dilation of a blood vessel brought about by brain centers when an abnormal increase in blood pressure is detected

Short Answer

33. List some benefits of endurance exercise. [p.164]

_____

_____

_____

True/False

If the statement is true, write a "T" in the blank. If the statement is false, make it correct by writing the word(s) in the blank that should take the place of the underlined word(s).

_____ 34. "Bulk flow" out of the capillary occurs when blood pressure inside a capillary is lower than fluid pressure outside. [p.165]

_____ 35. Because capillaries come within 0.01 millimeter of nearly all your living cells, most solutes move from bloodstream to cells by diffusion. [p.165]

_____ 36. Vesicles containing protein enter or leave the heart by endocytosis or exocytosis. [p.165]

_____ 37. Certain ions as well as white blood cells probably pass through pores in the capillary walls. [p.165]

_____ 38. Ultrafiltration and reabsorption are both examples of pressure-driven movements. [p.165]

_____ 39. The excess fluid that moves from capillaries into surrounding tissues is returned to the blood by the cardiovascular system. [p.165]

Elimination

Scratch out the incorrect answer(s) in parentheses so that each statement reads correctly. There may be more than one correct answer for a question.

40. In the United States, the most common cardiovascular disorders are (hypertension/angina pectoralis/atherosclerosis). [p.166]

41. Sustained high blood pressure is called (angioplasty/hypotension/hypertension). [p.166]

42. Thickening of the arterial wall and narrowing of the lumen is called (atherosclerosis/angioplasty/angina pectoralis). [p.166]

43. Hypertension and atherosclerosis cause most (strokes/organ failure/heart attacks). [p.166]

44. The nation's most costly health problem is (hypertension/heart failure/stroke). [p.166]

45. A risk factor(s) for heart disease is (smoking/obesity/drinking). [p.166]

46. An individual who knows that hypertension runs in his family should be careful about consuming too much (sugar/starch/salt). [p.166]

47. Arteriosclerosis (precedes/comes after/has nothing to do with) atherosclerosis. [p.166]

48. Mild chest pain associated with narrowed arteries is called (angioplasty/angina pectoralis/heart failure—HF). [p.166]

49. The underlying problem leading to a heart attack may be cured by (coronary bypass surgery/angioplasty/neither of these). [p.166]

50. In the bloodstream, proteins bind excess cholesterol and triglycerides, forming (CDLs/LDLs/HDLs), which are taken in by cells. [p.167]

51. HDLs are called "good cholesterol" because they remove excess (cholesterol/triglycerides/steroids) from the body. [p.167]

52. LDLs that are not removed from the bloodstream accumulate within (the liver/the heart/artery walls). [p.167]

53. Deposits in arteries result in atherosclerotic plaques, which often lead to (inflammation/clots/arrhythmia). [p.167]

54. A clot that breaks loose can cause a life-threatening (inflammation/arrhythmia/embolism). [p.167]

55. An electrocardiogram (ECG) may reveal an *abnormal* condition called (arrhythmia/bradycardia/tachycardia). [p.167]

56. Haphazard contraction of the cardiac muscle in parts of the ventricles is called (bradycardia/ventricular fibrillation/angina pectoralis). [p.167]

Identification

Identify each numbered part in the accompanying figure.

57. _____ [p.168]

58. _____ _____ [p.168]

59. _____ _____ [p.168]

60. _____ [p.168]

61. _____ _____ [p.168]

62. _____ _____ [p.168]

63. organized arrays of _____ [p.169]

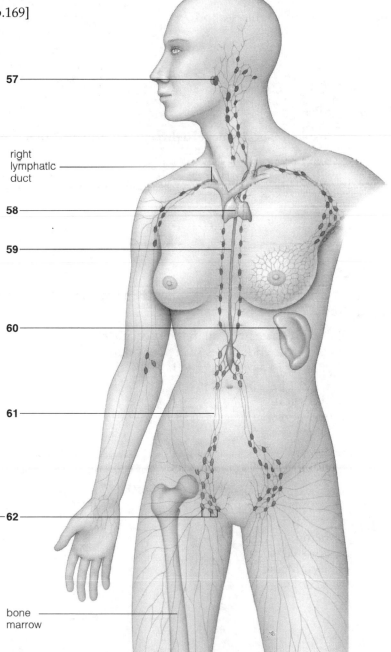

right lymphatic duct

valve (prevents backflow)

bone marrow

Matching

Choose the most appropriate description for each term.

64. _____ red pulp [p.169]

65. _____ lymph nodes [p.169]

66. _____ thymus [p.169]

67. _____ white pulp [p.169]

68. _____ spleen [p.169]

69. _____ lymph vessels [p.169]

70. _____ lymph capillaries [p.169]

A. Located at intervals along lymph vessels; battlegrounds of lymphocyte armies and foreign agents
B. Largest lymphoid organ; a filtering station for blood and a holding station for lymphocytes
C. Fluid enters their tips at flaplike "valves"
D. Have a larger diameter than lymph capillaries, smooth muscle in the wall, and valves that prevent backflow
E. Site where lymphocytes multiply, differentiate, and mature into fighters of specific disease agents; also produces hormones that influence these events
F. Portion of the spleen filled with lymphocytes to destroy specific invaders
G. Reservoir of red blood cells and macrophages inside the spleen

Short Answer

71. Briefly explain the "drainage, delivery, and disposal" functions of the lymph vascular system. [p.169]

_____

_____

_____

Self-Quiz

Are you ready for the exam? Test yourself on key concepts by taking the additional tests linked with your BiologyNow CD-ROM.

____ 1. Blood pressure in capillaries, especially at the arteriole end, forces water out of the capillaries in a process called _____. [p.155]
   a. diffusion
   b. reabsorption
   c. exocytosis
   d. endocytosis
   e. ultrafiltration

____ 2. _____ valves may be either bicuspid or tricuspid. [p.156]
   a. Sinoatrial
   b. Semilunar
   c. Atrioventricular
   d. Lymphatic
   e. Venous

____ 3. During systole, _____. [p.157]
   a. the aorta contracts
   b. both ventricles contract
   c. the entire heart relaxes
   d. only the left atrium and left ventricle contract
   e. only the right atrium and right ventricle contract

____ 4. The "lub-dup" sound of the heart comes from _____. [p.157]
   a. the SA valves closing followed by the AV valves closing
   b. the AV valves closing followed by the semilunar valves closing
   c. the semilunar valves closing followed by the SA valves closing
   d. the contraction of the atria followed by the contraction of the ventricles
   e. electrical impulses from the SA node

_____ 5. Begin with a red blood cell located in the superior vena cava and travel with it in proper sequence as it goes through the following structures. Which will be last in sequence? [pp.158–159]
   a. aorta
   b. left atrium
   c. pulmonary artery
   d. right atrium
   e. right ventricle

_____ 6. Arterioles and venules are connected by _____. [p.159]
   a. thoroughfare channels
   b. interstitial passageways
   c. hepatic portals
   d. true capillaries
   e. renal portals

_____ 7. The pacemaker of the human heart is the _____. [p.160]
   a. sinoatrial node
   b. semilunar valve
   c. inferior vena cava
   d. superior vena cava
   e. atrioventricular node

_____ 8. _____ are blood reservoirs in which resistance to flow is low. [p.163]
   a. Arteries
   b. Arterioles
   c. Capillaries
   d. Venules
   e. Veins

_____ 9. The lymph vascular system collects water and solutes from _____ fluid and returns them to the cardiovascular system. [p.169]
   a. vascular
   b. cellular
   c. pulmonary
   d. digestive
   e. interstitial

_____ 10. Fluid carried by lymph vessels is returned to collecting ducts that drain into veins in the _____. [p.169]
   a. kidneys
   b. lower neck
   c. right atrium
   d. left atrium
   e. liver

Chapter Objectives/Review Questions

This section lists general and detailed chapter objectives that can be used as review questions. You can make maximum use of these items by writing answers on a separate sheet of paper. Fill in answers where blanks are provided. To check for accuracy, compare your answers with information given in the chapter or glossary.

1. Describe the functional links that the circulatory system has with the lymphatic system. [p.155]
2. Describe the structure of the heart, including linings, chambers, and valves. [p.156]
3. The contraction phase of the cardiac cycle is called _____ [p.157] and the relaxation phase is called _____ [p.157]
4. Trace the path of blood in the human body. Begin with the aorta and name all major components of the circulatory system through which all blood passes before it returns to the aorta. [p.158]
5. The _____ [p.158] circuit receives blood from the body tissues and circulates it through the lungs for gas exchange; the _____ [p.159] circuit transports blood to and from tissues.
6. Describe the role of capillary beds in the cardiovascular system. Where are these structures found in the body? [pp.154, 159–160]
7. Explain what causes a heart to beat. [p.160]
8. Explain a blood pressure such as 118/76 in terms of systolic and diastolic pressure. Is this a healthy blood pressure? What medical terms refer to elevated blood pressure and low blood pressure? Which is more dangerous and why? [p.161]
9. Describe how the structures of arteries, arterioles, and capillaries differ; describe how the structures of venules and veins differ. [pp.162–163]
10. List and briefly describe the four routes by which substances enter and leave capillaries. [pp.164–165]
11. Distinguish between hypertension and atherosclerosis. What factors increase the risk of these cardiovascular disorders? What can either lead to? [p.166]
12. State the significance of high- and low-density lipoproteins to cardiovascular disorders. [p.167]
13. Describe the components and function of the lymphatic system. [pp.168–169]

Media Menu Review Questions

Questions 1–3 are drawn from the following InfoTrac College Edition article: "Drano for the Heart: An Experimental Drug No One Expected to Work Is Surprisingly Effective at Rooting Out Cholesterol." John Travis. *Science News*, March 15, 2003.

1. The experimental drug tested at the Cleveland Clinic reduced by 4.2 percent the _____ that triggers most heart attacks.
2. The research around this drug began with the discovery of a population in Italy that has a very rare type of _____, the "good" blood cholesterol, which seems to reverse plaque buildup.
3. Even if this drug is effective in reducing plaque buildup, it must still be proven that reducing plaque will always _____ the risk of heart disease.

Questions 4–6 are drawn from the following InfoTrac College Edition article: "Gene Mutation Causes Heart Problems." John Travis. *Science News*, January 11, 2003.

4. Atrial fibrillation is a heart condition that accounts for _____ of all strokes in people over the age of 65.
5. The genetic defect involved in atrial fibrillation is a mutation on chromosome _____.
6. The mutant gene produces an ion channel that stays open longer than usual, resulting in an abnormal inrush of _____ that disrupts the heart's normal rhythm.

Questions 7–9 are drawn from the following InfoTrac College Edition article: "Cardiovascular Benefits of Long-term Fruit and Vegetable Consumption." *FDA Consumer*, September/October, 2002.

7. Eating at least _____ servings of fruits and vegetables each day over an extended period may help against stroke and cardiovascular disease.
8. This study by the National Health and Nutrition Examination Survey involved _____ adults from 25–74 over an average time period of 19 years.
9. On average, people who had eaten at least three servings of fruits and vegetables daily had a _____ percent lower incidence of stroke and a 42 percent lower stroke mortality rate.

Integrating and Applying Key Concepts

You are observing that some people appear as though fluid has accumulated in their lower legs and feet. Their lower extremities resemble those of elephants. You inquire about what is wrong and are told that the condition is caused by a parasite injected by the bite of a mosquito. Construct a testable hypothesis that would explain (1) why the fluid was not being returned to the torso as normal, and (2) how the parasite might prevent the return of fluid.

10

IMMUNITY

Interactive Exercises

Impacts, Issues: Viral Villains [p.173]

10.1. THREE LINES OF DEFENSE [p.174]

10.2. COMPLEMENT PROTEINS: "DEFENSE TEAM" PARTNERS [p.175]

10.3. INFLAMMATION—RESPONSES TO TISSUE DAMAGE [pp.176–177]

For additional practice, use the interactive vocabulary exercises linked with your BiologyNow CD-ROM.

Selected Words: *athlete's foot* [p.174], *Lactobacillus* [p.174], *nonspecific response* [p.174], *specific* response [p.174], *chemotaxins* [p.177], *lactoferrin* [p.177], *endogenous pyrogen* [p.177]

Boldfaced, Page-Referenced Terms

[p.174] pathogen _____

_____

[p.174] lysozyme _____

_____

[p.174] immune response _____

_____

[p.175] complement system _____

_____

[p.175] membrane attack complexes _____

_____

[p.175] lysis _____

_____

[p.176] neutrophils _____

_____

[p.176] eosinophils _____

_____

[p.176] basophils _____

_____

[p.176] macrophages _____

_____

[p.176] inflammation _____

_____

[p.176] mast cells _____

_____

[p.176] histamine _____

_____

[p.176] interleukin _____

_____

[p.176] fever _____

_____

Choice

For questions 1–9, choose from the following possible answers:

        a. body surface defense   b. nonspecific response   c. immune response

1. _____ lysozyme in tears, saliva, and gastric fluid [p.174]
2. _____ T cell and B cell lymphocytes [p.174]

3. _____ occurs when a pathogen breaches surface barriers [p.174]

4. _____ dense populations of harmless bacteria that exclude pathogens [p.174]

5. _____ inflammation [p.174]

6. _____ first line of defense [p.174]

7. _____ third line of defense [p.174]

8. _____ involves recognition of a specific pathogen [p.174]

9. _____ general response to tissue damage [p.174]

Outlining

Complete this outline of section 10.3 of your textbook concerning inflammation.

I. Four kinds of (10) _____ [p.176] blood cells mount an initial response to tissue damage.

   A. Three of these respond in a swift and general (nonspecific) manner.

      1. (11) _____ [p.176] ingest, kill, and digest bacteria

      2. (12) _____ [p.176] attack parasitic worms, phagocytize foreign proteins

      3. (13) _____ [p.176] secrete substances (histamine) to sustain inflammation

   B. The fourth type is the slower (14) _____. [p.176]

      1. engulf and ingest any (15) _____ [p.176] agent

      2. help clean up (16) _____ [p.176] tissue

II. Inflammation

   A. Inflammation—a response to damaged (17) _____. [p.176]

   B. Phagocytes, complement, etc., escape from blood (18) _____ into damaged tissue. [p.176]

   C. (19) _____ [p.176] cells release histamine.

      1. causes arterioles to (20) _____ [p.176] (leads to redness and warmth)

      2. increases the (21) _____ [p.177] of capillaries, which leak plasma proteins (leads to edema)

   D. Within hours, (22) _____ [p.177] squeeze out across capillary walls.

   E. Macrophages arrive later.

      1. engulf pathogens and (23) _____ [p.177]

      2. secrete chemical communication signals

         a. (24) _____ [p.177] "alarm" molecules that attract more phagocytes

         b. (25) _____ [p.177] carries signals between B and T lymphocytes

         c. lactoferrin directly kills (26) _____ [p.177]

         d. endogenous pyrogen triggers (27) _____ [p.177] release, which resets the body's thermostat, resulting in a (28) _____ [p.177]

   F. The plasma proteins that leak into the tissue include (29) _____ [p.177] proteins and (30) _____ [p.177] factors. The resulting blood (31) _____ [p.177] wall off the inflamed area to protect surrounding tissue.

Matching

Write the letter of the event shown on the figure next to its description below.

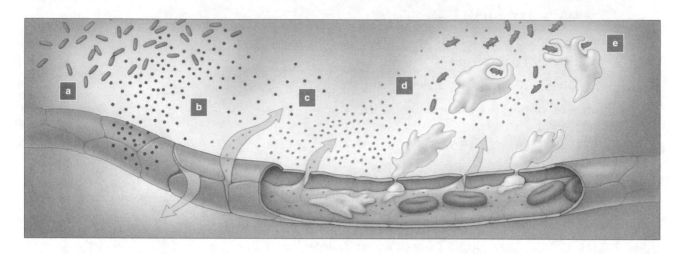

32. _____ Leakage from capillaries causes edema and pain. [p.177]

33. _____ Mast cells release histamine, causing vasodilation of arteries and capillary permeability. [p.177]

34. _____ Neutrophils, macrophages, and other phagocytes engulf invaders. Activated complement attracts phagocytes and directly kills invaders. [p.177]

35. _____ Tissue damage occurs due to bacterial invasion. [p.177]

36. _____ Plasma proteins attack bacteria; clotting factors wall off the inflamed area. [p.177]

10.4. AN IMMUNE SYSTEM OVERVIEW [pp.178–179]

10.5. HOW LYMPHOCYTES FORM AND DO BATTLE [pp.180–181]

10.6. A CLOSER LOOK AT ANTIBODIES IN ACTION [pp.182–183]

10.7. CELL-MEDIATED RESPONSES—COUNTERING THREATS INSIDE CELLS [pp.184–185]

10.8. ORGAN TRANSPLANTS: BEATING THE (IMMUNE) SYSTEM [p.185]

Selected Words: "nonself" markers [p.178], *effector cells* [p.178], *cell-mediated* responses [p.178], *antibody-mediated* responses [p.178], *variable regions* [p.180], *clonal selection* [p.180], *immunological memory* [p.180], *IgM* [p.182], *IgD* [p.183], *IgG* [p.183], *IgA* [p.183], *IgE* [p.183], "touch-kill" [p.185], *xenotransplantation* [p.185]

Boldfaced, Page-Referenced Terms

[p.178] immune system _____

_____

[p.178] self/nonself recognition _____

_____

[p.178] specificity _____

_____

[p.178] diversity _____

_____

[p.178] memory _____

_____

[p.178] antigen _____

_____

[p.178] memory cells _____

_____

[p.178] MHC markers _____

_____

[p.178] antigen-presenting cells _____

_____

[p.178] helper T cells _____

_____

[p.178] cytotoxic T cells _____

_____

[p.178] natural killer (NK) cells _____

_____

[p.178] B cells _____

_____

[p.178] antibody _____

_____

[p.179] suppressor T cells _____

_____

[p.180] TCRs (*T cell receptors*) _____

_____

[p.182] immunoglobulins (Igs) _____

_____

[p.184] interferons _____

_____

[p.185] apoptosis _____

_____

Short Answer

1. Which line of defense are B and T cells a part of? When does the body call upon this line of defense? [p.18]

   _____

   _____

   _____

2. List and briefly define the four features of the immune system. [p.178]

   _____

   _____

   _____

   _____

3. How do B and T cells distinguish between self and nonself? [p.178]

   _____

   _____

   _____

Sequencing

Sequence the events listed and write their letters in the appropriate blanks to show the order in which they occur. [p.178]

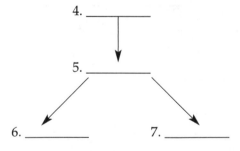

A. Repeated division of activated B and T cells

B. Specialization of B and T lymphocytes as memory cells for future attacks of the same invader

C. Specialization of B and T cell lymphocytes as effector cells to fight the invader

D. Recognition of "nonself" marker by B and T lymphocytes

4. _____

5. _____

6. _____   7. _____

Choice

Choose one of the white blood cell types (a–e) for each of the following functions.

     a. effector cytotoxic T cells and natural killer cells   d. helper T cells
     b. B cells   e. suppressor T cells
     c. antigen-presenting cells

8. _____ Prompt the specialization of B and T cells into effector and memory cells [p.178]

9. _____ Attack infected body cells, tumor cells, and organ transplant cells [p.178]

10. _____ Bind to antigen-MHC complexes and secrete chemical signals [p.178]

11. _____ Secrete huge numbers of antibodies that can bind to a specific invader that will then be destroyed by a phagocytic white blood cell [p.178]

12. _____ Produce chemical signals that help shut down an immune response [p.179]

13. _____ Ingest antigen, then link some antigen fragments to MHC markers for display [p.178]

14. _____ Take part in a cell-mediated response [p.178]

15. _____ Types include macrophages, B cells, and dendritic cells [p.178]

16. _____ Initiate repeated cell division to produce sensitized B and T cells producing effector and memory cells [p.178]

17. _____ Destroy cells by releasing chemicals that form lethal pores in a target's plasma membrane [p.178]

18. _____ Take part in an antibody-mediated response [p.178]

True/False

If the statement is true, write a "T" in the blank. If the statement is false, make it correct by writing the word(s) in the blank that should take the place of the underlined word(s).

_____ 19. Cytotoxic T cells attack tumor cells and cells of organ transplants. [p.178]

_____ 20. Helper T cells produce chemical signals that will cause cell division resulting in armies of sensitized T and B cells. [p.178]

_____ 21. Cytotoxic T cells and natural killer cells attack invader cells as part of an antibody-mediated response. [p.178]

_____ 22. Antibodies are released from NK cells. [p.178]

_____ 23. Antibodies bind to specific antigens. [p.178]

_____ 24. NK cells, macrophages, and neutrophils will destroy cells with antibodies attached to them. [p.178]

_____ 25. Antigen-presenting cells display ingested antibody fragments linked with MHC markers. [p.178]

Fill-in-the-Blanks

All of the antigen receptors on a particular B or T cell are identical, consisting of polypeptide chains. Certain parts of each chain are called (26) _____ _____ [p.180]. In each new T and B cell, genetic instructions for building these antigen-binding regions of receptor molecules are (27) _____ [p.180] around, producing unique polypeptides. As a B cell matures, it makes copies of its genetically unique, typically Y-shaped (28) _____ [p.180] molecule. The copies move to the B cell plasma membrane and the antigen-binding arms (29) _____ [p.180] above the lipid bilayer. This is repeated until the B cell is armed with multiple copies of its antibody. This cell is a (30) "_____" [p.180] B cell, as it has not yet encountered the specific antigen it is programmed to detect. As T cells form in the bone marrow, they migrate to the (31) _____ [p.180], where they mature and acquire unique antigen-binding receptors called (32) _____ [p.180]. Once released into the bloodstream, these molecules can recognize only (33) _____ -_____ [p.180] complexes. In the armies of B and T cells that form during an immune response, each one bears only a single type of (34) _____ [p.180], all specific for a detected invader. These armies arise from a

mechanism called (35)_____ _____ [p.180], in which an antigen binds only to the B or

T cell that is displaying a receptor specific for it. Once this first cell is activated, repeated rounds of cell

division produce a (36) _____ [p.180] of genetically identical cells. (37)_____

_____ [p.180] refers to the body's capacity to make a secondary immune response to a later

encounter with the same type of antigen that provoked the primary response. (38)_____

_____ [p.180] that form during the primary response circulate for years or decades. Because

there are many of these already on patrol, an infection by a previously encountered invader is

(39)_____ [p.180] before a person gets sick. A skin test for tuberculosis is a good example of

immunological memory.

In the skin, mucous membranes, and internal organs, the (40) _____ [p.181] and dendritic

cells pick up antigens by phagocytosis or endocytosis. They then migrate to the (41) _____ [p.181]

nodes all over the body. There, they present the antigen to (42) _____ and _____ [p.181]

cells. The antigen is presented to (43) _____ [p.181] B and T cells to start an immune response.

Any antigens that have made their way into tissue fluid enter the (44) _____ [p.181] vascular

system. Some may attempt to enter the bloodstream, but defending cells in the (45) _____

_____ [p.181] trap most of them. Antigens that so enter the bloodstream are intercepted by

the (46) _____ [p.181]. In the lymph nodes, antigens move through the region occupied by

B cells, macrophages, and dendritic cells. The invaders are captured, processed, and presented to

(47) _____ _____ [p.181] cells, thereby activating them. At this point, a full-blown

(48) _____ _____ [p.181] gets under way.

Sequencing

Write the letter of the first of the following events to occur beside number 49, with the other letters
following in sequence.

49. _____

50. _____

51. _____

52. _____

53. _____

54. _____

55. _____

A. Effector B cells produce huge numbers of antibodies that flag invaders for destruction by phagocytes. [p.182]

B. The clonal descendants of the activated B cell specialize as effector and memory B cells. [p.182]

C. An antigen binds with antibodies on the B cell and links them together. [p.182]

D. The TCRs of a helper T cell bind to the antigen-MHC complex displayed on a B cell, allowing an exchange of signals after which the cells disengage. [p.182]

E. The antibodies of a B cell bind to a second unprocessed antigen. This, along with interleukins secreted from helper T cells, causes B cell cloning. [p.182]

F. A naïve B cell is produced, bristling with many copies of a unique antigen. [p.182]

G. Endocytosis occurs and antigen-MHC complexes form that are displayed on the surface of the B cell. [p.182]

Matching

56. _____ IgG [p.183]
57. _____ IgA [p.183]
58. _____ IgM [p.182]
59. _____ IgE [p.183]
60. _____ IgD [p.183]

A. Binds to basophils and mast cells, prompting histamine release
B. Ten binding sites; first antibody made by newborns
C. Commonly found bound to naïve B cells; helps activate T helper cells
D. Found in tears, saliva, breast milk, and mucus
E. Most efficient at turning on complement proteins; neutralizes many toxins

Fill-in-the-Blanks

Pathogens such as viruses, some bacteria, some fungi, protozoans, and tumor cells hide inside body

(61) _____ [p.185]. The body's weapons against such dangers are (62) _____-

_____ [p.185] immune responses. Helper T cells and cytotoxic T cells respond to particular

(63) _____ [p.185]. NK cells, macrophages, neutrophils, and eosinophils make more

(64) _____ [p.185] responses. Chemical signals like interleukins and defensive proteins called

(65) _____ [p.185] attract defenders and stimulate them to divide, specialize, and attack.

Interferons also spur the making of (66) _____ [p.185] molecules, so some cells display

antigens more effectively.

Cytotoxic T cells produce molecules used to (67) "_____ _____" [p.185] infected and

abnormal body cells. A type called (68) _____ [p.185] form pores in a target's plasma membrane.

Cytotoxic T cells also secrete chemicals that cause the programmed death, or (69) _____ [p.185], of

a target cell, as well as contributing to the (70) _____ [p.185] of transplanted tissues and organs.

Natural killer (NK) cells act when infections trigger a flood of interferons and (71) _____ [p.185].

They are the (72) _____ [p.185] immune cells to act, and buy time while others are mobilizing.

Organ transplants are risky because the organ recipient's immune system will perceive donated tissues

as (73) _____ [p.185] and attempt to reject them. Transplants usually succeed only when the

donor and recipient share at least (74) _____ [p.185] percent of their MHC markers. Because of

this, the best donors are close (75) _____ [p.185] of the recipient. After surgery, the organ

recipient receives drugs that (76) _____ [p.185] the immune system. This means that the patient

must take large doses of (77) _____ [p.185] to control infections. Some researchers are trying to

genetically alter (78) _____ [p.185] to create varieties bearing common human MHC markers.

Transplantation of organs from one species to another is called (79) _____ [p.185]. Tissues of the

(80) _____ [p.185] and (81) _____ [p.185] are examples of exceptions to the "rule" that

transplanted tissues evoke immune defenses, thus making transplants in structures like the cornea simpler.

10.9. PRACTICAL APPLICATIONS OF IMMUNOLOGY [pp.186–187]

10.10. A CAN'T WIN PROPOSITION? [p.187]

10.11. DISORDERS OF THE IMMUNE SYSTEM [pp.188–189]

Selected Words: active immunization [p.186], "booster shot" [p.186], "attenuated" pathogens [p.186], "transgenic" viruses [p.186], *passive* immunization [p.186], "plantibody" [p.187], *gamma* interferon [p.187], *beta* interferon [p.187], *multiple sclerosis* [p.187], *lymphokines* [p.187], *influenza* virus [p.187], "gene swapping" [p.187], *allergens* [p.188], *hay fever* [p.188], *rheumatoid arthritis* [p.189], *diabetes mellitus* [p.189], *systemic lupus erythematosus* (SLE) [p.189], *severe combined immune deficiency* (SCID) [p.189]

Boldfaced, Page-Referenced Terms

[p.186] vaccine _____

_____

[p.186] monoclonal antibodies _____

_____

[p.184] allergy _____

_____

[p.188] anaphylactic shock _____

_____

[p.189] autoimmune response _____

_____

[p.189] immunodeficiency _____

_____

[p.189] AIDS (acquired immunodeficiency syndrome) _____

_____

Matching

1. _____ monoclonal antibodies
2. _____ gamma interferon
3. _____ active immunization
4. _____ passive immunization
5. _____ LAK cells
6. _____ attenuated
7. _____ "plantibody"
8. _____ booster shot
9. _____ beta interferon
10. _____ booster shot
11. _____ transgenic
12. _____ vaccine

A. Vaccine is injected or taken orally [p.186]
B. Elicits a secondary response that results in more effector and memory cells being made [p.186]
C. Being used to treat a type of multiple sclerosis [p.187]
D. Genetically engineered viruses used to make vaccines [p.186]
E. Injections of purified antibody molecules; for people who are already infected with pathogens [p.186]
F. Genetically engineered plant being used to make cost-effective and safe antibodies [pp.186–187]
G. B cells cloned from a single antibody-producing cell; used in research to find very small quantities of a substance [pp.186–187]
H. Produced by T cells, calls NK cells into action and boosts the activity of macrophages [p.187]
I. Killed or extremely weakened pathogens used in making vaccines [p.186]
J. Second administration of a vaccine; elicits a secondary immune response [p.186]
K. Tumor-infiltrating lymphocytes produced by research [p.187]
L. Prepared substance containing an antigen; elicits a primary immune response [p.186]

Short Answer

13. How do the three strains of influenza viruses differ from each other? In addition to this variation, name one way in which a strain can mutate to become a new variant of the flu. [p.187]

_____

_____

_____

Fill-in-the-Blanks

A(n) (14) _____ [p.188] is an immune response to a normally harmless substance. Such a substance is called a(n) (15) _____ [p.188]. Some people are genetically (16) _____ [p.188] to develop allergies. When an allergic person is first exposed to certain antigens, IgE (17) _____ [p.188] are secreted and bind to (18) _____ [p.188] cells. These secrete prostaglandins, (19) _____ [p.188], and other substances that cause inflammation. They also cause the person's airways to (20) _____ [p.188].

Food allergies are skewed responses of the immune system in which a particular food is interpreted as a(n) (21) "_____" [p.188]. If a whole-body response to an allergen occurs, the potentially lethal event called (22) _____ _____ [p.188] results. This may also result from an allergy to insect venom. Air passageways constrict, sever edema occurs, and blood pressure plummets, which can lead to complete collapse of the (23) _____ [pp.188–189] system. The emergency treatment is an injection of the hormone (24) _____ [p.189].

(25) _____ [p.189] are anti-inflammatory drugs that are used to relieve short-term allergy symptoms. In a desensitization program, doses of allergens are administered so that a person's body produces more circulating (26) _____ [p.189] molecules and memory cells, thus blocking allergic inflammation.

A(n) (27) _____ [p.189] response is a disorder in which the body mobilizes its forces against normal body cells or proteins. (28) _____ _____ [p.189] is an example of this kind of disorder, in which skeletal joints are chronically inflamed. Another example is type 1 (29) _____ [p.189], in which the immune system destroys the insulin-secreting cells of the pancreas. In the autoimmune disease (30) _____ _____ _____ [p.189], the affected person develops antibodies to her or his own DNA and other "self" components. Immunodeficiency refers to disorders in which the body does not have enough functioning (31) _____ [p.189]. Both T and B cells are in short supply in (32) _____ _____ _____ _____ [p.189] (SCID), an inherited life-threatening disorder. Infection by the (33) _____ _____ _____ [p.189] (HIV) causes AIDS—(34) _____ _____ _____ [p.189]. HIV is transmitted when (35) _____ _____ [p.189] of an infected person enter another person's tissues. The virus cripples the immune system by attacking (36) _____ _____ [p.189] cells and (37) _____ [p.189]. The body then becomes dangerously susceptible to opportunistic infections and to some otherwise rare forms of (38) _____ [p.189].

Self-Quiz

Are you ready for the exam? Test yourself on key concepts by taking the additional tests linked with your BiologyNow CD-ROM.

Multiple Choice

_____ 1. All the body's white blood cells are derived from stem cells in the _____. [p.176]
a. spleen
b. liver
c. thymus
d. bone marrow
e. thyroid

_____ 2. The plasma proteins that are activated when they contact a microorganism or virus are collectively known as the _____ system. [p.175]
a. shield
b. complement
c. IgG
d. MHC
e. HIV

_____ 3. The lymphocytes known as _____ attack only specific nonself cells or substances. [p.178]
a. macrophages
b. B and T cells
c. mast cells
d. basophils
e. interleukins

_____ 4. _____ produce and secrete antibodies that set up bacterial invaders for subsequent destruction by macrophages. [p.178]
a. B cells
b. Phagocytes
c. T cells
d. Mast cells
e. Thymus cells

_____ 5. Antibody molecules are shaped like the letter _____. [p.180]
a. Y
b. W
c. Z
d. H
e. E

_____ 6. The markers for every cell in the human body are referred to by the letters _____. [p.178]
a. HIV
b. MBC
c. RNA
d. DNA
e. MHC

_____ 7. B and T lymphocytes that enter a resting stage are called _____. [p.178]
a. effector cells
b. macrophages
c. mast cells
d. helper cells
e. memory cells

_____ 8. The great variety of naïve B and T cells differ in _____ regions of polypeptide chains. [p.180]
a. shuffled
b. constant
c. clonal
d. mutated
e. variable

____ 9. When an allergic person is first exposed to certain antigens, IgE antibodies cause _____. [p.188]
a. inflammation
b. clonal cells to be produced
c. B cell division
d. the immune response to be suppressed
e. an autoimmune disorder to develop

____ 10. The clonal selection theory explains _____. [p.180]
a. how all the B or T cells responding to an invasion bear a single type of antigen receptor
b. how B cells differ from T cells
c. how so many different kinds of antigen-specific receptors can be produced by lymphocytes
d. how memory cells are set aside from effector cells
e. how antigens differ from antibodies

Matching

Choose the most appropriate description for each term.

11. _____ autoimmune response [p.189]

12. _____ antibody [p.178]

13. _____ antigen [p.178]

14. _____ macrophage [p.176]

15. _____ clone [p.180]

16. _____ complement [p.175]

17. _____ histamine [pp.176–177]

18. _____ MHC marker [p.178]

19. _____ effector cells [p.178]

20. _____ T cell [p.180]

A. Begins its development in bone marrow, but matures in the thymus gland
B. A population of genetically identical cells that descended from a selected T or B cell
C. A chemical that causes blood vessels to dilate and let plasma proteins leak through the vessel walls
D. Y-shaped immunoglobulin produced by a B cell
E. A nonself marker that triggers the formation of lymphocytes
F. Produced by activated B and T cells to destroy a specific invader
G. A group of about twenty proteins that participate in nonspecific and specific defenses
H. A disorder in which the body's immune system attacks its own cells and proteins
I. Proteins sticking out of the plasma membrane of body cells that identify "self"
J. "Big eater" that lives for months and phagocytizes foreign agents

Chapter Objectives/Review Questions

This section lists general and detailed chapter objectives that can be used as review questions. You can make maximum use of these items by writing answers on a separate sheet of paper. To check for accuracy, compare your answers with information given in the chapter or glossary.

1. Describe typical external barriers that organisms such as humans present to invading organisms. [p.174]
2. List and discuss nonspecific defense responses that serve to exclude microbes from the body. [p.174]
3. Explain how the complement system is involved in destroying invaders and fanning inflammation. [p.175]
4. Describe what occurs in the course of an acute inflammation, and how these events stop the spread of a pathogen. [pp.176–177]
5. List the four features that define the immune system. [p.178]
6. List the three basic steps of an immune response. [p.178]
7. Distinguish between the cell-mediated response and the antibody-mediated response. [p.178]

8. Describe the clonal selection theory, and relate this concept to immunological memory. [p.180]
9. Explain how helper T cells and cytotoxic T cells are involved in countering threats inside cells. [pp.184–185]
10. Contrast active and passive immunization in the areas of procedure and effectiveness. [p.186]
11. Distinguish allergy from autoimmune disease. [pp.188–189]
12. Describe how AIDS specifically interferes with the human immune system. [p.189]

Media Menu Review Questions

Questions 1–10 are drawn from the following InfoTrac College Edition article: "Remembrance of Pathogens Past: A Physician Ponders Suggestive Evidence that Periodic Infections Are Needed to Foster a Normal Immune System." T. V. Rajan. *Natural History*, February 2002.

1. It is difficult to appreciate the wonders of modern _____ and sewage control if one has never been exposed to the realities experienced in most of the world.
2. Of all the accomplishments of modern science, the conquest of _____ illnesses must surely rank among the greatest.
3. Evidence from NOD (non-obese diabetic) mice demonstrates that diabetes is an autoimmune disease caused by malfunctioning _____.
4. Of the traits shared by diabetic mice, among the most important are the genes for the _____ (MHC). The result in these mice is that T cells attack _____ "self" cells instead of pathogens.
5. Research with the NOD mice suggests that exposure to and recovery from _____ infection prevented the development of autoimmune disease (diabetes in this case).
6. Rafi Ahmed's findings, at the Emory University School of Medicine, suggest that for most mammals, infection-specific _____ cells accumulate with each subsequent illness, becoming more abundant in the _____ of aging animals than "naïve" white cells.
7. Lack of exposure, the environmental factor working with genetic predisposition, may lead to greater vulnerability to _____ disorders.
8. Most biologists believe that infectious agents are major players in the _____ of the human genome, as first stated by Haldane in 1949.
9. Haldane's hypothesis is supported by the observation that certain genes, such as for _____ and _____ disease, are found almost exclusively in ethnic groups originating where a particular infectious disease is widespread.
10. It is fascinating to think that the human immune system is geared to accommodate a certain number of infectious agents that may play the _____ role of regulating the immune system.

Integrating and Applying Key Concepts

Vaccines can be developed against many pathogens, but not all. Diseases like the common cold and AIDS cannot be prevented by vaccination. Others, like influenza, must be vaccinated against every year. Humans are not capable of developing lasting immunity to any of the pathogens causing these diseases. Offer a possible genetic explanation for this variation among pathogens.

11

THE RESPIRATORY SYSTEM

Interactive Exercises

Impacts, Issues: Up in Smoke [p.193]

11.1. THE RESPIRATORY SYSTEM—BUILT FOR GAS EXCHANGE [pp.194–195]

For additional practice, use the interactive vocabulary exercises linked with your BiologyNow CD-ROM.

Selected Words: "respiration" [p.194], *nasal cavity* [p.194], "Adam's apple" [p.195], *epiglottis* [p.195], *glottis* [p.195], *laryngitis* [p.195], *pleural sac* [p.195], *intrapleural space* [p.195], *intrapleural fluid* [p.195], *pleurisy* [p.195], "bronchial trees" [p.195], *alveolar sac* [p.195]

Boldfaced, Page-Referenced Terms

[p.194] respiratory system _____

_____

[p.194] pharynx _____

_____

[p.194] larynx _____

_____

[p.195] trachea _____

_____

[p.195] bronchus _____

_____

[p.195] vocal cords _____

_____

[p.195] lungs _____

_____

[p.195] diaphragm _____

_____

[p.195] pleurae _____

_____

[p.195] bronchioles _____

_____

[p.195] respiratory bronchioles _____

_____

[p.195] alveolus (plural: alveoli) _____

_____

Fill-in-the-Blanks

Smoke immobilizes (1) _____ [p.193] in the passageways to the lungs. It also kills (2) _____ _____ _____ [p.193] that defend the respiratory tract. Microbes may start living there, leading to colds, (3) _____ [p.193], and asthma. Cigarette smoke contains (4) _____ [p.193] that can help stoke cancer in organs throughout the body. Other effects of smoking are increased (5) _____ [p.193], higher levels of the (6) _____ _____ [p.193], and lower levels of the (7) _____ _____ [p.193].

Labeling-Matching

Identify each of the components of the human respiratory system in the accompanying illustration by entering the correct names in the numbered blanks provided. Complete the exercise by matching and entering the letter of the correct function of each component in the parentheses that follow most of the labels. [p.194]

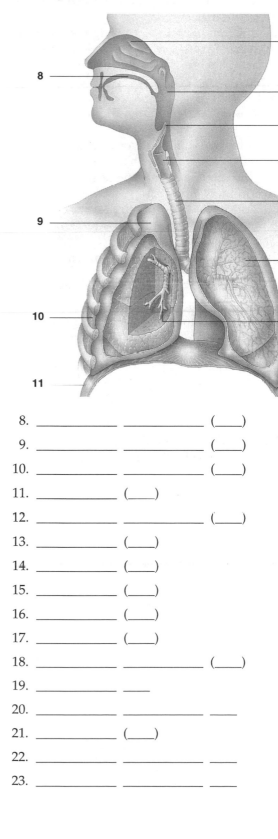

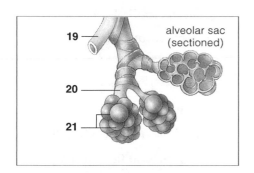

alveolar sac (sectioned)

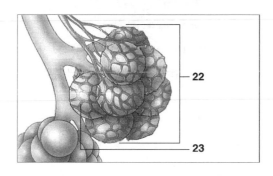

8. _____ _____ (___)

9. _____ _____ (___)

10. _____ _____ (___)

11. _____ (___)

12. _____ _____ (___)

13. _____ (___)

14. _____ (___)

15. _____ (___)

16. _____ (___)

17. _____ (___)

18. _____ _____ (___)

19. _____ ___

20. _____ _____ ___

21. _____ (___)

22. _____ _____ ___

23. _____ _____ ___

A. Airway where sound is produced; closed off while swallowing
B. Muscle sheet between chest cavity and abdominal cavity with roles in breathing
C. Increasingly branched airways between two bronchi and alveoli
D. Closes off larynx during swallowing
E. Chamber in which air is warmed, moistened, and filtered, and in which sounds resonate
F. Rib cage muscles with roles in breathing
G. Airway that connects larynx with two bronchi
H. Supplemental airway when breathing is labored
I. Thin-walled air sacs where gases are exchanged between lungs and pulmonary capillaries
J. Membranes that separate lungs from other organs; fluid-filled cavity between has roles in breathing
K. Airway that connects nasal cavity and mouth with larynx; enhances sounds; also connects with esophagus
L. Lobed, elastic organ of breathing that enhances gas exchange between the internal environment and the outside air

Matching

Choose the most appropriate description for each term.

24. _____ nose [p.194]

25. _____ pleural sac [p.191]

26. _____ alveoli [p.195]

27. _____ larynx [pp.194–195]

28. _____ vocal cords [p.195]

29. _____ epiglottis [p.195]

30. _____ bronchus [p.195]

31. _____ laryngitis [p.195]

32. _____ intrapleural space [p.195]

33. _____ intrapleural fluid [p.195]

34. _____ pleurisy [p.195]

35. _____ diaphragm [p.195]

36. _____ glottis [p.195]

37. _____ lung [p.195]

38. _____ trachea [p.195]

A. A sheet of muscle between the thoracic and abdominal cavities
B. Where air is filtered, warmed, and moisturized before entering the respiratory system
C. Painful inflammation of pleurae, causing membranes to dry out and rub against each other, or oversecrete fluid, which hampers breathing movements
D. A gap between the vocal cords that is forced open with each exhalation
E. Where gas diffusion between lungs and lung capillaries takes place
F. Inflammation of mucus lining of vocal cords
G. Left one has two lobes, right one has three
H. Formed by a thin, double membrane of epithelium covering each lung
I. Covers glottis during swallowing to prevent choking
J. Consists of horizontal folds of mucus membrane near entrance to larynx
K. Formed of nine pieces of cartilage, one of which is the "Adam's apple"
L. "Windpipe" supported by rings of cartilage
M. Lubricating substance filling the intrapleural space
N. Lined by cilia and cells that secrete mucus to trap bacteria and particles
O. A very narrow space between the two pleural membranes

Labeling-Matching

Arrange the following parts of the respiratory system in the order in which air would enter them after leaving the nasal cavity. Place the letter of the first by the number 39, the next by the number 40, etc.

39. _____

40. _____

41. _____

42. _____

43. _____

44. _____

45. _____

A. alveolus [p.195]
B. larynx [p.194]
C. bronchus [p.195]
D. pharynx [p.194]
E. glottis [p.195]
F. bronchiole [p.195]
G. trachea [p.195]

11.2. THE "RULES" OF GAS EXCHANGE [p.196]

11.3. BREATHING AS A HEALTH HAZARD [p.197]

11.4. BREATHING—AIR IN, AIR OUT [pp.198–199]

Selected Words: pressure gradients [p.196], *hypoxic* [p.196], *hyperventilation* [p.196], *carbon monoxide poisoning* [p.196], *bronchitis* [p.197], *emphysema* [p.197], *"second-hand smoke"* [p.197], *asthma* [p.197], *intrapulmonary pressure* [p.198], *negative pressure gradient* [p.199], *inspiratory reserve volume* [p.199], *expiratory reserve volume* [p.199], *residual volume* [p.199]

Boldfaced, Page-Referenced Terms

[p.196] respiratory surface _____

_____

[p.198] respiratory cycle _____

_____

[p.198] inspiration _____

_____

[p.198] expiration _____

_____

[p.199] tidal volume _____

_____

[p.199] vital capacity _____

_____

Fill-in-the-Blanks

(1) _____ _____ [p.196] relies on the tendency of oxygen and carbon dioxide to diffuse down their concentration gradients, or, as we say in the case of gases, their (2) _____ [p.196] gradients. The partial pressure of oxygen at sea level is (3) _____ [p.196] percent of 760 mm Hg (atmospheric pressure at sea level), which equals 160 mm Hg. The partial pressure of (4) _____ _____ [p.196] is about 0.3 mm Hg. Large animals such as humans must be capable of efficient gas exchange. Gases enter and leave the body by crossing a thin, moist (5) _____ [p.196] surface of epithelium. The surface must be moist because gases cannot diffuse across membranes unless they are (6) _____ [p.196] in fluid. The larger the surface area and the steeper the partial (7) _____ _____ [p.196], the faster diffusion occurs. In healthy human lungs, millions of thin-walled (8) _____ [p.196] provide a huge surface area for gas exchange. Gas exchange is facilitated by the (9) _____ [p.196] in red blood cells. In the lungs, each hemoglobin molecule binds with up to (10) _____ [p.196] oxygen molecules. When blood carries red blood cells into tissues where

oxygen concentration is low, hemoglobin (11) _____ [p.196] oxygen. By carrying oxygen away

from the respiratory surface, hemoglobin helps maintain the required pressure (12) _____ [p.196]

that helps draw oxygen into the lungs—and into blood in lung capillaries.

Dichotomous Choice

Circle one of two possible answers given between parentheses in each statement.

13. Partial pressure of oxygen (decreases/increases) as altitude increases. [p.196]

14. Higher than 2,400 meters, or about 8,000 feet, brain respiratory centers compensate for oxygen deficiency by triggering faster and deeper breathing called (hypoxia/hyperventilation). [p.196]

15. In carbon monoxide poisoning, tiny amounts of the gas can tie up half of the body's hemoglobin, resulting in tissues not receiving enough oxygen, a condition called (hypoxia/hyperventilation). [p.196]

Choice

For questions 16–24, choose from the following:

        a. bronchitis   b. emphysema   c. asthma

16. _____ A chronic, incurable condition of the respiratory tract in which breathing can become so difficult so quickly that the victim feels in imminent danger of suffocating [p.197]

17. _____ Scar tissue from chronic bronchitis builds up, the bronchi become clogged with mucus, and a reduced number of enlarged alveoli skew the balance between air flow and blood flow [p.197]

18. _____ Can be brought on by air pollution that increases mucus secretions and interferes with ciliary action in the lungs [p.197]

19. _____ Many sufferers lack a functional gene coding for antitrypsin, a protein that inhibits tissue-destroying enzymes produced by bacteria [p.197]

20. _____ Factors that trigger an attack actually cause the smooth muscle to contract in strong spasms, so that the bronchi and bronchioles suddenly narrow [p.197]

21. _____ Cilia are lost from the lining, and mucus-secreting cells multiply as the body fights against the accumulating debris [p.197]

22. _____ The disease can develop over 20 or 30 years, but by the time it is detected, lung tissue is permanently damaged [p.197]

23. _____ Triggering substances can include pollen, dairy products, shellfish, flavorings and other foods, pet hairs or dandruff—even mite dung in house dust [p.197]

24. _____ With this disorder, the lungs are so distended and inelastic that gases cannot be exchanged efficiently; running, walking, even exhaling can be difficult [p.197]

Analyzing Diagrams

To better understand the mechanisms of ventilation, study each indicated part of the following illustration. Then answer the accompanying questions. As you do the exercise, it aids understanding if you are conscious of the same parts of your own anatomy presently undergoing the ventilation cycle. [p.198]

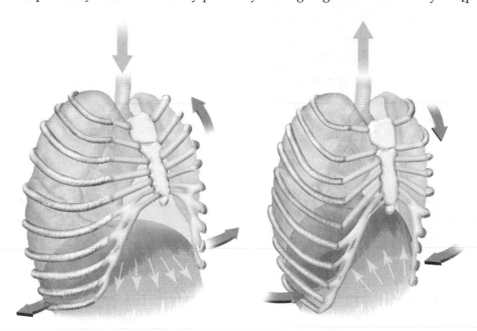

_____ 25. Which diagram, right or left, shows inspiration?

_____ 26. Which diagram, right or left, shows expiration?

_____ 27. Which muscles contract and relax to cause the rib cage movement indicated by the three larger arrows on each diagram?

_____ 28. The small arrows on each diagram show the contraction and relaxation of which muscle?

_____ 29. On the left-hand diagram, which direction does the diaphragm move?

_____ 30. On the left-hand diagram, which directions does the rib cage move?

_____ 31. What effect do the movements on the left-hand diagram have on lung volume?

_____ 32. On the right-hand diagram, which direction does the diaphragm move?

_____ 33. On the right-hand diagram, which directions does the rib cage move?

_____ 34. What effect do the movements on the right-hand diagram have on the lungs?

Dichotomous Choice

Circle one of two possibilities given between parentheses in each statement.

35. The air movements of ventilation result from rhythmic increases and decreases in the volume of the (lungs/thoracic cavity). [p.198]

36. When the air pressure in alveolar sacs is lower than atmospheric pressure, air flows down its gradient and (enters/exits) the alveoli. [p.199]

37. The normally passive—does not require energy use—part of the respiratory cycle is (inspiration/ expiration). [p.199]

38. When the muscles that cause inhalation relax, air flows (into/out of) the lungs. [p.199]

39. The negative pressure gradient outside the lungs keeps the lungs snug against the wall of the (pleural/thoracic) cavity even during exhalation. [p.199]

40. Pneumothorax, or collapsed lung, is a condition in which air can enter the (intercostals/pleural) cavity, preventing normal expansion of the lungs. [p.199]

41. The cohesiveness of water molecules in the fluid inside the (pleural sac/thoracic cavity) helps keep the lungs close to the thoracic wall. [p.199]

42. The amount of air taken in by a person in a normal breath, about 500 ml, is known as the (vital capacity/tidal volume). [p.199]

43. In addition to the air taken in as tidal volume, a person can forcibly inhale roughly 3,100 ml of air, called the (expiratory/inspiratory) reserve volume. [p.199]

44. Forcibly exhaling, you can expel an additional (expiratory/inspiratory) reserve volume of about 1,200 ml. [p.199]

45. The maximum volume of air that can move out of the lung after a person inhales as deeply as possible is called the (vital capacity/tidal volume). [p.199]

46. People rarely take more than (one-half/three-fourths) of their vital capacity, even when they breathe deeply during strenuous exercise. [p.199]

47. Even at the end of your deepest exhalation, your lungs still cannot be completely emptied of air; roughly another 1,200 ml of (expiratory reserve/residual) volume remains. [p.199]

48. About (150/350) ml of inhaled air is actually available for gas exchange. [p.199]

11.5. HOW GASES ARE EXCHANGED AND TRANSPORTED [pp.200–201]

11.6. HOMEOSTASIS DEPENDS ON CONTROLS OVER GAS EXCHANGE [pp.202–203]

11.7. TOBACCO AND OTHER THREATS TO THE RESPIRATORY SYSTEM [pp.204–205]

Selected Words: *external* respiration [p.200], *internal* respiration [p.200], *respiratory membrane* [p.200], *pulmonary surfactant* [p.200], *infant respiratory distress syndrome* [p.200], *carbaminohemoglobin* ($HbCO_2$) [p.201], breathing *rhythm* [p.202], breathing *magnitude* [p.202], *cerebrospinal fluid* [p.202], *apnea* [p.203], *sleep apnea* [p.203], "secondhand smoke" [p.204]

Boldfaced, Page-Referenced Terms

[p.200] oxyhemoglobin (HbO_2) _____

_____

[p.201] carbonic anhydrase _____

_____

[p.203] carotid bodies _____

_____

[p.203] aortic bodies _____

_____

Matching

Choose the most appropriate description for each term.

1. _____ aortic bodies [p.203]
2. _____ carbaminohemoglobin ($HbCO_2$) [p.201]
3. _____ carbonic anhydrase [p.201]
4. _____ carotid bodies [p.203]
5. _____ external respiration [p.200]
6. _____ infant respiratory distress syndrome [p.200]
7. _____ internal respiration [p.200]
8. _____ oxyhemoglobin (HbO_2) [p.200]
9. _____ respiration [p.200]
10. _____ respiratory centers [p.203]
11. _____ pulmonary surfactant [p.200]
12. _____ ventilation [p.200]

A. Hemoglobin with bound oxygen
B. Enzyme in red blood cells mediating the chemical reactions that form and dissociate carbonic acid
C. Phase of respiration in which oxygen moves from blood into tissues, and carbon dioxide moves from tissues into blood
D. Sensory receptors located where carotid arteries branch to the brain; detect arterial changes in carbon dioxide and oxygen levels as well as pH
E. Provides the body as a whole with oxygen for aerobic respiration in cells and disposes of carbon dioxide
F. Stimulated by decreasing pH of cerebrospinal fluid
G. Life-threatening breathing disorder in premature infants whose partially developed lungs do not yet have functional surfactant-secreting cells
H. Sensory receptors in arterial walls near the heart that detect changes in carbon dioxide, oxygen, and pH levels
I. Refers only to movement of gases into and out of the lungs
J. Carbon dioxide bound with hemoglobin
K. Phase of respiration in which oxygen moves from alveoli to blood and carbon dioxide moves from blood to alveoli
L. Secreted by certain cells of the alveolar epithelium; reduces surface tension of the watery fluid film between alveoli

Dichotomous Choice

Circle one of the two possibilities given between parentheses in each statement.

13. The higher the partial pressure of oxygen around the alveoli, the (less/more) oxygen will be picked up by hemoglobin. [p.200]

14. Oxygen-binding to hemoglobin weakens as temperature rises, or as pH (increases/decreases). [p.200]

15. The (more/less) DPG bound to hemoglobin, the less affinity hemoglobin has for binding oxygen and thus more oxygen can be available to tissues. [p.200]

16. About 70 percent of the body's carbon dioxide is transported in plasma in the form of (carbonic acid/bicarbonate). [p.201]

17. Bicarbonate forms after carbon dioxide combines with (water/carbonic acid) in plasma. [p.201]

18. Carbonic anhydrase mediates reactions that convert unbound carbon dioxide to carbonic acid and its dissociation products; the blood level of carbon dioxide then (rises/falls) rapidly. [p.201]

19. In the alveoli, the partial pressure of carbon dioxide is (higher/lower) than it is in the surrounding capillaries, so that it diffuses into the sacs and is exhaled. [p.201]

20. The H^+ formed in red blood cells along with bicarbonate binds to hemoglobin, which acts as a (buffer/plasma protein) to minimize pH change. [p.201]

Fill-in-the-Blanks

Gas exchange is most efficient when breathing is regulated so that the rate of air flow is matched with the rate of blood flow. The (21) _____ [p.202] system acts to balance the flow rates. It controls breathing (22) _____ [p.202], which is the rate of breathing, and breathing (23) _____ [p.202], which is the depth of breathing. Breathing rhythm involves nervous system control of the (24) _____ [p.202] and rib cage muscles. One group of cells of the reticular formation in the brain stem coordinates signals calling for (25) _____ [p.202]; another cell cluster coordinates the signals for (26) _____ [p.202]. The resulting rhythmic contractions are fine-tuned by another part of the brain stem, the (27) _____ [p.202], which can stimulate or inhibit both cell clusters.

The body's control over breathing magnitude involves monitoring the level of (28) _____ [p.202] in the blood. However, the nervous system is more sensitive to levels of (29) _____ _____ [p.202]. Both oxygen and carbon dioxide levels are monitored in blood flowing through (30) _____ [p.202]. When conditions warrant, nervous system signals adjust contractions of the (31) _____ [p.202] and muscles in the chest wall, and so adjust the rate and depth of breathing. Sensory receptors in the (32) _____ [p.202] of the brain detect rising carbon dioxide levels. The receptors detect hydrogen ions in the (33) _____ [p.202] fluid, which bathes the medulla. The shift in pH stimulates the receptors that signal changes to the brain's (34) _____ [p.203] centers. The rate and depth of breathing fall, followed by a drop in the blood level of (35) _____ [p.203]. In addition, the brain receives input from sensory receptors such as the (36) _____ [p.203] bodies located where the carotid arteries branch to the brain, and the (37) _____ [p.203] bodies in arterial walls near the heart. Both types of receptors can detect changes in carbon dioxide and oxygen levels in arterial blood as well as changes in blood pH. The brain responds by increasing the (38) _____ [p.203] rate, so more oxygen can be delivered to tissues.

Dichotomous Choice

Circle one of the two possibilities between parentheses in each statement.

39. If blood flow is too fast compared to ventilation rate, there is an a(n) (increase/decrease) in the blood level of carbon dioxide. [p.203]

40. An increase in the blood level of carbon dioxide affects smooth muscle in the bronchiole walls so that the bronchioles (constrict/dilate), increasing air flow. [p.203]

41. A decrease in the blood level of carbon dioxide causes the bronchiole walls to (constrict/dilate), so air flow decreases. [p.203]

42. The mechanisms by which the nervous system regulates respiration normally operate under (voluntary/involuntary) control. [p.203]

43. Breathing that stops briefly and then resumes is called (hiccupping/apnea). [p.203]

44. Sleep apnea is a common problem in (very young/elderly) people. [p.203]

Fill-in-the-Blanks

Cigarette smoke—including (45) "_____ [p.204] smoke" inhaled by a nonsmoker—causes lung cancer and contributes to various other ills. Cigarette smoking causes at least (46) _____ [p.204] percent of all lung cancer deaths.

Consider Your Own Health

47. A considerable amount of scientific research has been reported that firmly establishes the devastating effects of smoking on human beings. Study the Focus on Your Health essay and table on page 200 closely, and then complete the following table by first checking in the left column only those risks associated with smoking that *you personally are willing to take* as a smoker. Be prepared to seriously discuss with your friends, relatives, and classmates the reasons you have for accepting the risks you checked. Then, in the right column, check which risks you are willing to subject nonsmokers among your family and close friends to through secondhand smoke. [p.204]

Personal Risks	Risks Associated with Smoking	Risks to Others
	a. shortened life expectancy	
	b. chronic bronchitis, emphysema	
	c. lung cancer	
	d. cancer of the mouth	
	e. cancer of the larynx	
	f. cancer of the esophagus	
	g. cancer of the pancreas	
	h. cancer of the bladder	
	i. coronary heart diseases	
	j. effects on your offspring	
	k. impaired immune system	
	l. slow bone healing (about 30% longer)	

Self-Quiz

Are you ready for the exam? Test yourself on key concepts by taking the additional tests linked with your BiologyNow CD-ROM.

_____ 1. Most forms of life depend on
_____ [p.196] down
concentration gradients to obtain oxygen
and eliminate carbon dioxide.
a. active transport
b. bulk flow
c. diffusion
d. osmosis
e. muscular contractions

_____ 2. _____ [p.196] is the most
abundant gas in Earth's atmosphere.
a. Water vapor
b. Oxygen
c. Carbon dioxide
d. Hydrogen
e. Nitrogen

_____ 3. _____ [p.203] are involved in
local chemical controls over air flow
operation in the lungs.
a. Smooth muscles in bronchiole walls
b. Nerve cell clusters in the pons and
medulla
c. The vagus nerve and its associated
stretch receptors
d. Carotid bodies
e. Aortic bodies

_____ 4. The amount of a gas diffusing across a
respiratory surface does _not_ depend on the
_____ [p.196].
a. amount of surface area of the
membrane involved
b. level of glucose in the blood
c. differences in partial pressures
of a gas across the membrane
involved
d. presence and amount of hemoglobin
e. whether or not a respiratory membrane
is moist

_____ 5. Immediately before reaching the alveoli,
air passes through the _____
[p.195].
a. bronchioles
b. glottis
c. larynx
d. pharynx
e. trachea

_____ 6. During inhalation, _____ [p.198].
a. air pressure in the alveolar sacs is lower
than atmospheric pressure
b. fresh air follows a gradient from the
alveoli up to the trachea
c. the diaphragm moves upward and
becomes more curved
d. the thoracic cavity volume decreases
e. all of the above happen

_____ 7. Hemoglobin _____ [p.200].
a. releases oxygen more readily in active
tissues
b. tends to release oxygen in places where
the temperature is lower
c. tends to hold on to oxygen when the
pH of the blood drops
d. tends to give up oxygen in regions
where partial pressure of oxygen is
higher than in the blood
e. does all of the above

_____ 8. Because mechanisms for sensing carbon
dioxide and oxygen levels become less
effective, along with loss of lung elasticity
with age, older people are more affected
by _____ [p.203].
a. sleep apnea
b. emphysema
c. asthma
d. bronchitis
e. edema

_____ 9. The wall of an alveolus separated from
lung capillaries by a thin film of interstitial
fluid constitutes a _____ [p.200].
a. bronchial tree
b. pleural sac
c. respiratory membrane
d. pulmonary surfactant
e. respiratory center

_____ 10. Inflammation of the mucous lining of the
vocal cords results in _____
[p.195].
a. emphysema
b. bronchitis
c. hypoxia
d. laryngitis
e. asthma

Chapter Objectives/Review Questions

This section lists general and detailed chapter objectives that can be used as review questions. You can make maximum use of these items by writing answers on a separate sheet of paper. Fill in answers where blanks are provided. To check for accuracy, compare your answers with information given in the chapter or glossary.

1. List all the principal parts of the human respiratory system and explain how each structure contributes to transporting oxygen from the external world to the bloodstream. [pp.194–195]
2. Describe the behavior of gases and the type of respiratory surface that participates in gas exchange in humans. [p.196]
3. Describe the functional relationship of the human lung to the pleural sac and to the thoracic cavity. [p.199]
4. Distinguish bronchitis from emphysema. [p.197]
5. The two phases of ventilation are _____ and _____. [p.198]
6. For the human lung, distinguish tidal volume from vital capacity. [p.199]
7. Distinguish respiration from ventilation. [p.200]
8. Explain why oxygen diffuses from the bloodstream into the tissues far from the lungs. Then explain why carbon dioxide diffuses into the bloodstream from the same tissues. [pp.200–201]
9. Explain why oxygen diffuses from alveolar air spaces, through interstitial fluid, and across capillary epithelium. Then explain why carbon dioxide diffuses in the reverse direction. [pp.200–201]
10. Hemoglobin with oxygen bound to it is called _____ [p.200]; when carbon dioxide binds to hemoglobin, it is called _____ [p.201].
11. Describe what happens to carbon dioxide when it dissolves in water under conditions normally present in the human body. [p.201]
12. List the structures involved in detecting carbon dioxide levels in the blood and in regulating the rate of breathing. Name the location of each structure. [pp.202–203]
13. List some of the things that go awry with the respiratory system, and describe the characteristics of the breakdown. [p.203]
14. How long after quitting smoking does it take smokers to return to the life expectancy of nonsmokers? [p.204]

Media Menu Review Questions

Questions 1–5 are drawn from the following InfoTrac College Edition article: "Breathing on the Edge." Jessica Gorman. *Science News*, March 31, 2001.

1. One of the biggest challenges for people who travel to high altitudes is _____, a condition in which the body doesn't get enough oxygen.
2. What differs between low and high altitudes is not the percentage of air that is oxygen, but that there are _____ total air molecules at higher elevations.
3. Many studies have shown that high-altitude climbers have cognitive impairments for 2 to 10 _____ after returning to low altitude.
4. Cynthia Beall's research showed that Andean highlanders had higher _____ concentrations the higher they lived. She found that Tibetans have a higher resting _____ than people at sea level.
5. To her surprise, Beall found that both Andean and Tibetan highlanders release high amounts of _____ _____, suggesting that this may play a role in the efficient uptake of oxygen.

Questions 6–10 are drawn from the following InfoTrac College Edition article: "More than a Kick." Kendall Morgan. *Science News*, March 22, 2003.

6. Scientists are concerned about the effects of nicotine outside the nervous system because the neurotransmitter it mimics, _____, has effects on many cells, including those in the lungs and skin.

7. Nicotine prevents a cellular form of suicide called _____, a mechanism that usually causes mutated cells, such as those that become cancerous, to kill themselves.

8. Nicotine has been shown to cause tumor-nurturing _____ _____ to grow to tumor cells rapidly.

9. Nicotine _____ up normal cellular activity, so that cells might go through the same life stages in 10 days that would normally take 10 weeks.

10. Nornicotine, a minor metabolite of nicotine, has been shown to severely alter _____ molecules in ways implicated in diabetes, cancer, and aging.

Integrating and Applying Key Concepts

Explain the need for an extensive and efficient circulatory system to go along with the human respiratory system, including the huge number of capillaries, efficient heart, and presence of hemoglobin.

12

THE URINARY SYSTEM

Interactive Exercises

Impacts, Issues: Double-Edged Sword [p.209]

12.1. THE CHALLENGE: SHIFTS IN EXTRACELLULAR FLUID [pp.210–211]

12.2. THE URINARY SYSTEM—BUILT FOR FILTERING AND WASTE DISPOSAL [pp.212–213]

For additional practice, use the interactive vocabulary exercises linked with your BiologyNow CD-ROM.

Selected Words: "insensible" water losses [p.210], *uric acid* [p.211], *gout* [p.211], *ammonia* [p.211], "deamination" reactions [p.211], *urea* [p.211], *cortex* [p.212], *medulla* [p.212], *renal capsule* [p.212], *collecting ducts* [p.212], *renal pelvis* [p.212], *afferent arteriole* [p.213], *efferent arteriole* [p.213]

Boldfaced, Page-Referenced Terms

[p.210] extracellular fluid (ECF) _____

_____

[p.210] intracellular fluid _____

_____

[p.210] urinary system _____

_____

[p.210] urinary excretion _____

_____

[p.211] electrolytes _____

_____

[p.212] kidney _____

_____

[p.212] nephrons _____

_____

[p.212] ureter _____

_____

[p.212] urinary bladder _____

_____

[p.212] urethra _____

_____

[p.213] renal corpuscle _____

_____

[p.213] glomerulus _____

_____

[p.213] Bowman's (glomerular) capsule _____

_____

[p.213] proximal tubule _____

_____

[p.213] loop of Henle _____

_____

[p.213] distal tubule _____

_____

[p.213] peritubular capillaries _____

_____

Short Answer

1. What is the basic task of the urinary system? Use the terms extracellular fluid and intracellular fluid in your answer. [p.210]

_____

_____

_____

Complete the Table

2. Complete the table by categorizing the movements of water and solutes. [pp.210–211]

Water Gain	Water Loss	Solute Gain	Solute Loss
a.	c.	g.	k.
b.	d.	h.	l.
_____	e.	i.	m.
_____	f.	j.	_____

True/False

If the statement is true, write a "T" in the blank. If the statement is false, make it correct by writing the word(s) in the blank that should take the place of the underlined word(s).

_____ 3. When there is a water deficit in body tissues, the <u>kidneys</u> compel us to seek out water. [p.210]

_____ 4. Of the ways that water is lost, the body exerts the most control over <u>sweating</u>. [p.210]

_____ 5. Water loss that a person is not aware of, such as through evaporation from lungs or skin, is called "<u>insensible</u>." [p.210]

_____ 6. The <u>excretory</u> system brings oxygen into the blood, and respiring cells add carbon dioxide to it. [p.211]

_____ 7. The most abundant metabolic waste is <u>carbon dioxide</u>. [p.211]

_____ 8. All metabolic wastes besides carbon dioxide leave in the <u>feces</u>. [p.211]

_____ 9. <u>Uric acid</u> forms when cells break down nucleic acids. [p.211]

_____ 10. Ammonia is a product of <u>carbohydrate</u> metabolism. [p.211]

_____ 11. The liver combines ammonia and carbon dioxide to make <u>urine</u>. [p.211]

_____ 12. The kidneys regulate the balance of ions such as sodium, potassium, and calcium that are called <u>electrolytes</u>. [p.211]

Labeling

Identify each indicated part of the accompanying illustrations. [pp.212–213]

13. _____

14. _____

15. _____ _____

16. _____

17. _____ _____

18. _____ _____

19. _____

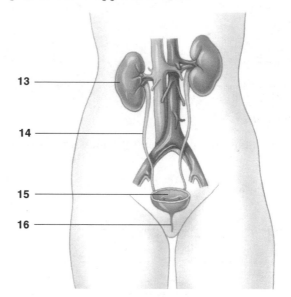

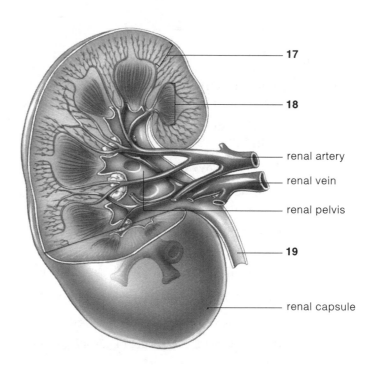

renal artery

renal vein

renal pelvis

renal capsule

Fill-in-the-Blanks

Urine formation occurs in a pair of (20) _____ [p.208], the central components of the urinary system. Each kidney contains blood vessels and slender tubes called (21) _____ [p.208] that filter water and (22) _____ [p.208] from the (23) _____ [p.208]. More than a (24) _____ [p.208] nephrons are packed inside each kidney. A(n) (25) _____ [p.209] arteriole delivers blood to each nephron, and filtration starts at the (26) _____ _____ [p.209]. Here, the nephron balloons around a cluster of capillaries called the (27) _____ [p.209]. Its wall region, the (28) _____ [p.209] capsule, receives water and solutes filtered from the blood. This filtrate moves into the (29) _____ [p.209] tubule, then through the loop of (30) _____ [p.209] and then into the (31) _____ [p.209] tubule that empties into a collecting duct.

The capillaries inside a glomerular (Bowman's) capsule converge to form a(n) (32) _____ [p.209] arteriole. This branches into a set of (33) _____ [p.209] capillaries that weave around the nephron's tubules. Water and essential solutes (34) _____ [p.209] from the tubule system enter these capillaries and are returned to the bloodstream.

Labeling

Identify each indicated part of the accompanying illustration. [p.209]

35. _____ _____
36. _____ _____
37. _____ _____
38. _____ _____
39. _____ _____
40. _____
41. _____ _____
42. _____ _____

12.3. HOW URINE FORMS [pp.214–215]

12.4. KEEPING WATER AND SODIUM IN BALANCE [p.216]

12.5. HERE'S TO THE HEALTH OF YOUR URINARY TRACT! [p.217]

12.6. ADJUSTING REABSORPTION: HORMONES ARE THE KEY [p.214]

Selected Words: urination [p.215], internal urethral sphincter [p.215], external urethral sphincter [p.215], stress incontinence [p.215], kidney stones [p.215], lithotripsy [p.215], sodium "pumps" [p.216], cystitis [p.217], pyelonephritis [p.217], chlamydia [p.217], urinalysis [p.217], diuretic [p.218], renin [p.219]

Boldfaced, Page-Referenced Terms

[p.214] urine _____

_____

[p.214] filtration _____

_____

[p.214] reabsorption _____

_____

[p.214] secretion _____

_____

[p.219] juxtaglomerular apparatus _____

_____

[p.219] thirst center _____

_____

Complete the Table

1. Complete the table below by categorizing events that occur in the kidneys into the three steps of urine formation. [p.214]

Step 1: Filtration	Step 2: Reabsorption	Step 3: Secretion
a. occurs at the _____	d. occurs along tubes of _____	g. occurs along tubes of _____
b. moves _____ and small _____ out of blood	e. most of filtrate moves _____ of nephron	h. wastes and ions enter nephron by diffusion out of _____
c. forces substances into _____ capsule, then into _____ tubule of nephron	f. returns most of filtrate into neighboring _____	i. substances are secreted into forming _____

Dichotomous Choice

Circle one of the two possible answers given between parentheses in each statement.

2. Drug testing relies on the use of (urinalysis/dialysis), which shows which substances have been secreted in the urine. [p.215]

3. Urination is a reflex response signaled by tension across the walls of the (kidney/urinary bladder). [p.215]

4. During urination, the internal urethral sphincter (relaxes/contracts) while the bladder walls force urine through the urethra. [p.215]

5. A person can exert control over the (internal/external) urethral sphincter. [p.215]

6. (Gall/Kidney) stones are deposits of substances that have settled out of the urine. [p.215]

7. The kidneys handle more blood flow than any other organ except the (heart/lungs). [p.215]

8. Because the afferent arterioles provide reduced resistance to blood flow and the efferent arterioles provide high resistance, blood pressure in the glomerular capillaries is (higher/lower) than in other capillaries. [p.215]

9. Because the glomerular capillaries are (impermeable/very permeable) to water and small solutes, an adult's kidney can filter an average of 45 gallons per day. [p.215]

10. When blood pressure falls, cells in the walls of arterioles leading to glomeruli secrete chemicals that trigger (vasodilation/vasoconstriction). [p.215]

Fill-in-the-Blanks

Of all the water and sodium the kidneys filter, about two-thirds is quickly reabsorbed at the part of the nephron nearest the glomerulus called the (11) _____ _____ [p.216]. The epithelial cells here have transport proteins at their outer surface that function as (12) _____ "_____" [p.216]. Sodium ions are actively transported from the inside of the tubule into the (13) _____ _____ [p.216] outside. Glucose and (14) _____ _____ [p.216] are reabsorbed by active transport linked to sodium reabsorption. The wall of the proximal tubule is quite permeable to water, so water follows its gradient and flows (15) _____ [p.216] of the tubule into the (16) _____ _____ [p.216], returning to the bloodstream.

The hairpin-shaped loop of Henle descends into the kidney (17) _____ [p.216]. The (18) _____ [p.216] limb of the loop is permeable to water, so water moves (19) _____ [p.216] and is reabsorbed. In the ascending limb, water cannot cross the tubule wall, but (20) _____ [p.217] ions are actively transported out of the nephron. The filtrate that finally reaches the distal tubule in the kidney cortex is quite (21) _____ [p.217], with a low sodium concentration.

Matching

Choose the most appropriate description for each term.

22. _____ glucose

23. _____ pyelonephritis

24. _____ bile pigments

25. _____ red blood cells

26. _____ white blood cells

27. _____ prostate gland

28. _____ cystitis

29. _____ albumin and other proteins

A. High levels may indicate a urinary tract infection [p.217]

B. High levels may indicate kidney disease or severe hypertension [p.217]

C. High levels may indicate liver problems [p.217]

D. High levels may indicate diabetes [p.217]

E. High levels may indicate infection, kidney stones, cancer, or an injury [p.217]

F. Swelling narrows urethra, preventing effective urination

G. Inflammation of the bladder [p.217]

H. Inflammation of the kidneys [p.217]

Dichotomous Choice

Circle one of the two possible answers given between parentheses in each statement.

30. Antidiuretic hormone (ADH) is controlled by the (kidney/hypothalamus). [p.218]

31. ADH is made in the (hypothalamus/pituitary gland). [p.218]

32. ADH acts on (proximal/distal) tubules and collecting ducts in the kidney cortex. [p.218]

33. ADH causes (less/more) water to be reabsorbed from urine. [p.218]

34. A diuretic promotes the (loss/retention) of water in the urine. [p.218]

35. Alcohol is a diuretic because it suppresses the release of (urine/ADH). [p.218]

36. While ADH affects water balance, aldosterone affects the rate at which (nitrogen/sodium) is reabsorbed. [p.219]

37. When the volume of extracellular fluid falls, cells of the juxtaglomerular apparatus secrete the enzyme (ADH/renin). [p.219]

38. Angiotensin I is converted to (sodium/angiotensin II). [p.219]

39. Aldosterone causes cells of the distal tubules to reabsorb sodium (faster/slower). [p.219]

40. Excess sodium causes a (drop/rise) in blood pressure. [p.219]

41. A thirst center in the hypothalamus can inhibit the production of (ADH/saliva), making you feel thirsty. [p.219]

12.7. MAINTAINING THE BODY'S ACID–BASE BALANCE [p.220]

12.8. KIDNEY DISORDERS [p.221]

12.9. MAINTAINING THE BODY'S CORE TEMPERATURE [pp.222–223]

Selected Words: polycystic kidney disease [p.221], *nephritis* [p.221], *glomerulonephritis* [p.221], "dialysis" [p.221], *hemodialysis* [p.221], *peritoneal dialysis* [p.221], *peripheral vasoconstriction* [p.222], *pilomotor response* [p.223], *nonshivering heat production* [p.223], "brown fat" [p.223], *hypothermia* [p.223], *frostbite* [p.223], *peripheral vasodilation* [p.223], *hyperthermia* [p.223], *heat exhaustion* [p.223], *heat stroke* [p.223]

Boldfaced, Page-Referenced Terms

[p.220] acid–base balance _____

_____

[p.220] bicarbonate–carbon dioxide buffer system _____

_____

[p.222] endotherms _____

_____

[p.222] core temperature _____

_____

Fill-in-the-Blanks

Normal pH of extracellular body fluids ranges from (1) _____ to _____ [p.220].

Buffer systems have a temporary effect in neutralizing excess H^+, but only the urinary system can

(2) _____ [p.220] excess H^+. Depending on changes in the blood's acid–base balance, the kidneys

can either excrete (3) _____ [p.220] or form new bicarbonate and add it to (4) _____

[p.220]. The chemical reactions involved occur in the cells of (5) _____ _____ [p.220] walls.

Matching

Choose the most appropriate description for each term.

6. _____ polycystic kidney disease
7. _____ peritoneal dialysis
8. _____ nephritis
9. _____ glomerulonephritis
10. _____ hemodialysis

A. Fluid is put into the abdominal cavity and later drained out [p.221]
B. Inflammation of the kidneys; interferes with blood filtering [p.221]
C. Damage to kidneys from hypertension, diabetes, etc. [p.221]
D. Blood pumped from bloodstream is cleaned and returned [p.221]
E. Masses form in and destroy kidney tissue [p.221]

Matching

Choose the most appropriate description for each term.

11. _____ heat stroke
12. _____ shivering
13. _____ pilomotor response
14. _____ core temperature
15. _____ hyperthermia
16. _____ hypothermia
17. _____ peripheral vasoconstriction
18. _____ peripheral vasodilation
19. _____ heat exhaustion
20. _____ nonshivering heat production
21. _____ frostbite

A. Freezing of cells, often destroying tissue [p.223]
B. Severe heat stress involving dry skin and potentially lethal body temperature [p.223]
C. Reduction of blood flow to capillaries near body surface so that body retains heat [p.222]
D. Temperature of the head and torso [p.222]
E. Body hair "stands on end" to trap warm air and reduce heat loss [p.223]
F. Blood pressure drops due to vasodilation and water loss; skin is cold and clammy [p.223]
G. Core temperature rises above normal [p.223]
H. Low-level contractions of skeletal muscle that produce heat
I. Core temperature falls below normal [p.223]
J. Heat production associated with breakdown of "brown fat" [p.223]
K. Increase of blood flow to capillaries near surface so that body loses excess heat [p.223]

22. What is the effect of high body temperature (over 105.8°F) on enzymes? What is the effect of very low body temperature on enzymes? [p.222]

_____

_____

_____

23. What is the value of sweating during exercise? What is the intended purpose of sports drinks in this context? [p.223]

_____

_____

_____

24. Explain why shivering occurs when a fever "breaks." [p.223]

_____

_____

_____

Self-Quiz

Are you ready for the exam? Test yourself on key concepts by taking the additional tests linked with your BiologyNow CD-ROM.

____ 1. The most toxic waste product of metabolism is _____. [p.211]
   a. water
   b. uric acid
   c. urea
   d. ammonia
   e. carbon dioxide

____ 2. The functional subunit of a kidney that filters blood, and restores solute and water balance is called a _____. [p.212]
   a. glomerulus
   b. loop of Henle
   c. nephron
   d. ureter
   e. none of the above

____ 3. In humans, the thirst center is located in the _____. [p.219]
   a. adrenal cortex
   b. thymus
   c. heart
   d. adrenal medulla
   e. hypothalamus

____ 4. Of all the water and sodium that leaves the bloodstream at the glomerulus, about two-thirds is promptly reabsorbed across the _____. [p.216]
   a. loop of Henle
   b. proximal tubule
   c. ureter
   d. Bowman's capsule
   e. collecting duct

____ 5. During reabsorption, sodium ions cross the proximal tubule walls into the interstitial fluid principally by means of diffusion or _____. [p.214]
   a. phagocytosis
   b. countercurrent multiplication
   c. bulk flow
   d. active transport
   e. all of the above

____ 6. Filtration of the blood in the kidney takes place in the _____. [p.214]
a. loop of Henle
b. proximal tubule
c. distal tubule
d. glomerulus
e. all of the above

____ 7. _____ primarily controls the concentration of sodium in urine. [p.219]
a. Insulin
b. Glucagon
c. Antidiuretic hormone
d. Aldosterone
e. Epinephrine

____ 8. Hormonal control over excretion primarily affects the _____. [pp.218–219]
a. Bowman's capsule
b. distal tubules
c. proximal tubules
d. urinary bladder
e. loops of Henle

____ 9. The last portion of the excretory system that urine passes through before it is eliminated from the body is the _____. [p.212,215]
a. renal pelvis
b. bladder
c. ureter
d. collecting ducts
e. urethra

____ 10. _____ is the principal waste product of protein breakdown, a combination of ammonia and carbon dioxide produced by the liver. [p.211]
a. Urea
b. Uric acid
c. Creatine
d. Amino acid
e. ADH

Chapter Objectives/Review Questions

This section lists general and detailed chapter objectives that can be used as review questions. You can make maximum use of these items by writing answers on a separate sheet of paper. To check for accuracy, compare your answers with information given in the chapter or glossary.

1. List some of the factors that can change the composition and volume of body fluids. [pp.210–211]
2. List three electrolytes that the kidneys help to maintain correct levels of in the body. [p.211]
3. List successively the parts of the human urinary system that constitute the path of urine formation and excretion. [pp.212–213]
4. Locate the processes of filtration, reabsorption, and secretion along a nephron, and tell what makes each process happen. [pp.214–215]
5. Discuss the factors that allow the kidneys to effectively filter large amounts of blood fairly quickly. [p.215]
6. State explicitly how the hypothalamus, pituitary, and tubules of the nephrons are interrelated in regulating water and solute levels in body fluids. [pp.216–217]
7. Why are urinary tract infections more common in women than in men? Why are they more of a problem for men as they get older? [p.217]
8. Describe the role of the kidneys in maintaining the pH of the extracellular fluids between 7.37 and 7.43. [p.220]
9. List three kidney disorders and explain what can be done if kidneys become too diseased to work properly. [p.221]
10. Explain how humans and other endotherms maintain their body temperature when environmental temperatures fall. [p.222]
11. Define *hypothermia* and state the situations in which a human might experience the disorder. [p.223]

12. Explain how humans and other endotherms maintain their body temperature when environmental temperatures rise 3 to 4 degrees Fahrenheit above standard body temperature. [pp.222–223]
13. Discuss what can happen as a result of hyperthermia. [p.223]
14. What happens in the body at the onset of a fever, as well as when a fever "breaks"? [p.223]

Media Menu Review Questions

Questions 1–4 are drawn from the following InfoTrac College Edition article: "American Kidney Fund Warns about Impact of High-Protein Diets on Kidney Health." *Obesity, Fitness and Wellness Week*, June 29, 2002.

1. Popular high-protein diets place such a strain on the kidneys that even conditioned athletes can become _____.
2. Dehydration forces the kidneys to work harder to clean toxins from the _____.
3. Increased protein intake leads to a buildup of _____ that must be diluted in urine.
4. Hyperfiltration can produce _____ in the kidneys, reducing kidney function.

Integrating and Applying Key Concepts

The hemodialysis machine used in hospitals is expensive and time consuming. So far, artificial kidneys capable of allowing people who have nonfunctional kidneys to purify their blood by themselves, without having to go to a hospital or clinic, have not been developed. Which aspects of the hemodialysis procedure do you think have presented the most problems for developing a method of home self-care? If you had an unlimited budget and were appointed head of a team to develop such a procedure and its instrumentation, what strategy would you pursue?

13

THE NERVOUS SYSTEM

Interactive Exercises

Impacts, Issues: In Pursuit of Ecstasy [p.227]

13.1. NEURONS—THE COMMUNICATION SPECIALISTS [pp.228–229]

13.2. A CLOSER LOOK AT ACTION POTENTIALS [pp.230–231]

13.3. CHEMICAL SYNAPSES: COMMUNICATION JUNCTIONS [pp.232–233]

For additional practice, use the interactive vocabulary exercises linked with your BiologyNow CD-ROM.

Selected Words: "ecstasy" [p.227], "speed" [p.227], *stimulus* [p.228], "glia" [p.228], *input zones* [p.228], *axon hillock* [p.228], "output zones" [p.228], *conducting zone* [p.228], *excitable* [p.228], "nerve impulse" [p.229], channels that "leak" or have "gates" [p.229], *electric* gradient [p.230], *all-or-nothing* events [p.231], *presynaptic cell* [p.231], *post*synaptic cell [p.231], *excitatory* signals [p.232], *inhibitory* signals [p.232], *serotonin* [p.232], *nitric* oxide [p.232], *endorphins* [p.233], *depolarize* [p.233], *hyperpolarize* [p.233], *summation* [p.233]

Boldfaced, Page-Referenced Terms

[p.228] neurons _____

_____

[p.228] sensory neurons _____

_____

[p.228] interneurons _____

_____

[p.228] motor neurons _____

_____

[p.228] neuroglia _____

_____

[p.228] dendrites _____

_____

[p.228] trigger zone _____

_____

[p.228] axon _____

_____

[p.229] resting membrane potential _____

_____

[p.229] threshold _____

_____

[p.229] action potential _____

_____

[p.231] sodium–potassium pumps _____

_____

[p.232] neurotransmitters _____

_____

[p.232] chemical synapses _____

_____

[p.232] acetylcholine (ACh) _____

[p.232] neuromodulators _____

[p.232] synaptic integration _____

Matching

Choose the most appropriate description for each term.

1. _____ threshold level
2. _____ axon
3. _____ action potential
4. _____ interneurons
5. _____ sodium–potassium pump
6. _____ sensory neurons
7. _____ neuroglia
8. _____ resting membrane potential
9. _____ dendrite
10. _____ motor neurons

A. Physically support and protect neurons [p.228]
B. A "nerve impulse" [p.229]
C. Carrier protein; moves Na^+ and K^+ across membrane [p.231]
D. Collect information about stimuli and relay it to the brain [p.228]
E. Minimum shift in voltage difference required for an action potential [p.229]
F. Steady charge difference across neuron cell membrane [p.229]
G. Receive and integrate input, then signal other neurons [p.228]
H. Neuron's "input zone"; receives incoming signals [p.228]
I. Relay messages from interneurons to muscles and glands [p.228]
J. Neuron's "conducting zone"; carries outgoing signals [p.228]

Short Answer

11. What does it mean to say that a *neuron* is excitable? [p.228]

_____

_____

_____

Labeling [p.228]

12. _____
13. _____ _____
14. _____ _____
15. _____
16. _____ _____

12 ⎫
13 ⎬ INPUT ZONE
14

CONDUCTING ZONE

15

16

OUTPUT ZONE

Dichotomous Choice

Circle one of the two possible answers given between parentheses in each statement.

17. Shifts in the (charges/concentrations) of sodium and potassium set the stage for nervous system signals. [p.229]

18. At rest, a neuron's gated sodium channels are (open/closed). [p.229]

19. Of sodium and potassium ions, the neuron's plasma membrane allows more (sodium/potassium) to leak through. [p.229]

20. Based on the rules of diffusion, sodium tends to move (in/out) and potassium tends to move (in/out). [p.229]

21. Ion concentrations result in the cytoplasm just inside the membrane being (positively/negatively) charged compared to the fluid just outside the membrane. [p.229]

22. The steady difference in charge across the membrane of –70 millivolts is called the resting membrane (differential/potential), indicating that the difference has the ability to do work. [p.229]

23. A strong enough signal arrives; it causes the voltage difference to (increase/reverse). [p.229]

24. A strong enough signal causes sodium gates to (open/close). [p.229]

25. The threshold level of stimulation for a neuron will result in a (positive/negative) feedback situation in which more and more sodium channels open in response to previous ones. [p.229]

True/False

If the statement is true, write a "T" in the blank. If the statement is false, make it correct by writing the word(s) in the blank that should take the place of the underlined word(s).

_____ 26. Propagation of action potentials must begin at a neuron's <u>dendrites</u>. [p.230]

_____ 27. Action potentials always propagate <u>toward</u> a trigger zone. [p.230]

_____ 28. A neuron can't respond to an incoming signal unless concentration and electric <u>gradients</u> are in place across its membranes. [p.230]

_____ 29. A resting neuron uses energy to power a <u>facilitated diffusion</u> mechanism to maintain sodium and potassium gradients. [p.231]

_____ 30. Sodium–potassium pumps move potassium <u>out of</u> the cell and sodium in the opposite direction. [p.231]

_____ 31. Every action potential spikes to the same level as an <u>all-or-nothing</u> event. [p.231]

_____ 32. When a spike ends, the gated sodium channels close, while <u>chloride</u> channels are open so that ions flow out and restore the original voltage difference across the membrane. [p.231]

Labeling

Label the parts of the accompanying illustration. [p.231]

33. _____ _____

34. _____ _____

35. _____ _____

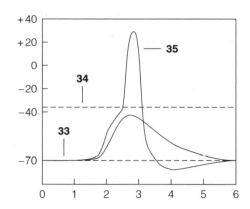

Fill-in-the-Blanks

The arrival of a signal may prompt the release of one or more chemical messengers called (36) _____ [p.232]. These molecules diffuse across a(n) (37) _____ _____ [p.232], a gap between the neuron's output zone and the input zone of the neighboring neuron. The (38) _____ [p.232] cell stores neurotransmitter molecules. When an action potential arrives, (39) _____ [p.232] ions flow from outside the cell down their gradient through gated channels into the cell, resulting in synaptic (40) _____ [p.232] fusing with the plasma membrane and releasing neurotransmitters. These diffuse across the synapse and bind with receptor proteins on the receiving, or (41) _____ [p.232], cell. This binding causes a membrane channel to open so that ions can flow through and enter the receiving cell.

Some neurotransmitters are excitatory and help drive the membrane to a(n) (42) _____ _____ [p.232], while others are (43) _____ [p.232] signals. Examples of neurotransmitters include (44) _____ [p.232], or ACh, and serotonin. Neuromodulators can magnify or impede the effects of a(n) (45) _____ [p.232]. An example is the group of natural painkillers called (46) _____ [p.233]. These inhibit nerves from releasing substance P, which conveys information about (47) _____ [p.233].

Between 1,000 and 10,000 communication lines form (48) _____ [p.233] with a typical neuron in the brain, and the brain has at least 100 (49) _____ [p.233] neurons. At any moment, many signals are washing over the input zone of a receiving neuron. All are (50) _____ [p.233] potentials. Signals called (51) _____ [p.233] depolarize the membrane, bringing it closer to threshold. (52) _____ [p.232] will hyperpolarize the membrane and drive it away from threshold. (53) _____ _____ [p.233] tallies up competing signals in a process called (54) _____ [p.233]. (55) _____ [p.233] occurs when neurotransmitter molecules from more than one presynaptic cell reach a neuron's input zone at the same time.

The flow of signals through the nervous system depends on the rapid, controlled (56) _____ [p.233] of neurotransmitters from synapses. Some diffuse out, others are cleaved by (57) _____ [p.233] in the synapse such as acetylcholinesterase. Some are actively (58) _____ [p.233] back into the presynaptic cells. Some drugs, such as cocaine, inhibit the (59) _____ [p.233] of neurotransmitters. Antidepressants like Prozac alter mood by specifically blocking the reuptake of (60) _____ [p.233].

13.4. INFORMATION PATHWAYS [pp.234–235]

13.5. THE NERVOUS SYSTEM: AN OVERVIEW [pp.236–237]

13.6. AN ENVIRONMENTAL ASSAULT ON THE NERVOUS SYSTEM [p.237]

Selected Words: "reverberating" circuits [p.235], *somatic* subdivision [p.236], *autonomic* subdivision [p.236], *ganglia* (singular: ganglion) [p.236]

Boldfaced, Page-Referenced Terms

[p.234] nerve _____

_____

[p.234] myelin sheath _____

_____

[p.234] Schwann cells _____

_____

[p.234] reflex _____

_____

[p.235] reflex arcs _____

_____

[p.236] central nervous system (CNS) _____

_____

[p.236] peripheral nervous system (PNS) _____

_____

Fill-in-the-Blanks

A(n) (1) _____ [p.234] consists of the long axons of sensory neurons, motor neurons, or both. Each axon has a(n) (2) _____ _____ [p.234] that speeds the rate at which (3) _____ _____ [p.234] propagate. The sheath consists of glia called (4) _____ [p.234] cells. An exposed (5) _____ [p.234] separates each Schwann cell from the next one. Action potentials (6) _____ [p.234] from node to node in what is called saltatory conduction. In a large, sheathed axon, action potentials (7) _____ [p.234] at the rate of 120 meters per second! The (8) _____

_____ _____ [p.234] has no Schwann cells. There, glia called (9) _____ [p.234]

sheath myelinated axons.

Sensory and motor neurons take part in a path known as the (10) _____ _____

[p.234]. In the simplest (11) _____ _____ [p.235], sensory neurons synapse directly on

motor neurons. The stretch reflex (12) _____ [p.235] a muscle involuntarily when gravity or some

other load has caused the muscle to stretch. In most reflex pathways, the sensory neurons interact with sev-

eral (13) _____ [p.235], which excite or inhibit motor neurons. This allows a coordinated response.

Blocks of hundreds or thousands of neurons in the brain and spinal cord are parts of (14) _____

[p.235] within which integration of signals occurs. Some blocks fan out to many others or

(15) _____ [p.235], while others funnel many signals to a few. Others repeat signals among

themselves and are called (16) "_____" [p.235] signals.

Labeling

Label the parts of the illustrated nerve. [p.234]

17. _____

18. _____ _____

19. _____ _____

20. _____ _____

the nerve's
outer wrapping

17

18

19

20

Matching

Match the following choices with the correct number in the diagram. [p.235]

21. _____   A. Muscle cell plasma membrane stimulated to contract (response)

22. _____   B. Action potentials generated in motor neuron and propagated along its axon toward muscle

23. _____   C. Axon endings of motor neuron synapse with muscle cells, release neurotransmitter

24. _____   D. Spinal cord

25. _____   E. Receptor endings of stimulated sensory neurons generate action potential toward spinal cord

26. _____   F. Muscle spindle stretches (stimulus)

27. _____   G. Axon endings of sensory neurons synapse on input zone of motor neuron, release neurotransmitter

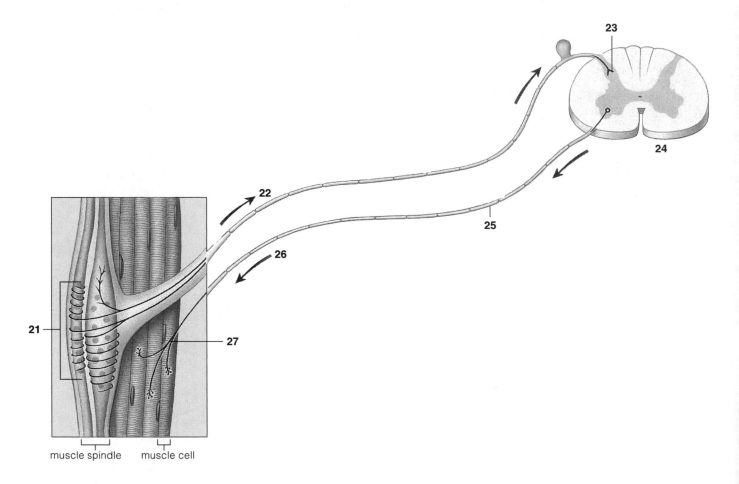

muscle spindle   muscle cell

Fill-in-the-Blanks

The nervous system has two main regions. The brain and spinal cord, including all interneurons, comprise the (28) _____ _____ _____ [p.236], or CNS. The motor and sensory nerves throughout the rest of the body are the (29) _____ _____ _____ [p.236], or PNS. The PNS is organized into (30) _____ [p.236] and (31) _____ [p.236] subdivisions, with the latter being subdivided again. The PNS has 31 pairs of (32) _____ [p.236] nerves, and 12 pairs of (33) _____ [p.236] nerves. Places where the cell bodies of several neurons cluster together are called (34) _____ [p.236].

Labeling

Label each numbered part of the accompanying illustration. [p.236]

35. _____ nerves

36. _____ _____

37. _____ nerves

38. _____ nerves

39. _____ nerves

40. _____ nerves

41. _____ nerves

brain

35

36

37

38

sciatic nerve

ulnar nerve

39

40

41

Short Answer

42. What food has the Food and Drug Administration warned pregnant women to avoid? Why? Is this problem diminishing or increasing? [p.237]

_____

_____

_____

13.7. MAJOR EXPRESSWAYS: PERIPHERAL NERVES AND THE SPINAL CORD [pp.238–239]

13.8. THE BRAIN—COMMAND CENTRAL [pp.240–241]

13.9. A CLOSER LOOK AT THE CEREBRUM [pp.242–243]

13.10. MEMORY AND STATES OF CONSCIOUSNESS [pp.244–245]

Selected Words: parasympathetic nerves [p.238], *sympathetic* nerves [p.238], "rebound effect" [p.239], *white matter* [p.239], *gray matter* [p.239], *meninges* [p.239], *spinal reflexes* [p.239], *autonomic reflexes* [p.239], *dura mater* [p.240], *olfactory bulbs* [p.240], *nuclei* [p.240], *basal nuclei* [p.240], *meningitis* [p.241], *motor* areas [p.242], *sensory* areas [p.242], *association* areas [p.242], "emotional-visceral brain" [p.243], *short-term* storage [p.244], *long-term* storage [p.244], *facts* [p.244], *skills* [p.244], *amnesia* [p.244]

Boldfaced, Page-Referenced Terms

[p.238] somatic nerves _____

_____

[p.238] autonomic nerves _____

_____

[p.239] parasympathetic nerves _____

_____

[p.239] sympathetic nerves _____

_____

[p.239] fight-flight response _____

_____

[p.239] spinal cord _____

_____

[p.240] brain _____

_____

[p.240] meninges _____

_____

[p.236] cerebrospinal fluid _____

_____

[p.240] brain stem _____

_____

[p.240] medulla oblongata _____

_____

[p.240] cerebellum _____

_____

[p.240] pons _____

_____

[p.241] cerebrum _____

_____

[p.241] thalamus _____

_____

[p.241] hypothalamus _____

_____

[p.241] reticular formation _____

_____

[p.241] cerebrospinal fluid _____

_____

[p.241] blood-brain barrier _____

_____

[p.242] cerebral hemispheres _____

_____

[p.242] cerebral cortex _____

_____

[p.243] limbic system _____

_____

[p.244] memory _____

_____

Choice

Choose A or S to match each statement about the peripheral nervous system.

A. autonomic nerves   S. somatic nerves

1. _____ Signals travel to and from internal organs and other structures [p.238]
2. _____ Sensory axons carry information from receptors in skin, skeletal muscles, and tendons [p.238]
3. _____ Signals concern moving the head, trunk, and limbs [p.238]
4. _____ Includes preganglionic and postganglionic neurons [p.238]
5. _____ Motor axons carry messages to smooth muscle, cardiac muscle, and glands [p.238]
6. _____ Motor axons deliver commands to skeletal muscles [p.238]
7. _____ Includes parasympathetic and sympathetic nerves [p.238]

Dichotomous Choice

Circle one of the two possible answers given between parentheses in each statement.

8. (Sympathetic/Parasympathetic) nerves cause the pupils to constrict and heart rate to decrease, as well as increasing stomach and intestinal movements. [p.238]

9. (Sympathetic/Parasympathetic) nerves cause glandular secretions in the airways to decrease and salivary gland secretions to thicken. [p.238]

10. (Sympathetic/Parasympathetic) nerves tend to slow down the body when there is not much outside stimulation. [p.239]

11. (Sympathetic/Parasympathetic) nerves tend to speed up the body during heightened awareness, excitement, or danger. [p.239]

12. The release of (endorphins/norepinephrine) primes the body in what is called the fight-flight response. [p.239]

13. The (gray matter/white matter) is found on the inside of the spinal cord. [p.239]

14. The (gray matter/white matter) contains dendrites, cell bodies of neurons, interneurons, and neuroglial cells. [p.239]

15. The spinal cord is protected by the vertebral column, as well as three layers of coverings called the (intervertebral disks/meninges). [p.239]

16. Spinal reflexes (do/do not) require direct input from the brain. [p.239]

17. The spinal cord contributes to (somatic/autonomic) reflexes that deal with internal organ functions such as emptying the bladder. [p.239]

Labeling

Identify the numbered parts of the accompanying illustration. [p.239]

18. _____ _____

19. _____

20. _____

21. _____

22. _____ _____

23. _____

24. _____ _____

25. _____ _____

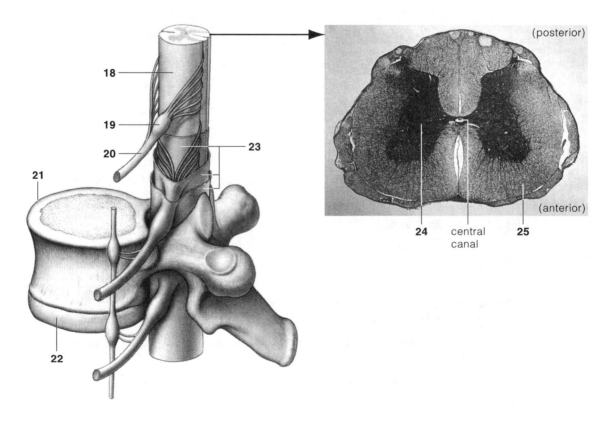

Fill-in-the-Blanks

The spinal cord merges with the (26) _____ [p.240], a master control center. The brain is protected by the bones of the (27) _____ [p.240] and by the three (28) _____ [p.240]. Folds in the tough, outermost (29) _____ [p.240] mater separate the brain into right and left (30) _____ [p.240]. The meninges also enclose spaces filled with fluid that (31) _____ [p.240] and helps nourish the brain. The tissue of the brain that controls basic reflexes is the (32) _____ _____ [p.240]. Through evolution, expanded layers of (33) _____ _____ [p.240] developed over this structure. The most recent layers are correlated with humans' increasing reliance on the nose, ears, and eyes.

Matching

To the left of each number, indicate what division of the brain includes the structure by writing H for hindbrain, M for midbrain, or F for forebrain.

Then, match the named part with the letter of the phrase that describes its function in the blank to the right of each number.

_____ 34. _____ cerebrum

_____ 35. _____ hypothalamus

_____ 36. _____ cerebellum

_____ 37. _____ medulla oblongata

_____ 38. _____ tectum

_____ 39. _____ olfactory bulbs

_____ 40. _____ pons

_____ 41. _____ thalamus

A. Monitors internal organs; influences behaviors related to thirst, hunger, sexual behavior, and emotional expression [p.241]

B. Information is processed, and sensory input and motor responses are integrated [p.241]

C. Relays and coordinates sensory signals to cerebrum through clustered neuron cell bodies (nuclei) [p.241]

D. Deal with sensory information about smell [p.241]

E. Directs signal traffic between cerebellum and forebrain [p.240]

F. Visual and auditory sensory input converges on gray matter roof before being sent to higher brain centers [p.241]

G. Coordinates sensory and motor nerve signals for movement and balance [p.240]

H. Site of reflex centers involved in respiration and blood circulation [p.240]

Fill-in-the-Blanks

The mesh of interneurons extending from the uppermost spinal cord, through the brain stem, and into the cerebral cortex is called the (42) _____ _____ [p.241]. It helps govern muscle activity associated with balance, (43) _____ [p.241], and muscle tone. It can also activate centers in the (44) _____ _____ [p.241] and so help govern the entire nervous system.

The brain and spinal cord are surrounded by transparent (45) _____ [p.241] fluid. It is secreted from specialized capillaries inside cavities called (46) _____ [p.241]. The fluid is also found in the central canal of the spinal cord, and in the space between the innermost layer of the (47) _____ [p.241] and the brain itself, helping to cushion the brain and spinal cord from jarring movements.

The structure of brain capillaries forces substances to pass through endothelial cells, not between them, in order to reach the brain. This (48) _____ - _____ _____ [p.241] helps control which substances reach the brain's neurons. The "loophole" in the system allows (49) _____ - soluble [p.241] substances through, including caffeine, nicotine, alcohol, and many illegal drugs.

True/False

If the statement is true, write a "T" in the blank. If the statement is false, make it correct by writing the word(s) in the blank that should take the place of the underlined word(s).

_____ 50. The cerebellum is divided into right and left hemispheres. [p.242]

_____ 51. The cerebral cortex is a thin, outer layer of gray matter. [p.242]

_____ 52. The left hemisphere deals with visual-spatial relationships, music, and other creative activities. [p.242]

_____ 53. The left hemisphere dominates the right hemisphere in most people. [p.242]

_____ 54. The corpus callosum is a band of connective tissue between the hemispheres. [p.242]

_____ 55. Each hemisphere is divided into four regions called lobes. [p.242]

_____ 56. Everything people comprehend, communicate, remember, and voluntarily act upon arises in the brain stem. [p.242]

_____ 57. Motor areas in the primary motor cortex of the frontal lobe control coordinated movements of skeletal muscles. [p.242]

_____ 58. The premotor cortex deals with instinctive behaviors. [p.242]

_____ 59. Broca's area, used in speech, as well as the eye field controlling voluntary eye movements, are in the temporal lobe of each hemisphere. [p.242]

_____ 60. The main receiving center for sensory input from the skin and joints is in the frontal lobe. [p.242]

_____ 61. Taste is perceived in the parietal lobe, sight in the occipital lobe, and sound and smell in the temporal lobes. [p.242]

_____ 62. Association areas in all parts of the cortex integrate, analyze, and respond to many inputs. [p.242]

_____ 63. The limbic system is located on the outside of the cerebral hemispheres. [p.243]

_____ 64. The limbic system governs emotions and has roles in memory. [p.243]

Labeling

Identify each numbered part of the accompanying illustration. [p.242]

65. _____

66. _____

67. _____ _____

68. _____

69. _____

70. _____ _____

71. _____

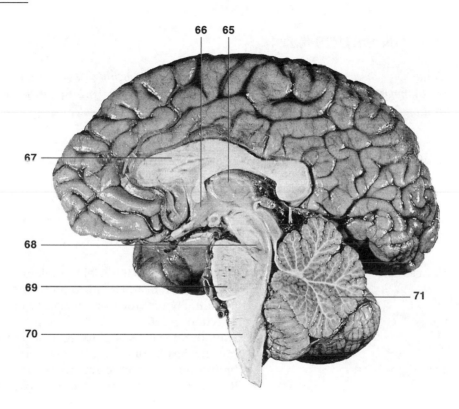

Fill-in-the-Blanks

The first stage of forming memories is (72) _____-_____ [p.244] storage, which lasts a few seconds to a few hours, and is limited to bits of sensory information. In (73) _____-_____ [p.244] storage, a great amount of information is kept more or less permanently. While facts are often forgotten or filed in long-term storage, (74) _____ [p.244] are gained by practicing specific motor activities. A memory circuit leading to fact memory flows from the sensory cortex to the (75) _____ [p.244] and (76) _____ [p.244] in the limbic system. From there, information flows to the (77) _____ [p.244] cortex and to the basal nuclei that send it back to the cortex for reinforcement. Skill memory flows from the sensory cortex to the (78) _____ _____ [p.245] that promotes motor responses. The circuit extends to the (79) _____ [p.245] that coordinates motor activity. (80) _____ [p.245] is the loss of fact memory, but does not affect a person's capacity to learn new (81) _____ [p.245]. (82) _____ [p.245] disease destroys basal nuclei and (83) _____ [p.245] ability, although skill memory remains. (84) _____ [p.245] disease results in people remembering long-standing information, but they have trouble remembering very recent events.

EEGs and (85) _____ [p.245] scans show brain activity. Part of the (86) _____ _____ [p.245] promotes chemical changes that influence whether you stay awake or fall asleep. One of its sleep centers releases (87) _____ [p.245] that triggers drowsiness and sleep.

13.11. DISORDERS OF THE NERVOUS SYSTEM [p.246]

13.12. THE BRAIN ON DRUGS [p.247]

Selected Words: *Parkinson's disease* [p.246], *Alzheimer's disease* [p.246], *meningitis* [p.246], *encephalitis* [p.246], *multiple sclerosis* [p.246], *myasthenia gravis* [p.246], *concussion* [p.246], *epilepsy* [p.246], *seizure disorders* [p.246], *blood alcohol concentration* (BAC) [p.247], *analgesic* [p.247], *hallucinogen* [p.247], *tolerance* [p.247], *habituation* [p.247]

Boldfaced, Page-Referenced Term

[p.247] stimulants _____

_____

Matching

1. _____ multiple sclerosis
2. _____ epilepsy (seizure disorders)
3. _____ meningitis
4. _____ concussion
5. _____ myasthenia gravis
6. _____ encephalitis
7. _____ Parkinson's disease
8. _____ Alzheimer's disease

A. Slow death of basal nuclei of thalamus; produces muscle tremors and balance problems [p.246]
B. Inflammation of the brain usually due to viral infection [p.246]
C. Autoimmune disease causing destruction of the myelin sheaths of the central nervous system [p.246]
D. Brain's normal electric activity becomes chaotic [p.246]
E. Results from violent blow to head or neck [p.246]
F. Degeneration of brain neurons and buildup of amyloid protein, leading to loss of memory and intellect [p.246]
G. Autoimmune disorder affecting ACh receptors; produces muscle weakness, drooping eyelids, and fatigue [p.246]
H. Often fatal inflammation of the meninges caused by viral or bacterial infection [p.246]

Matching

9. _____ amphetamine
10. _____ nicotine
11. _____ cocaine
12. _____ alcohol

A. Diminishes judgment and may produce disorientation and lack of coordination; depresses brain activity [p.247]
B. Slows motor activity, elicits a mild euphoria, skews performance of complex tasks [p.247]

13. _____ heroin

14. _____ Oxycontin

15. _____ marijuana

C. Cause a flood of dopamine and norepinephrine, thus stimulating the brain's pleasure center [p.247]
D. Synthetic prescription drug version of morphine; produces euphoria [p.247]
E. Mimics acetylcholine by stimulating sensory receptors; increases heart rate and blood pressure [p.247]
F. Blocks pain signals by binding with certain CNS receptors; from opium poppy seed pods; similar to heroin [p.247]
G. Blocks reabsorption of neurotransmitters

Short Answer

16. Tolerance develops when detoxifying liver enzymes increase to the point that all of the drug in the bloodstream is broken down before it can have an effect. What must an addict do to get past tolerance?

_____

_____

_____

17. What is habituation?

_____

_____

_____

Self-Quiz

Are you ready for the exam? Test yourself on key concepts by taking the additional tests linked with your BiologyNow CD-ROM.

_____ 1. The conducting zone of a neuron is the _____. [p.228]
a. axon
b. axon endings
c. cell body
d. dendrite

_____ 2. An action potential is brought about by _____. [p.229]
a. a sudden membrane impermeability
b. the movement of negatively charged proteins through the neuronal membrane
c. the movement of lipoproteins to the outer membrane
d. a local change in membrane permeability caused by a greater-than-threshold stimulus

_____ 3. The resting membrane potential _____. [p.229]
a. exists as long as a charge difference that could do physiological work exists across a membrane
b. occurs because there are more potassium ions outside the neuronal membrane than there are inside
c. occurs because of the unique distribution of receptor proteins located on the dendrite exterior
d. is brought about by a local change in membrane permeability caused by a greater-than-threshold stimulus

___ 4. The phrase "all-or-nothing" used in conjunction with the discussion about an action potential means that _____. [p.231]
   a. a resting membrane potential has been received by the cell
   b. it will always spike totally once stimulated past threshold
   c. the membrane either achieves total equilibrium or remains as far from equilibrium as possible
   d. propagation along the neuron is saltatory

___ 5. _____ are responsible for integration of sensory and motor signals within the central nervous system. [p.228]
   a. Interneurons
   b. Schwann cells
   c. Motor neurons
   d. Sensory neurons

___ 6. _____ nerves generally dominate internal events when environmental conditions permit normal body functioning. [p.239]
   a. Ganglia
   b. Pacemaker
   c. Sympathetic
   d. Parasympathetic

___ 7. The center of consciousness, memory, and intelligence is the _____. [p.241]
   a. hindbrain
   b. reticular formation
   c. midbrain
   d. brain stem
   e. forebrain

___ 8. The left cerebral hemisphere is generally responsible for _____. [p.242]
   a. music
   b. mathematics
   c. spatial relationships
   d. abstract abilities
   e. artistic ability

___ 9. The part of the brain that controls the basic responses necessary to maintain life processes (breathing, heartbeat) is the _____. [p.240]
   a. cerebral cortex
   b. cerebellum
   c. corpus callosum
   d. medulla oblongata

___ 10. The center for balance and coordination in the human brain is the _____. [p.240]
   a. cerebrum
   b. pons
   c. cerebellum
   d. hypothalamus
   e. thalamus

Chapter Objectives/Review Questions

This section lists general and detailed chapter objectives that can be used as review questions. You can make maximum use of these items by writing answers on a separate sheet of paper. To check for accuracy, compare your answers with information given in the chapter or glossary.

1. Draw a neuron and label it according to its three general zones, its specific structures, and the specific function(s) of each structure. [p.228]
2. Explain the chemical basis of the action potential. Look at Figure 13.6 in the text and determine which part of the curve represents the following: [pp.229–231]
   a. the point at which the stimulus was applied;
   b. the events prior to achievement of the threshold value;
   c. the opening of the ion gates and the diffusing of the ions;
   d. the change from net negative charge inside the neuron to net positive charge and back again to net negative charge; and
   e. the active transport of sodium ions out of and potassium ions into the neuron.

3. Explain what a reflex is by drawing and labeling a diagram and telling how it functions. [p.235]
4. Define and contrast the central and peripheral nervous systems. [p.236]

5. Explain how parasympathetic nerve activity balances sympathetic nerve activity. List activities of the sympathetic and parasympathetic nerves in regulating pupil diameter, rate of heartbeat, activities of the gut, and action of sphincter muscles. [p.238]
6. Compare the structures of the spinal cord and brain with respect to white matter and gray matter. [pp.239,242]
7. List the parts of the brain found in the hindbrain, midbrain, and forebrain, and tell the basic functions of each. [pp.240–241]
8. Explain what an electroencephalogram is and what EEGs can tell us about the levels of conscious experience. [p.245]
9. Describe six major disorders/diseases of the human nervous system by naming the causes and symptoms of each. [p.246]
10. Name six drugs that are commonly abused and state their effects. [p.247]

Media Menu Review Questions

Questions 1–5 are drawn from the following InfoTrac College Edition article: "Solving the MS Puzzle." Katherine Hobson. *U.S. News & World Report*, October 7, 2002.

1. In MS patients, T cells attack _____, the sheath covering nerves in the central nervous system, resulting in scarring that greatly diminishes the function of the nervous system.
2. Different responses of MS patients to the same treatment is leading doctors to think that MS may be _____ than one disease.
3. MRI brain scans reveal that lesions develop in the brain without a person knowing it; these incidents are called _____ attacks.
4. Because researchers once suspected a virus might cause MS, _____ has been used to slow down the progress of the disease, reducing annual attacks by about 30 percent.
5. There is strong evidence that _____ play a role in who will get MS, how severe the symptoms will be, and which medications are effective.

Questions 6–10 are drawn from the following InfoTrac College Edition article: "The Addicted Brain." *Harvard Mental Health Letter*, July 2004.

6. When an action satisfies a need or desire, dopamine is released into the nucleus accumbens just beneath the cerebral hemispheres. The resulting pleasure is a signal that the action promotes _____ or reproduction.
7. Addictive drugs, each in its own way, set in motion a biological process that floods the nucleus accumbens with _____.
8. Addicts who relapse after years of abstinence are victims of _____ learning, which creates habitual responses.
9. In the last few years, it has been shown that addiction alters the strength of connections at synapses of nerve cells, especially those that use the excitatory neurotransmitter _____.
10. Twin and adoption studies show that about 50 percent of individual variation in susceptibility to addiction is _____.

Integrating and Applying Key Concepts

Suppose that anger is eventually determined to be caused by excessive amounts of specific transmitter substances in the brains of angry people. Also suppose that an inexpensive antidote that neutralizes these anger-producing transmitter substances is readily available. Can convicted violent murderers now argue that they have been wrongfully punished because they were victimized by their brain's transmitter substances and could not have acted in any other way? Suppose an antidote is prescribed to curb violent temper in an easily angered person. Suppose also that the person forgets to take the pill and subsequently murders a family member. Can the murderer still claim to be victimized by transmitter substances?

14

SENSORY SYSTEMS

Interactive Exercises

Impacts, Issues: Private Eyes [p.251]

14.1. SENSORY RECEPTORS AND PATHWAYS: AN OVERVIEW [pp.252–253]

14.2. SOMATIC "BODY" SENSATIONS [pp.254–255]

For additional practice, use the interactive vocabulary exercises linked with your BiologyNow CD-ROM.

Selected Words: *compound* sensations [p.252], "special senses" [p.253], *somatic pain* [p.255], *visceral* pain [p.255], "referred pain" [p.255], *phantom pain* [p.255]

Boldfaced, Page-Referenced Terms

[p.252] sensory systems _____

_____

[p.252] sensation _____

_____

[p.252] perception _____

_____

[p.252] sensory receptors _____

_____

[p.252] mechanoreceptors _____

_____

[p.252] thermoreceptors _____

_____

[p.252] nociceptors _____

_____

[p.252] chemoreceptors _____

_____

[p.252] osmoreceptors _____

_____

[p.252] photoreceptors _____

_____

[p.253] sensory adaptation _____

_____

[p.254] somatic sensations _____

_____

[p.254] somatosensory cortex _____

_____

[p.254] free nerve endings _____

_____

[p.254] encapsulated receptors _____

_____

[p.250] pain _____

_____

Short Answer

1. Distinguish between a sensation and a perception. [p.252]

_____

_____

_____

2. How does a compound sensation differ from a simple sensation? [p.252]

_____

_____

_____

3. List the three ways in which the brain determines the nature of a given stimulus. [p.253]

_____

_____

_____

4. Name one sensation with which sensory adaptation occurs. Name one with which it does not. [p.253]

_____

_____

5. Distinguish between somatic sensations and special senses. [p.253]

_____

_____

_____

Matching

Choose the type of sensory receptor associated with each phrase. [p.252]

6. _____ detect visible light

7. _____ stimulated by tissue damage

8. _____ olfactory receptors in nose are an example

9. _____ includes auditory receptors

10. _____ detects CO_2 concentration in the blood

11. _____ detects heat energy of environment

12. _____ detects internal body temperature

13. _____ stimulated by mechanical pressure against body

14. _____ rods and cones are examples

15. _____ includes baroreceptors and balance

16. _____ located in hypothalamus

17. _____ stimulated by stretching

A. Chemoreceptors
B. Mechanoreceptors
C. Nociceptors
D. Photoreceptors
E. Thermoreceptors
F. Osmoreceptors

Dichotomous Choice

Circle one of two possible answers given between parentheses in each statement.

18. Somatic sensations travel from body receptors to the spinal cord, and then to the somatosensory cortex in the brain's (cerebellum/cerebrum). [p.254]

19. Interneurons in the somatosensory cortex form a "map" of the body's surface with the largest areas corresponding to the body parts where the density of sensory receptors is (least/greatest). [p.254]

20. You might feel a spider walking on your arm because of (chemoreceptors/mechanoreceptors). [p.254]

21. Encapsulated receptors that respond to low-frequency vibration are (Meissner's corpuscles/bulbs of Krause). [p.254]

22. Receptors that respond to steady touching and pressure, as well as to high temperature, are called (Pacinian corpuscles/Ruffini endings). [p.254]

23. Pain associated with internal organs is called (somatic pain/visceral pain). [p.255]

24. Damaged cells release chemicals such as bradykinins that (cause immediate pain/activate pain receptors). [p.255]

25. Sensing the pain of a heart attack along the left shoulder is an example of (referred pain/phantom pain). [p.255]

14.3. TASTE AND SMELL—CHEMICAL SENSES [pp.256–257]

14.4. A TASTY MORSEL OF SENSORY SCIENCE [p.257]

14.5. HEARING: DETECTING SOUND WAVES [pp.258–259]

14.6. BALANCE: SENSING THE BODY'S NATURAL POSITION [pp.260–261]

14.7. NOISE POLLUTION: AN ATTACK ON THE EARS [p.261]

Selected Words: *chemical* senses [p.256], *gustation* [p.256], *umami* [p.256], "*sexual nose*" [p.257], "*tastant*" molecules [p.257], *amplitude* [p.258], *frequency* [p.258], *outer ear* [p.258], *middle ear* [p.258], *inner ear* [p.258], *semicircular canals* [p.258], *malleus* [p.258], *incus* [p.258], *stapes* [p.258], *oval window* [p.258], *cochlear duct* [p.258], *scala vestibuli* [p.258], *scala tympani* [p.258], *basilar membrane* [p.259], "*pitch*" [p.259], *round window* [p.259], *eustachian tube* [p.259], "*equilibrium position*" [p.260], *cupula* [p.260], *static* equilibrium [p.261], *otolith organ* [p.261], *motion sickness* [p.261], *decibels* [p.261]

Boldfaced, Page-Referenced Terms

[p.256] taste receptors _____

_____

[p.256] olfactory receptors _____

_____

[p.258] cochlea _____

_____

[p.258] tympanic membrane _____

_____

[p.259] organ of Corti _____

[p.259] hair cells _____

[p.259] tectorial membrane _____

[p.260] vestibular apparatus _____

[p.260] semicircular canals _____

Fill-in-the-Blanks

With both taste and smell, (1) _____ [p.256] bind molecules that are dissolved in the fluid bathing them. Sensory information travels from the receptors through the (2) _____ [p.256] and on to the cerebral cortex. In the case of taste, these receptors are often part of sense organs called (3) _____ _____ [p.256] such as those scattered over the tongue. These distinguish only (4) _____ [p.256] basic types of flavors: sweet, sour, (5) _____ [p.256], bitter, and (6) _____ [p.256].

Animals smell substances by means of (7) _____ [p.256] receptors that detect water-soluble or volatile substances. Sensory nerve pathways lead from the nasal cavity to the region of the brain called the (8) _____ [p.256] bulbs. From there, other neurons forward the message to a center in the (9) _____ _____ [p.256], which interprets it as a particular smell. About half an inch inside the nose is the (10) _____ [p.257] organ, or "sexual nose." It detects signal molecules called (11) _____ [p.257]. These affect the behavior of other individuals.

Each of the five categories of taste is associated with particular (12) _____ [p.257]. Each taste bud has receptors that can respond to tastants in at least (13) _____ [p.257] and sometimes all five of the taste classes. Some taste receptors, such as "bitter," are extremely (14) _____ [p.257], while others require much more stimulation in order to release a neurotransmitter.

The loudness of a sound corresponds to the (15) _____ [p.258] of its wave form. The number of wave cycles per second is the sound's (16) _____ [p.258]. The sense of hearing starts with vibration-sensitive (17) _____ [p.258] deep in the ear that respond to fluid motion in the ear. Vibrations from sound waves cause this motion, which bends the tips of (18) _____ [p.258] on the mechanoreceptors. Enough bending results in a(n) (19) _____ _____ [p.258] that ultimately reaches the brain as a(n) (20) "_____" [p.258].

Matching

Choose the most appropriate description for each term.

21. _____ malleus

22. _____ incus

23. _____ stapes

24. _____ tympanic membrane

25. _____ cochlea

26. _____ basilar membrane

27. _____ Eustachian tube

28. _____ organ of Corti

29. _____ inner ear

30. _____ middle ear

31. _____ outer ear

32. _____ round window

A. Location of semicircular canals and cochlea [p.254]
B. "Anvil" of middle ear [p.254]
C. Pathway by which sound waves enter the ear [p.258]
D. "Hammer" of middle ear [p.258]
E. Its hair cells touch the tectorial membrane [p.259]
F. "Release valve" for force of sound waves [p.259]
G. Eardrum [p.258]
H. Equalizes air pressure of middle ear with outside pressure [p.259]
I. Location of three tiny bones that transmit sound waves [p.258]
J. "Stirrup" of middle ear [p.258]
K. Different sound frequencies cause different parts to vibrate [p.259]
L. Location of scala vestibuli, scala tympani, and cochlear duct [p.258]

Sequencing

In the blanks beside the numbers, write the letters of the ear parts to show the pathway of sound vibrations through the ear.

33. _____

34. _____

35. _____

36. _____

37. _____

38. _____

39. _____

40. _____

41. _____

42. _____

43. _____

A. Round window
B. Basilar membrane
C. Fluid of cochlear duct
D. Malleus
E. Organ of Corti hair cells pushing against tectorial membrane (trigger action potential)
F. Oval window
G. Incus
H. Tympanic membrane
I. Outer ear
J. Fluid inside scala vestibuli and scala tympani
K. Stapes

Labeling

Identify each indicated part of the accompanying illustrations.

44. _____ _____ [p.258]

45. _____ _____ _____ [p.258]

46. _____ _____ [p.258]

47. _____ [p.258]

48. _____ _____ [p.258]

49. _____ _____ [p.259]

50. _____ _____ [p.259]

51. _____ _____ [p.259]

52. _____ _____ [p.259]

53. _____ _____ [p.259]

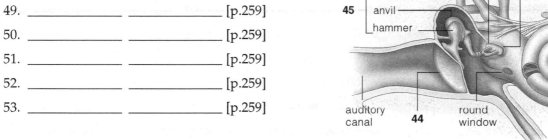

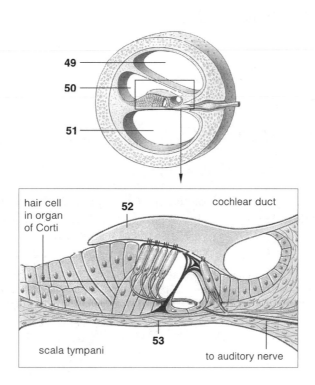

Fill-in-the-Blanks

The baseline for assessing any body displacement from its natural position is called the
(54) "_____ _____" [p.260]. Our sense of balance relies partly on messages from
(55) _____ [p.260] in the eyes, skin, and joints. Within the inner ear is the (56) _____
[p.256] apparatus, which consists of three (57) _____ _____ [p.260]. These are positioned
at right angles to one another. Sensory receptors inside monitor (58) _____ _____
[p.260] while other receptors in the apparatus monitor acceleration and (59) _____ [p.260].
Sensory hairs at the base of each semicircular canal project into a jellylike (60) _____ [p.260] and
respond to rotation of the head. The head's position in space tracks (61) _____ [p.261] equilibrium. The receptors that detect this are two fluid-filled sacs in the (62) _____ [p.261] apparatus,
called the utricle and the saccule. Each contains a(n) (63) _____ _____ [p.216] with hairs
embedded in a jellylike membrane. Movements of the membrane and bits of calcium carbonate called
(64) _____ [p.261] signal changes in the head's orientation. Action potentials from different parts
of the vestibular apparatus travel to reflex centers in the (65) _____ _____ [p.261]. The
brain integrates this information with other sensory information, then orders movements that help you
keep your (66) _____ [p.261]. (67) _____ _____ [p.261] develops when extreme
motion overstimulates hair cells in the balance organs, or when conflicting messages about motion or
position are received from the eyes and ears.

14.8. VISION: AN OVERVIEW [pp.262–263]

14.9. FROM VISUAL SIGNALS TO "SIGHT" [pp.264–265]

14.10. DISORDERS OF THE EYE [pp.266–267]

Selected Words: "tunics" [p.262], *sclera* [p.262], *choroid* [p.262], *pupil* [p.262], *ciliary body* [p.262], *aqueous humor* [p.262], *vitreous humor* [p.262], "blind spot" [p.262], *accommodation* [p.263], *bipolar* interneurons [p.265], *ganglion cells* [p.265], *horizontal* cells [p.265], *amacrine* cells [p.265], "receptive fields" [p.265], "visual field" [p.265], *red-green color blindness* [p.266], *astigmatism* [p.266], *nearsightedness* (myopia) [p.266], *farsightedness* (hyperopia) [p.266], *histoplasmosis* [p.266], *Herpes simplex* [p.266], *trachoma* [p.266], *cataracts* [p.267], *macular degeneration* [p.267], *glaucoma* [p.267], *retinal detachment* [p.267], *corneal transplant surgery* [p.267], "lasik" and "lasek" [p.267], *conductive keratoplasty* [p.267], *laser coagulation* [p.267]

Boldfaced, Page-Referenced Terms

[p.262] vision _____

_____

[p.262] eyes _____

_____

[p.262] cornea _____

_____

[p.262] iris _____

_____

[p.262] lens _____

_____

[p.262] retina _____

_____

[p.262] visual cortex _____

_____

[p.264] rod cells _____

_____

[p.264] cone cells _____

_____

[p.264] rhodopsin _____

_____

[p.264] fovea _____

_____

Fill-in-the-Blanks

Vision [p.262] requires a system of (1) _____ [p.262], (2) _____ [p.262] centers that can interpret images. The sense of (3) _____ [p.262] is an awareness of the position, shape, brightness, distance, and movement of visual stimuli. The (4) _____ [p.262] are sensory organs that contain a tissue with a dense array of photoreceptors. The outer layer of the eye consists of a sclera and transparent (5) _____ [p.262]. The middle layer includes a choroid, ciliary body, and (6) _____ [p.262]. The key feature of the inner layer is the (7) _____ [p.262].

Matching

8. _____ pupil
9. _____ sclera
10. _____ visual cortex
11. _____ lens
12. _____ vitreous humor
13. _____ ciliary body
14. _____ retina
15. _____ aqueous humor
16. _____ choroid
17. _____ cornea
18. _____ iris

A. Part of brain where signals are interpreted as sight [p.262]
B. Focuses incoming light onto the retina [p.262]
C. Pigmented ring behind cornea; adjusts amount of light entering eye [p.262]
D. Pigmented area beneath sclera; prevents light from scattering [p.262]
E. Jellylike substance in chamber behind lens [p.262]
F. Clear fluid bathing both sides of lens [p.262]
G. Smooth muscles that focuses light [p.262]
H. Entrance through which light enters eye [p.262]
I. Transparent layer covering iris and pupil [p.262]
J. Layer of neural tissue at back of eye [p.262]
K. Fibrous "white" of eye [p.262]

Labeling

Identify each indicated part of the accompanying illustration. [p.262]

19. _____

20. _____ _____

21. _____ _____

22. _____

23. _____

24. _____

25. _____

26. _____ _____

27. _____

28. _____

29. _____

30. _____ _____

31. _____ _____

Multiple Choice

____ 32. The correct path of light waves and/or electrochemical impulses used in photoreception and
vision is _____. [pp.262–263]
   a. cornea → lens → retina → optic nerve → thalamus → visual cortex
   b. sclera → iris → retina → optic nerve → visual cortex → thalamus
   c. cornea → lens → retina → thalamus → optic nerve → visual cortex
   d. sclera → lens → retina → optic nerve → visual cortex → thalamus
   e. cornea → retina → lens → optic nerve → thalamus → visual cortex

____ 33. What is the "blind spot" or optic disk of the retina? [p.262]
   a. where the optic nerve exits the eye
   b. an area with no photoreceptors
   c. area that can be scanned as a form of identification
   d. area most vulnerable to damage by strong, focused light
   e. both a and b

____ 34. What causes "upside-down and backwards" imaging that must be corrected in the brain?
[pp.262–263]
   a. density of the vitreous humor
   b. presence of a blind spot
   c. "double" focusing at both the lens and the retina
   d. curve of cornea causing light trajectories to bend
   e. contraction of ciliary muscles around iris

____ 35. Which of these is NOT true concerning focusing? [p.263]
   a. The lens is adjusted so that light strikes the retina very precisely.
   b. Adjustments of the lens are called accommodation.
   c. Focusing is necessary because light rays strike the cornea at different angles.
   d. Ciliary muscle changes the size of the pupil.
   e. Muscle contractions cause the lens to bulge.

True/False

If the statement is true, write a "T" in the blank. If the statement is false, make it correct by writing the word(s) in the blank that should take the place of the underlined word(s).

_____ 36. Daytime vision and color perception is the job of rods. [p.264]

_____ 37. Rods are better than cones at detecting light intensity. [p.264]

_____ 38. Vitamin A is used in making the pigment rhodopsin. [p.264]

_____ 39. To start an action potential to allow sight in dim surroundings, many photons must be absorbed by rhodopsin. [p.264]

_____ 40. In the part of the retina called the fovea, visual acuity is lower than in the rest of the retina. [p.264]

_____ 41. The optic nerves form from the axons of bipolar neurons. [p.265]

_____ 42. Horizontal cells and amacrine cells play a role in processing before visual information is sent to the brain. [p.265]

_____ 43. The organization of the retina into receptive fields leads to the brain being bombarded with signals that cause confusion. [p.265]

_____ 44. The part of the outside world that a person actually sees is his or her "receptive field." [p.265]

_____ 45. The optic nerve leading out of each eye delivers signals from both side(s) of a person's visual field. [p.265]

Matching

Choose the most appropriate description for each term.

46. _____ astigmatism

47. _____ cataracts

48. _____ farsightedness (hyperopia)

49. _____ glaucoma

50. _____ *Herpes simplex*

51. _____ nearsightedness (myopia)

52. _____ red green color blindness

53. _____ retinal detachment

54. _____ trachoma

55. _____ histoplasmosis

56. _____ macular degeneration

A. Caused by a physical blow to the head or an illness that separates the retina from the choroid [p.267]

B. Damaged eyeballs and conjunctiva caused by chlamydial bacteria; in North Africa and the Middle East [p.266]

C. Objects close to the eye are focused in front of the retina; eyeball too long from front to back [p.266]

D. Gradual clouding of the lens [p.267]

E. Inherited abnormality; retina lacks a particular type of cone cell [p.266]

F. Objects close to the eye are focused beyond the retina; eyeball too short from front to back [p.266]

G. Sometimes causes ulcerated cornea [p.266]

H. Excess aqueous humor accumulates inside the eyeball, causing neurons in the retina and optic nerve to die [p.267]

I. Uneven curvature of the cornea; cannot bend incoming light rays to the same focal point [p.266]

J. Portion of retina breaks down and is replaced by scar tissue [p.267]

K. Occurs when fungal infection of lungs moves to the eye [p.266]

Self-Quiz

Are you ready for the exam? Test yourself on key concepts by taking the additional tests linked with your BiologyNow CD-ROM.

Multiple Choice

___ 1. What type of mechanoreceptors help sense limb motions and the body's position in space? [p.255]
   a. stretch receptors
   b. baroreceptors
   c. touch receptors
   d. balance receptors
   e. auditory receptors

___ 2. The principal place in the human ear in which sound waves are amplified is the _____. [p.258]
   a. pinna
   b. ear canal
   c. middle ear
   d. organ of Corti
   e. none of the above

___ 3. The place in which vibrations are translated into patterns of nerve impulses is _____. [p.259]
   a. the pinna
   b. the ear canal
   c. the middle ear
   d. the organ of Corti
   e. none of the above

___ 4. Nearsightedness is caused by _____. [p.266]
   a. eye structure that focuses an image in front of the retina
   b. uneven curvature of the lens
   c. eye structure that focuses an image posterior to the retina
   d. uneven curvature of the cornea
   e. none of the above

Choice

For questions 5–9, choose from the following:

   a. fovea   b. cornea   c. iris   d. retina   e. sclera

5. The white protective fibrous tissue of the eye is the _____. [p.262]

6. Rods and cones are located in the _____. [p.264]

7. The highest concentration of cones is in the _____. [p.264]

8. The adjustable ring of contractile and connective tissues that controls the amount of light entering the eye is the _____. [p.262]

9. The outer transparent protective covering of part of the eyeball is the _____. [p.262]

Multiple Choice

10. Accommodation involves the ability to _____. [p.263]
   a. change the sensitivity of the rods and cones by means of transmitters
   b. change the width of the lens by relaxing or contracting certain muscles
   c. change the curvature of the cornea
   d. adapt to large changes in light intensity
   e. do all of the above

Chapter Objectives/Review Questions

This section lists general and detailed chapter objectives that can be used as review questions. You can make maximum use of these items by writing answers on a separate sheet of paper. To check for accuracy, compare your answers with information given in the chapter or glossary.

1. Define and distinguish among chemoreceptors, mechanoreceptors, photoreceptors, and thermoreceptors. Name one example of each type that appears in humans. [p.252]
2. Describe the function of nociceptors. [p.255]
3. Explain how a taste bud works. [p.256]
4. Follow a sound wave from the outer ear to the organ of Corti; mention the name of each structure it passes and state where the sound wave is amplified and where the pattern of pressure waves is translated into nervous impulses. [pp.258–259]
5. State how low- and high-frequency sounds affect the basilar membrane and the organ of Corti. [p.259]
6. Explain the roles of the oval and round windows in hearing. [pp.258–259]
7. Explain how the three semicircular canals of the human ear detect changes of position and acceleration in a variety of directions. [pp.260–261]
8. Describe the structure of the human eye. [p.262]
9. Describe how the human eye perceives color and black-and-white. [p.264]
10. Explain the general principles that affect how light is detected by photoreceptors and changed into electrochemical messages. [p.264]
11. Define nearsightedness and farsightedness and relate each to eyeball structure. [p.266]
12. Describe inherited eye problems. List and tell the cause of several eye diseases. [pp.266–267]

Media Menu Review Questions

Questions 1–4 are drawn from the following InfoTrac College Edition article: "How the Retina Works." Helga Kolb. *American Scientist*, January/February 2003.

1. Photoreceptors are found at the very back of the _____ so that light must pass through the entire retina before exciting pigment molecules.
2. Photoreceptors called _____ are used for low-light vision, and _____ for daylight, bright-colored vision.
3. Most fish, amphibian, reptile, and bird species have three to five types of cones and consequently very good _____ vision. (This is not true of most mammals.)
4. Primates and raptors have a _____, a tremendously cone-rich spot devoid of rods where images focus.

Questions 5–13 are drawn from the following InfoTrac College Edition article: "The New Science of Headaches." Christine Gorman and Alice Park. *Time*, October 7, 2002.

5. Many neurologists now believe that most severely disabling headaches are actually _____ in disguise and need to be so medicated.
6. Dr. Stephen Silberstein's findings with the anti-_____ drug topiramate offers hope that a whole class of this type of drugs could be useful against headaches.
7. The most common type of self-contained headache is the familiar _____ headache.
8. Cluster headache attacks typically _____ themselves, often daily, usually striking at certain times of year more often in men.
9. The pain from a _____ is a throbbing one, usually on one side of the head, with extreme sensitivity to both light and sound.
10. Unlike early ideas, it is now known that swollen blood vessels are the _____ of a migraine, not its cause.

11. Research is linking the _____ nerve to all types of primary headaches, with migraineurs having a low threshold for activating this nerve.
12. To prevent a migraine, avoid chocolate and _____ lights.
13. Getting older seems to _____ the blow of migraines.

Integrating and Applying Key Concepts

Discuss the benefits and problems associated with iris scans as a form of identification. Can you think of any other sensory structures or characteristics that are being or could be used for identification?

15

THE ENDOCRINE SYSTEM

Interactive Exercises

Impacts, Issues: Hormones in the Balance [p.271]

15.1. THE ENDOCRINE SYSTEM: HORMONES [pp.272–273]

15.2. HORMONE CATEGORIES AND SIGNALING [pp.274–275]

For additional practice, use the interactive vocabulary exercises linked with your BiologyNow CD-ROM.

Selected Words: "local signaling molecules" [p.272], "peptide hormones" [p.274], *testicular feminization syndrome* [p.274], "first messenger" [p.274]

Boldfaced, Page-Referenced Terms

[p.272] target cell _____

_____

_____

[p.272] hormones _____

_____

_____

[p.272] pheromones _____

_____

_____

[p.272] endocrine system _____

_____

_____

[p.272] opposing interaction _____

_____

_____

[p.272] synergistic interaction _____

_____

_____

[p.272] permissive interaction _____

_____

_____

[p.274] second messenger _____

_____

_____

Matching

Choose the most appropriate description for each term.

1. _____ hormones
2. _____ opposing interaction
3. _____ permissive interaction
4. _____ target cells
5. _____ synergistic interaction
6. _____ endocrine system
7. _____ pheromones

A. Sum total of the actions of two or more hormones necessary to produce the required effect on a target [p.272]
B. Group of glands that release hormones [p.272]
C. Exocrine gland secretions; signaling molecules that act on animals of the same species to help integrate social behavior [p.272]
D. Effect of one hormone works against the effect of another [p.272]
E. Secretions from endocrine glands or cells, and some neurons distributed by the bloodstream to target cells [p.272]
F. Cells that have receptors for a given type of signaling molecule [p.272]
G. One hormone exerts its effect only when a target cell has been "primed" to respond to that hormone [p.272]

Complete the Table

8. Complete the following table by identifying the numbered components of the endocrine system shown in the illustration as well as the hormones produced by each gland. [p.273]

Gland Name	Number	Hormone(s) Produced
a. parathyroids (four)		
b. adrenal cortex		1)
		2)
c. pineal		
d. pancreatic islets		
e. ovaries		1)
		2)
f. hypothalamus		1) *six*
		2)
		3)
g. pituitary, anterior lobe		*four*
h. thyroid		1)
		2)
i. thymus		
j. testes		
k. pituitary, posterior lobe		*two*
l. adrenal medulla		1)
		2)

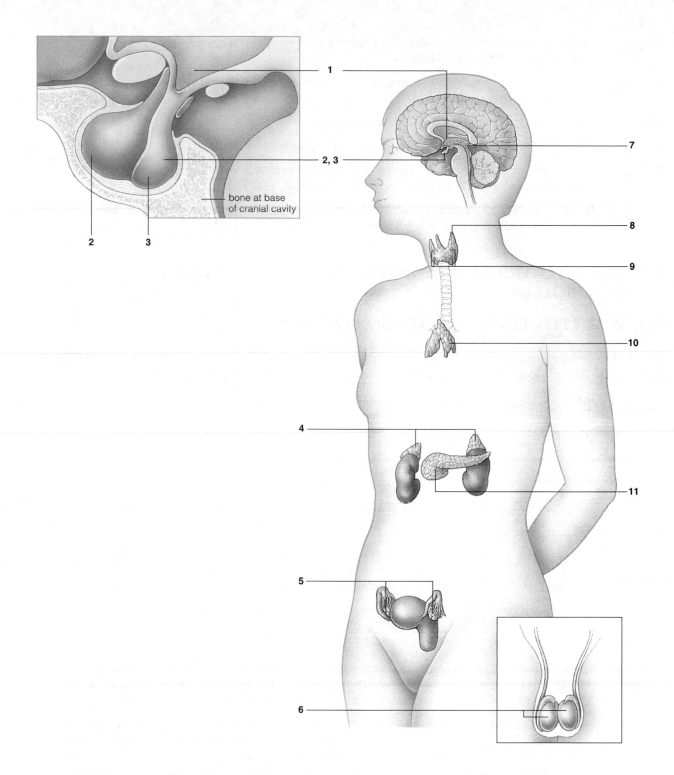

bone at base
of cranial cavity

Choice

For questions 9–16, choose from the following:

a. steroid hormones   b. nonsteroid hormones

9. _____ Lipid-soluble molecules synthesized from cholesterol; made in adrenal glands, ovaries, and testes [p.274]

10. _____ Includes amines, peptides, proteins, and glycoproteins [p.274]

11. _____ One example involves testosterone, defective receptors, and a condition called *testicular feminization syndrome* [p.274]

12. _____ Secreted by endoplasmic reticulum and mitochondrion of cells in the adrenal glands as well as in the ovaries and testes [p.274]

13. _____ Hormones that often activate second messengers [p.274]

14. _____ Peptide hormones that bind to receptors at the plasma membrane [p.274]

15. _____ Acts by forming a hormone-receptor complex that interacts with specific DNA regions to stimulate or inhibit transcription of mRNA [p.274]

16. _____ Involves molecules such as cyclic AMP that activate many enzymes in the cytoplasm that, in turn, cause alteration in some cell activity [pp.274–275]

15.3. THE HYPOTHALAMUS AND PITUITARY GLAND—MAJOR CONTROLLERS [pp.276–277]

15.4. WHEN PITUITARY SIGNALS GO AWRY [p.278]

Selected Words: *posterior* lobe and *anterior* lobe of the pituitary [p.276], "metabolic hormone" [p.277], *gigantism* [p.278], *pituitary dwarfism* [p.278], *acromegaly* [p.278], *diabetes insipidus* [p.278]

Boldfaced, Page-Referenced Terms

[p.276] hypothalamus _____

_____

[p.276] pituitary gland _____

_____

[p.277] releasers _____

_____

[p.277] inhibitors _____

_____

Choice-Match

Label each hormone given in the following list with an "A" if it is secreted by the anterior lobe of the pituitary and a "P" if it is released from the posterior pituitary. Complete the exercise by entering the letter of the corresponding target and hormone action in the parentheses following each label.

1. _____ (___) ACTH [p.276]

2. _____ (___) ADH [p.276]

3. _____ (___) FSH [pp.276–277]

4. _____ (___) GH (STH) [pp.276–277]

5. _____ (___) LH [pp.276–277]

A. Acts on ovaries and testes to produce gametes

B. Acts on mammary glands to stimulate and sustain milk production

C. Acts on ovaries and testes to release gametes; promotes testosterone secretion in males and formation of corpus luteum in females

D. Induces uterine contractions and milk movement into secretory ducts

6. _____ (___) oxytocin [p.276]

7. _____ (___) prolactin [pp.276–277]

8. _____ (___) TSH [p.276]

E. Acts on the thyroid gland to stimulate release of thyroid hormones

F. Acts on the kidneys to induce water conservation and control extracellular fluid volume

G. Acts on the adrenal cortex to stimulate release of adrenal steroid hormones

H. Acts on most cells to promote growth in young; induces protein synthesis and cell division; plays roles in glucose and protein metabolism

Dichotomous Choice

Circle one of two possibilities given between parentheses in each statement.

9. The (hypothalamus/pituitary gland) monitors internal organs and activities related to their functioning, such as eating, sexual behavior, and body temperature; it also secretes some hormones. [p.276]

10. The (posterior/anterior) lobe of the pituitary stores and secretes two hypothalamic hormones, ADH and oxytocin. [p.276]

11. The (posterior/anterior) lobe of the pituitary produces and secretes its own hormones, which govern the release of hormones from other endocrine glands. [p.276]

12. Most hypothalamic hormones acting in the anterior pituitary lobe are (releaser/inhibitor) hormones, and cause target cells to secrete their own hormones. [p.276]

13. Some hypothalamic hormones slow down hormone secretion from their targets, these are classed as (releaser/inhibitor) hormones. [p.277]

14. (ACTH/TSH) is an anterior pituitary hormone acting on the adrenal glands. [p.276]

15. In addition to LH, the anterior pituitary hormone having a role in reproduction is (FSH/TSH). [pp.276–277]

16. (Pituitary dwarfism/Gigantism) results when not enough somatotropin is produced during childhood. [p.278]

17. Production of excessive amounts of somatotropin during childhood results in (pituitary dwarfism/gigantism). [p.278]

18. Excess somatotropin production during adulthood results in thicker bone, cartilage, and connective tissues of hands, feet, jaws, and epithelia; this condition is known as (gigantism/acromegaly). [p.278]

15.5. SOURCES AND EFFECTS OF OTHER HORMONES [p.279]

15.6. HORMONES AND FEEDBACK CONTROLS—THE ADRENALS AND THYROID [pp.280–281]

15.7. FAST RESPONSES TO LOCAL CHANGES—PARATHYROIDS AND THE PANCREAS [pp.282–283]

15.8. SWEET TREACHERY—THE DIABETES EPIDEMIC [pp.284–285]

15.9. SOME FINAL EXAMPLES OF INTEGRATION AND CONTROL [pp.286–287]

Selected Words: *cortisol* [p.280], *gluconeogenesis* [p.280], "glucose sparing" [p.280], *hypoglycemia* [p.280], *aldosterone* [p.280], "fight-flight" response [p.281], *simple goiter* [p.281], *hypothyroidism* [p.281], *Graves disease* [p.281], *toxic goiters* [p.281], *hyperparathyroidism* [pp.281,282], *exocrine* cells [p.282], *endocrine* cells [p.282], *alpha cells* [p.282], *glucagons* [p.282], *beta cells* [p.282], *insulin* [p.282], *delta cells* [p.282], *somatostatin* [p.282], *diabetes mellitus* [p.282], *metabolic acidosis* [p.282], "insulin-dependent" diabetes [p.283], "juvenile-onset" diabetes [p.283], "prediabetes" [p.285], *seasonal affective disorder* (SAD) [p.286], *atrial natriuretic peptide* (ANP) [p.286], *local signaling molecules* [p.287], *epidermal growth factor* (EGF) [p.287], *nerve growth factor* (NGF) [p.287]

Boldfaced, Page-Referenced Terms

[p.280] adrenal cortex _____

_____

[p.280] glucocorticoids _____

_____

[p.280] mineralocorticoids _____

_____

[p.281] adrenal medulla _____

_____

[p.281] thyroid gland _____

_____

[p.282] parathyroid glands _____

_____

[p.282] pancreatic islet _____

_____

[p.286] pineal gland _____

_____

[p.286] biological clock _____

_____

[p.286] thymus _____

_____

[p.287] prostaglandins _____

_____

[p.287] growth factors _____

_____

Complete the Table

Complete the following table by matching the gland/organ and the hormone(s) produced by it to the descriptions of hormone action.

Gland/Organ

A. adrenal cortex [pp.279–280]
B. adrenal medulla [pp.279,281]
C. thyroid [pp.279,281]
D. parathyroids [pp.279,282]
E. testes [p.279]
F. ovaries [p.279]
G. pancreas (alpha cells) [p.282]
H. pancreas (beta cells) [p.282]
I. pancreas (delta cells) [p.282]
J. thymus [pp.279,286]
K. pineal [pp.279,286]

Hormone

a. thyroxine and triiodothyronine
b. glucagon
c. PTH
d. androgens (including testosterone)
e. somatostatin
f. thymosins
g. glucocorticoids (including cortisol)
h. estrogens
i. epinephrine
j. melatonin
k. insulin
l. progesterone
m. mineralocorticoids (including aldosterone)
n. calcitonin
o. norepinephrine

Gland/Organ	Hormone	Hormone Action
1.		Elevates calcium and phosphate levels in the bloodstream by stimulating bone cells to release these elements and the kidneys to convert them; also helps activate vitamin D
2.		Influences carbohydrate metabolism by control of food digestion; can block secretion of insulin and glucagon
3.		Required in egg maturation and release; prepares and maintains the uterine lining for pregnancy; influences growth and development
4.		Promote protein breakdown and conversion to glucose
5.		Lowers blood sugar level by stimulating glucose uptake by liver, muscle, and adipose cells; promotes protein and fat synthesis; inhibits protein conversion to glucose
6.		Required in sperm formation, genital development, and maintenance of sexual traits; influences growth and development
7.		Regulates metabolism; roles in growth and development
8.		Influences daily biorhythms; influences gonad development and reproductive cycles
9.		Raises blood level of sugar and fatty acids; increases heart rate and contraction force, the "fight-flight" response
10.		Plays roles in immune responses

Gland/Organ	Hormone	Hormone Action
11.		Raises blood sugar level by causing glycogen and amino acid conversion to glucose in liver
12.		Prepares and maintains uterine lining for pregnancy; stimulates breast development
13.		Promotes sodium reabsorption; controls salt, water balance
14.		Promotes constriction or dilation of blood vessels
15.		Lowers calcium levels in blood

Matching

16. _____ glucocorticoids

17. _____ cortisol

18. _____ gluconeogenesis

19. _____ hypoglycemia

20. _____ mineralocorticoids

21. _____ aldosterone

22. _____ fight-flight response

23. _____ adrenal medulla

24. _____ adrenal cortex

A. Ongoing low blood glucose level; can develop due to cortisol deficiency [p.280]
B. Type of adrenal hormones that adjust the concentrations of mineral salts in the extracellular fluid [p.280]
C. Outer part of each adrenal gland [p.280]
D. Effects of epinephrine and norepinephrine, including increased heart rate and enhanced respiration [p.281]
E. Type of adrenal hormones that raise the blood level of glucose [p.280]
F. Synthesis of glucose from amino acids [p.280]
G. Most abundant mineralocorticoid [p.280]
H. Body's main glucocorticoids [p.280]
I. Inner part of each adrenal gland [p.281]

Fill-in-the-Blanks

Thyroxine (T4) and triiodothyronine (T3) are the main hormones secreted by the human (25) _____ [p.281] gland. They affect a person's overall (26) _____ [p.281] rate, growth, and development. The thyroid also makes (27) _____ [p.281], a hormone that lowers the level of calcium (and phosphate) in the blood. The synthesis of thyroid hormones requires (28) _____ [p.281], which is obtained from the diet. In the absence of iodine, blood levels of these hormones decrease. The anterior pituitary responds by secreting (29) _____ [p.281]. Excess TSH overstimulates the thyroid gland and causes it to enlarge. This tissue enlargement leads to an enlargement of the gland called simple (30) _____ [p.281]. Insufficient thyroid output is called (31) _____ [p.281]. Hypothyroid adults tend to be (32) _____ [p.281], sluggish, intolerant of cold, and sometimes feel confused and depressed. Simple goiter is no longer common in areas where people use (33) _____ _____ [p.281]. When blood levels of thyroid hormones become too high, (34) _____ [p.281] disease or (35) _____ [p.281] goiter may develop. Conditions like these that are caused by elevated levels of thyroid hormones are called (36) _____ [p.281]. Symptoms include (37) _____ [p.281] heart rate, profuse sweating, and elevated blood pressure.

The (38) _____ [p.282] glands are four glands located on the back of the thyroid. They secrete (39) _____ [p.282] hormone (PTH), the main regulator of (40) _____ [p.282] level in blood. PTH is the hormone in charge of bone (41) _____ [p.282]. PTH stimulates the reabsorption of calcium from the filtrate flowing through (42) _____ [p.282] nephrons. In addition, it helps to activate vitamin (43) _____ [p.282], which improves the absorption of calcium from food in the GI tract. In vitamin D deficiency, too little calcium and phosphorus are absorbed, so bones develop improperly. This ailment is called (44) _____ [p.282]. In hyperparathyroidism, PTH causes too much calcium to be withdrawn from the (45) _____ [p.282], which weakens bone tissue, and causes kidney stones and improper functioning of muscles and the nervous system.

Dichotomous Choice

Circle one of the two possibilities given between parentheses in each statement.

46. The pancreas has the (exocrine/endocrine) function of secreting digestive enzymes, and the (exocrine/endocrine) function of secreting hormones. [p.282]

47. The endocrine cells of a pancreas are found in clusters called pancreatic (islets/patches). [p.282]

48. Glucagon is secreted by (alpha/beta) cells. [p.282]

49. Insulin is secreted by (alpha/beta) cells. [p.282]

50. Somatostatin secreted by delta cells acts on beta and alpha cells to (stimulate/inhibit) the secretion of insulin and glucagon. [p.282]

51. Insulin deficiency can lead to diabetes mellitus, a disorder in which the blood glucose level (rises/decreases) and glucose accumulates in the urine. [p.282]

52. In a person with diabetes mellitus, water loss through urination is (reduced/excessive), so the body's water–solute balance becomes disrupted. [p.282]

53. Lacking a steady glucose supply, body cells of a person with diabetes mellitus begin breaking down fats and proteins for (energy/water). [p.282]

54. In diabetes mellitus, metabolic acidosis may develop in which blood pH is dangerously (increased/decreased). [p.282]

55. Glucose released after a meal causes (insulin/glucagons) to be released, which enable cells to use or store glucose. [p.283]

56. Blood glucose levels decrease between meals. (Glucagon/Insulin) converts glycogen back to glucose to provide glucose for circulation. [p.283]

57. In (type 1 diabetes/type 2 diabetes) the body mounts an immune response against its own insulin-secreting beta cells and destroys them. [p.283]

58. Juvenile-onset diabetes is also known as (type 1 diabetes/type 2 diabetes). [p.283]

59. In (type 1 diabetes/type 2 diabetes), insulin levels are close to or above normal, but target cells fail to respond to insulin. [p.283]

60. (Type 1 diabetes/Type 2 diabetes) is appearing more commonly in young adults, and is related to obesity. [p.283]

61. The trigger for type 1 diabetes may be (obesity/a viral infection). [p.284]

62. A huge risk factor for type 2 diabetes is (obesity/a viral infection). [p.284]

63. There (is/is not) a genetic predisposition to type 2 diabetes. [p.284]

64. Elevated blood sugar damages (white blood cells/capillaries). [p.285]

65. Poor circulation results in tissue (edema/death), leading to blindness, amputation, and severe kidney problems. [p.285]

66. "Metabolic syndrome" refers to a group of characteristics that indicate a person (has/is at risk for) type 2 diabetes. [p.285]

67. Once a person is "prediabetic," proper diet and exercise (can/cannot) prevent worsening of the situation. [p.285]

Choice

For questions 68–81, choose from the following integration and control examples:

a. pineal gland   d. prostaglandins   f. pheromones
b. thymus gland   e. growth factors   g. local signaling molecules
c. heart

68. _____ Produces ANP, a hormone with various effects that include regulating blood pressure [p.286]

69. _____ Produced by many animals; released outside of the individual and then pass through air or water to reach another individual [p.287]

70. _____ An ancient photosensitive organ in the brain [p.286]

71. _____ Actions are confined to the immediate vicinity of change [p.287]

72. _____ Located behind the breastbone, between the lungs [p.286]

73. _____ EGF influences many cell types [p.287]

74. _____ Hormones known as *thymosins* [p.286]

75. _____ Menstrual cramping is one effect [p.287]

76. _____ Examples are prostaglandins and growth factors [p.287]

77. _____ Secretes melatonin into the blood and cerebrospinal fluid [p.286]

78. _____ EGF, IGF, and NGF [p.287]

79. _____ T lymphocytes multiply, differentiate, and mature in this gland [p.286]

80. _____ Evidence shows that human behavior may be affected by these [p.287]

81. _____ May influence biological clock as evidenced by its effect on SAD [p.286]

Self-Quiz

Are you ready for the exam? Test yourself on key concepts by taking the additional tests linked with your BiologyNow CD-ROM.

Multiple Choice

_____ 1. The anterior lobe of the _____ governs the release of hormones from other endocrine glands, while the posterior lobe stores hormones secreted by the _____. [p.276]
a. pituitary; hypothalamus
b. pancreas; hypothalamus
c. thyroid; parathyroid glands
d. hypothalamus; pituitary
e. pituitary; thalamus

_____ 2. ADH is sometimes called vasopressin because it increases _____. [p.276]
a. heart rate
b. sweat production
c. water excretion
d. blood vessel diameter
e. blood pressure

___ 3. The anterior lobe of the pituitary secretes
_____ different hormones, while
the posterior lobe secretes _____
hormones. [p.276]
a. two; six
b. six; six
c. six; no
d. six; two
e. two; two

Choice

For questions 4–6, choose from the following answers:

a. PTH   b. cortisol   c. aldosterone   d. calcitonin   e. melatonin

4. _____ Lowers the level of calcium and phosphate in the blood [p.281]

5. _____ Stimulates kidney to reabsorb sodium ions and excrete potassium ions [p.280]

6. _____ Affects sleep/wake cycles [p.286]

For questions 7–9, choose from the following answers:

a. adrenal medulla   b. adrenal cortex   c. thyroid   d. anterior pituitary   e. posterior pituitary

7. _____ The _____ produces glucocorticoids that help maintain the blood level of glucose and
suppress inflammatory responses. [p.280]

8. _____ The gland that is most closely associated with emergency situations is the _____. [p.281]

9. _____ The _____ gland regulates the basic metabolic rate. [p.281]

Multiple Choice

___ 10. If all sources of calcium were eliminated
from your diet, your body would secrete
more _____ in an effort to
release calcium stored in your body and
send it to the tissues that require it. [p.282]
a. parathyroid hormone
b. aldosterone
c. calcitonin
d. mineralocorticoids
e. none of the above

Matching

Choose the most appropriate description for each term.

11. _____ ACTH [p.276]

12. _____ ADH [p.276]

13. _____ calcitonin [p.279]

14. _____ cortisol [p.279]

15. _____ epinephrine and norepinephrine [p.279]

16. _____ estrogen [p.279]

17. _____ GH (STH) [p.276]

A. Raises the glucose level in the blood
B. Influences daily biorhythms, gonad
development, and reproductive cycles
C. Affects development of male sexual traits,
required for sperm formation
D. Increases heart rate and controls blood
volume; the "emergency hormones"
E. Essential for egg maturation and maintenance
of secondary sexual characteristics in the
female

18. _____ glucagon [p.279]

19. _____ insulin [p.279]

20. _____ melatonin [p.279]

21. _____ oxytocin [p.276]

22. _____ parathyroid hormone [p.279]

23. _____ progesterone [p.279]

24. _____ testosterone [p.279]

25. _____ thymosins [p.279]

26. _____ thyroxine [p.279]

27. _____ TSH [p.276]

F. The water-conservation hormone; released from posterior pituitary

G. Lowers blood sugar by signaling cells to take in glucose; promotes synthesis of proteins and fats

H. Stimulates adrenal cortex to secrete steroid hormones

I. Causes increase of calcium in the blood

J. Influences overall metabolic rate, growth, and development

K. Roles in immunity

L. Triggers uterine contractions during labor and causes milk release during nursing

M. Prepares and maintains uterine lining for pregnancy; stimulates breast development

N. Promotes conversion of protein to glucose; primary glucocorticoids

O. Lowers calcium levels in blood

P. Secreted by anterior pituitary; stimulates release of thyroid hormones

Q. Secreted by anterior pituitary; enhances growth in young animals, especially cartilage and bone

Chapter Objectives / Review Questions

This section lists general and detailed chapter objectives that can be used as review questions. You can make maximum use of these items by writing answers on a separate sheet of paper. Fill in answers where blanks are provided. To check for accuracy, compare your answers with information given in the chapter or glossary.

1. _____ cells have receptors for a specific signaling molecule and that may alter their behavior in response to it. [p.272]

2. Where is a hormone made, how does it travel to its target, and what happens when it arrives there? [p.272]

3. Collectively, sources of hormones are referred to as the _____ system. [p.272]

4. Name and define the three kinds of hormonal interactions. [p.272]

5. Locate and name the components of the human endocrine system on a diagram such as Figure 15.1 of the main text. [p.273]

6. Contrast the proposed mechanisms of hormonal action on target cell activities by (a) steroid hormones and (b) nonsteroid hormones. [pp.274–275]

7. State the relationship between both the anterior and posterior pituitary lobes and the hypothalamus. [p.276]

8. Identify the hormones released from the posterior lobe of the pituitary and state their target tissues. [p.276]

9. Identify the hormones produced by the anterior lobe of the pituitary and tell which target tissues or organs each acts on. [pp.276–277]

10. Most hypothalamic hormones acting in the anterior lobe are _____ that cause target cells to secrete hormones of their own. Some are _____ that slow down secretion from their targets. [p.277]

11. Pituitary dwarfism, gigantism, and acromegaly are all associated with abnormal secretion of _____ by the pituitary gland. [p.278]

12. Describe the major human hormone sources, their secretions, main targets, and primary actions as shown on Table 15.3 of the main text. [p.279]
13. The adrenal _____ secretes glucocorticoids. [p.280]
14. Define *hypoglycemia*; describe the roles of the hypothalamus and the anterior pituitary in this condition. [p.280]
15. The most abundantly produced mineralocorticoid is _____; cite its function. [p.280]
16. Thymosins are secreted by the _____ gland; they appear to be related to the proper functioning of the _____ system. [p.280]
17. The _____ _____ is the part of the adrenal gland that is involved in response to stress. [p.281]
18. List the features of the "fight-flight" response. [p.281]
19. Describe the characteristics of hypothyroidism and hyperthyroidism. [p.281]
20. Name the glands that secrete PTH, and give the function of this hormone. [p.282]
21. Describe the ailment called *rickets* and cite its cause. [p.282]
22. Name the hormones secreted by alpha, beta, and delta pancreatic cells; list the effect of each. [p.282]
23. Describe the symptoms of diabetes mellitus and distinguish between type 1 and type 2. [pp.282–283]
24. Name factors that put a person at risk for diabetes type 2. [p.284]
25. What is "metabolic syndrome"? List its characteristics. [p.285]
26. The pineal gland secretes the hormone _____; give two examples of the action of this hormone. [p.286]
27. Give two examples that illustrate the effects of local signaling molecules. [p.287]
28. More than sixteen different kinds of the fatty acids called _____ have been identified in tissues throughout the body; list their major effects. [p.287]
29. List three examples of growth factors in humans, and state their functions. [p.287]
30. Define *pheromone*; cite a possible function in humans. [p.287]

Media Menu Review Questions

Questions 1–3 are drawn from the following InfoTrac College Edition article: "Incidence of Diabetes in Developing Nations May Double." *Diabetes Week*, December 8, 2003.

1. According to the World Health Organization, the number of diabetes cases in developing countries could double over the next 30 years because of increasingly unhealthy _____ and less _____.

2. Over 90% of the world's estimated 171 million people with the disease have type _____ diabetes, which is associated with obesity and lack of exercise.

3. At least one death in every _____ worldwide is due to diabetes.

Questions 4–12 are drawn from the following InfoTrac College Edition article: "Hormone Disruptors: A Clue to Understanding the Environmental Causes of Disease." Sheldon Krimsky. *Environment*, June 2001.

4. Scientists have postulated a relationship between _____ in the environment and abnormalities/diseases in humans.

5. According to some laboratory studies on animals, observable effects of a(n) _____-disrupting chemical on fetal development are detectable at concentrations as low as parts per trillion.

6. In 1991, a scientific conference dubbed "Wingspread Work Session" brought together wildlife biologists who displayed data on _____ and developmental abnormalities of birds and marine species as related to environmental _____.

7. The primary evidence of the effects of endocrine-disrupting chemicals on wildlife is not disputed, although there is uncertainty about the _____ and dose-responses in specific species.

8. Scientists have postulated a link between endocrine-disrupting chemicals and three areas of human abnormalities: _____ count declines, _____ (breast, testicular, prostate), and _____ disorders.

9. The strongest evidence linking nonoccupational exposure of endocrine-disrupting chemicals with adverse human health consequences is in the areas of _____ and neurodevelopmental effects.
10. In 1996, Congress passed the Food Quality Protection Act but did not require businessmen to agree to pesticide-free food (the _____ clause).
11. The EPA has targeted the year _____ as the deadline for more food testing.
12. One study reported that approximately _____ percent of the herbicides used in the United States are endocrine disruptors.

Integrating and Applying Key Concepts

Suppose you suddenly quadruple your already high daily consumption of calcium. State which organs would be affected and tell how they would be affected. Name two hormones whose levels would most probably be affected, and tell whether your body's production of them would increase or decrease. Suppose you continue this high rate of calcium consumption for ten years. Can you predict the organs that would be subject to the most stress as a result?

16

REPRODUCTIVE SYSTEMS

Interactive Exercises

Impacts, Issues: Sperm with a Nose for Home? [p.291]

16.1. THE MALE REPRODUCTIVE SYSTEM [pp.292–293]

16.2. HOW SPERM FORM [pp.294–295]

For additional practice, use the interactive vocabulary exercises linked with your BiologyNow CD-ROM.

Selected Words: epididymides (singular: epididymis) [p.292], "germ cells" [p. 294], "haploid" [p.294], *gamete* [p.294], *spermatogonia* [p.294], *mitosis* [p.294], *meiosis* [p.294], *primary spermatocytes* [p.294], *secondary spermatocytes* [p.294], *spermatids* [p.294], *spermatozoa* [p.294], *spermatogenesis* [p.295]

Boldfaced, Page-Referenced Terms

[p.291] testes (singular: testis) _____

_____

[p.291] ovaries _____

_____

[p.292] seminiferous tubules _____

_____

[p.292] vas deferentia (singular: vas deferens) _____

_____

[p.292] semen _____

_____

[p.292] seminal vesicles _____

_____

[p.293] prostate gland _____

_____

[p.293] bulbourethral glands _____

_____

[p.294] sperm _____

_____

[p.295] Sertoli cells _____

_____

[p.295] acrosome _____

_____

[p.295] Leydig cells _____

_____

[p.295] testosterone _____

_____

[p.295] LH (luteinizing hormone)_____

_____

[p.295] FSH (follicle-stimulating hormone)_____

_____

True/False

If the statement is true, write a "T" in the blank. If the statement is false, make it correct by changing the underlined word(s) and writing the correct word(s) in the answer blank.

_____ 1. Testes and ovaries are the male and female gonads. [p.291]

_____ 2. In a male, the testes descend into the scrotum after birth. [p.292]

_____ 3. For sperm to develop properly, the temperature inside the scrotum must be a few degrees warmer than inside the body. [p.292]

_____ 4. When human sperm leave the testes, they are not mature. [p.292]

_____ 5. Until sperm leave the body, they are stored in the scrotum. [p.292]

_____ 6. Semen consists of <u>sperm only</u>. [p.292]

_____ 7. Seminal vesicles secrete <u>fructose</u> and prostaglandins. [p.293]

_____ 8. Secretions from the <u>bulbourethral</u> gland buffer the acidic environment of the female reproductive tract. [p.293]

_____ 9. There is <u>a pair of</u> prostate gland(s). [p.293]

_____ 10. Germ cells in testes and ovaries are <u>diploid</u>. [p.294]

_____ 11. Sperm and eggs are <u>diploid</u>, having only 23 chromosomes. [p.294]

_____ 12. Sperm and eggs are both called <u>gametes</u>. [p.294]

_____ 13. When two gametes unite at <u>fertilization</u>, the diploid number of 46 chromosomes is restored. [p.294]

Sequence and Label

First, write the letter of each structure listed in the order that the sperm travel through them. Put the first location of sperm by number 14, and so on. As you do this, label the diagram below.

a. epididymis   b. urethra   c. ejaculatory duct   d. vas deferens   e. testis

14. _____

15. _____

16. _____

17. _____

18. _____

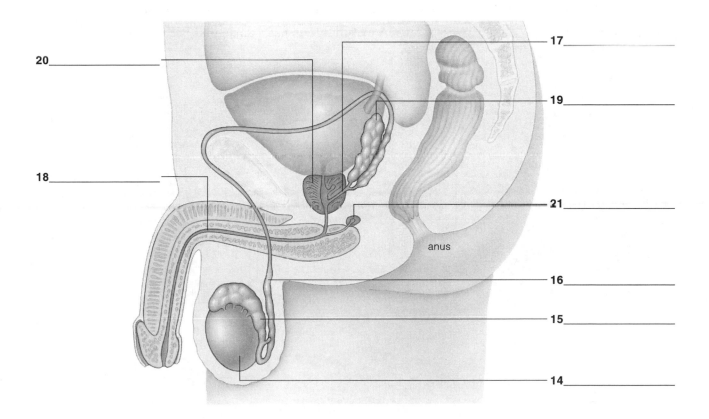

Now, place the letter of each gland that contributes to the semen in the order that the sperm passes through them prior to ejaculation. As you do this label the diagram on the previous page.

a. bulbourethral gland   b. seminal vesicle   c. prostate gland

19. _____

20. _____

21. _____

Fill-in-the-Blanks

The numbered items in the accompanying illustrations represent missing information; complete the numbered blanks in the following narrative to supply the missing information on each illustration. Some illustrated structures are numbered more than once to aid identification.

Each testis has over 400 feet of (22) _____ _____ [p.294] packed inside. Inside their walls are cells called (23) _____ [p.294]. These cells undergo divisions, including a type called (24) _____ [p.294] and a type called (25) _____ [p.294]. This process results in the specialized haploid male reproductive cells called *sperm*. Spermatogonia develop into (26) _____ _____ [p.294], which after a meiotic division known as (27) _____ _____ [p.294] are termed (28) _____ _____ [p.294]. A second division known as (29) _____ _____ [p.294] results in immature sperm known as (30) _____ [p.294], which gradually develop into spermatozoa, or simply (31) _____ [p.294]—the male gametes. The (32) "_____" [p.295], or flagellum, of each sperm arises at the very end of the process, which takes 9 to 10 weeks. These developing cells receive nourishment and chemical signals from adjacent (33) _____ [p.295] cells. A mature sperm has a tail, a midpiece, and a(n)

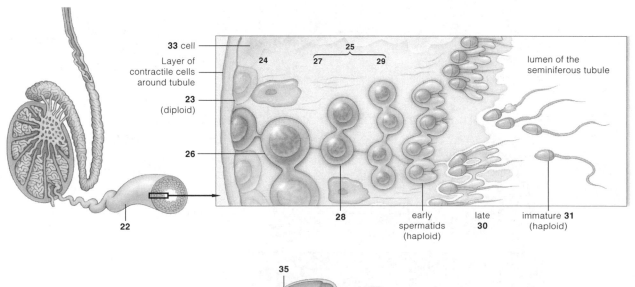

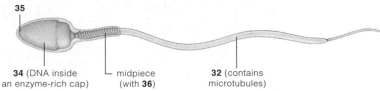

(34) _____ [p.295]. Within the head, a nucleus contains DNA organized into chromosomes. An enzyme-containing cap, the (35) _____ [p.295], covers most of the head. Its enzymes help sperm penetrate the extracellular material around an egg at fertilization. In the midpiece, (36) _____ [p.295] supply energy for the tail's whiplike movements.

Dichotomous Choice

Circle one of two possibilities given between parentheses in each statement.

37. Testosterone is secreted by (Leydig cells/the hypothalamus). [p.295]

38. (Testosterone/FSH) governs the growth, form, and functions of the male reproductive tract. [p.295]

39. Sexual behavior and secondary sexual traits are associated with (LH/testosterone). [p.295]

40. LH and FSH are secreted by the (anterior/posterior) lobe of the pituitary gland. [p.295]

41. The (testes/hypothalamus) govern(s) sperm production by controlling interactions among testosterone, LH, and FSH. [p.295]

42. When blood levels of testosterone (increase/decrease) beyond a certain set point, the hypothalamus secretes GnRH, which stimulates the anterior pituitary lobe to release LH and FSH, which the bloodstream distributes to the testes. [p.295]

43. Within the testes, (LH/FSH) acts on Leydig cells; they secrete testosterone, which stimulates diploid germ cells to become sperm. [p.295]

44. Sertoli cells have (LH/FSH) receptors and this compound is crucial to establishing spermatogenesis at the time of puberty. [p.295]

45. When blood testosterone levels (increase/decrease) past a set point, feedback loops to the hypothalamus slow down testosterone secretion and sperm formation. [p.295]

16.3. THE FEMALE REPRODUCTIVE SYSTEM [pp.296–297]

16.4. THE OVARIAN CYCLE: OOCYTES DEVELOP [pp.298–299]

16.5. VISUAL SUMMARY OF THE MENSTRUAL AND OVARIAN CYCLES [p.300]

Selected Words: *fallopian tube* [p.296], *cervix* [p.296], *vagina* [p.296], *vulva* [p.296], *labia majora* [p.296], *labia minora* [p.296], *clitoris* [p.296], *primary* oocyte [p.296], *secondary* oocyte [p.296], *menstrual phase* [p.296], *proliferative phase* [p.296], *progestational phase* [p.296], *menarche* [p.297], *menopause* [p.297], "hot flashes" [p.297], *endometriosis* [p.297], *granulose cells* [p.298], *first polar body* [p.299], *fimbriae* [p.299]

Boldfaced, Page-Referenced Terms

[p.296] oocytes _____

_____

[p.296] oviduct _____

_____

[p.296] uterus _____

_____

[p.296] endometrium _____

_____

[p.296] menstrual cycle _____

_____

[p.296] menstruation _____

_____

[p.296] progesterone _____

_____

[p. 296] estrogen _____

_____

[p.298] ovarian cycle _____

_____

[p.298] follicle _____

_____

[p.298] zona pellucida _____

_____

[p.299] secondary oocyte _____

_____

[p.299] corpus luteum _____

_____

Fill-in-the-Blanks

The numbered items on the accompanying illustration represent missing information; complete the numbered blanks in the following narrative to supply the missing information on the illustration.

An oocyte (immature egg) is released from a(n) (1) _____ [p.296]. When the oocyte is released from either ovary, it moves into a(n) (2) _____ [p.296] and is transported to the (3) _____ [p.296], a hollow, pear-shaped organ in which a baby can grow and develop. The wall of the uterus consists of a thick layer of smooth muscle, the (4) _____ [p.296], and an interior lining, the (5) _____ [p.296], which includes epithelial tissue, connective tissue, glands, and blood vessels. The lower portion of the uterus is the (6) _____ [p.296]. A muscular tube, the (7) _____ [p.296], extends from the cervix to the body surface and receives the penis and sperm and functions as part of the birth canal. Outermost is a pair of fat-padded skin folds, the (8) _____ [p.296]. Those folds enclose a smaller pair of skin folds, the (9) _____ _____ [p.296], that are highly vascularized but have no fatty tissue. The smaller folds partly enclose the (10) _____ [p.296], a small organ sensitive to stimulation that is developmentally analogous to the penis. The opening of the (11) _____ [p.296] is about midway between the clitoris and the vaginal opening.

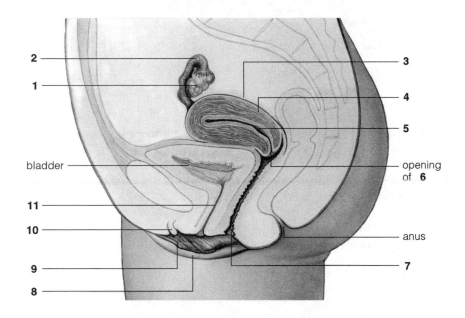

bladder

opening
of **6**

anus

Outlining

Fill in the outline below covering the reproductive cycle of female humans.

I. The (12) _____ -phase menstrual cycle [pp.296–297]

   A. Menstrual phase

      1. (13) _____ marks the first day of a new cycle

      2. (14) _____ disintegration

      3. oocyte maturation

   B. (15) _____ phase

      1. (16) _____ begins to thicken again

      2. (17) _____: release of an oocyte from an ovary

   C. (18) _____ phase

      1. (19) _____ _____ forms

      2. (20) _____ is primed for pregnancy by progesterone and estrogen

   D. Terms related to the menstrual cycle

      1. (21) _____: first menstruation, between ages ten and sixteen

      2. (22) _____: menstrual cycles stop; occurs in late 40s or early 50s

      3. (23) _____: disorder in which endometrial tissue spreads outside of uterus

II. The (24) _____ cycle: steps leading to ovulation [pp.298–299]

   A. Primary oocyte maturation

      1. About (25) _____ exist at age seven

      2. Found near the surface of a(n) (26) _____

      3. Surrounded and nourished by the (27) _____ cells

      4. Follicle grows due to anterior pituitary secretions of (28) _____ and LH

5. Glycoprotein coating called the (29) _____ _____ forms around oocyte

6. FSH and LH stimulate cells outside zona pellucida to secrete (30) _____

7. Primary oocyte completes meiosis I, giving rise to a (31) _____ oocyte and the first (32) _____ body

B. Ovulation

1. Surge of LH causes the (33) _____ to swell and rupture

2. (34) _____ _____ and first polar body are released

3. Secondary oocyte released into abdominal cavity and is drawn into a(n) (35) _____ by the beating of the cilia or fimbriae

4. (36) _____ typically occurs in the oviduct

5. At fertilization, the oocyte completes meiosis II and is a mature (37) _____

III. Hormones prepare the uterus for pregnancy [p.293]

A. Before ovulation

1. Estrogens stimulate growth of the (38) _____ and its glands

2. Cells of follicle wall secrete (39) _____ as well as estrogens

B. At ovulation, estrogens cause the (40) _____ to secrete a thin mucus for sperm to swim through

C. After ovulation

1. LH surge leads to development of the yellowish glandular (41) _____ _____

2. Corpus luteum secretes progesterone, which maintains the (42) _____ during a pregnancy

3. While corpus luteum persists, a pituitary-ordered decrease in FSH prevents other (43) _____ from developing

4. Corpus luteum breaks down after about twelve days if an embryo doesn't (44) _____ in the endometrium

5. When progesterone and estrogen levels drop, the (45) _____ breaks down and menstruation occurs

6. The cycle begins again as rising levels of (46) _____ stimulate the repair and growth of the endometrium

Analyzing Diagrams

Study the diagram on the next page to correctly complete the following dichotomous choice statements. [p.300]

47. With increasing levels of FSH and LH secreted by the pituitary, a follicle (shrinks/grows).

48. Ovulation occurs when there is a sharp surge of (FSH/LH) levels.

49. FSH and LH levels (decrease/increase) after ovulation.

50. As a follicle develops (prior to ovulation), estrogen levels (decrease/increase).

51. After ovulation, estrogen levels drop slightly, while progesterone levels (decrease/increase).

52. As long as the corpus luteum remains, estrogen and progesterone levels (remain stable/continue to increase).

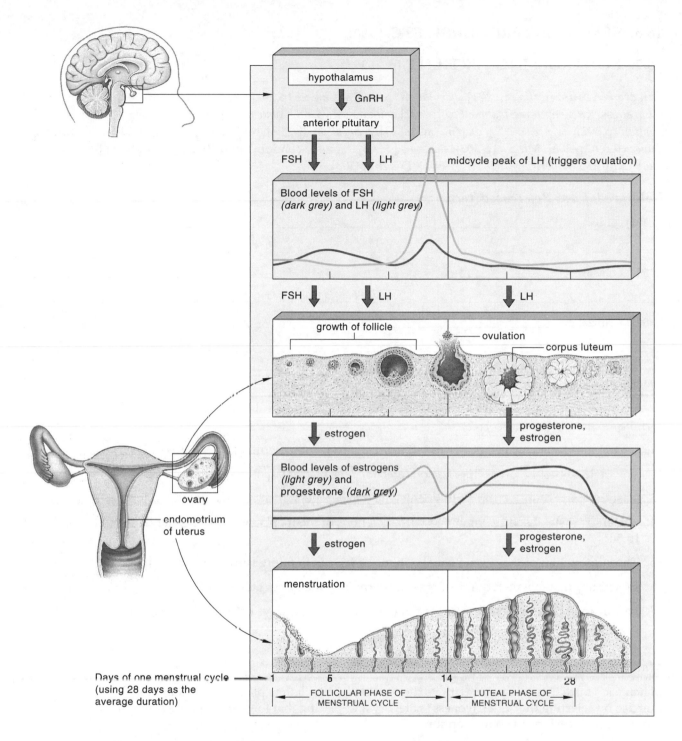

53. As long as the corpus luteum remains, the endometrium is (very thin/fully developed).

54. The endometrium is fully developed during the (follicular/luteal) stage of the menstrual cycle.

55. The menstrual cycle begins with (ovulation/menstruation).

56. Progesterone appears to cause the endometrium to (deteriorate/be maintained).

57. Estrogen production continues throughout the existence of a follicle, while progesterone is only made after the formation of the (corpus luteum/endometrium).

58. Fertilization is possible (throughout/around the middle of) the menstrual cycle.

16.6. SEXUAL INTERCOURSE, ETC. [p.301]

16.7. CONTROLLING FERTILITY [p.302]

Selected Words: erection [p.301], *ejaculation* [p.301], *abstinence* [p.302], *rhythm method* [p.302], "fertility awareness" or *sympto-thermal method* [p.302], *withdrawal* [p.302], *douching* [p.302], *vasectomy* [p.302], *tubal ligation* [p.302], *Essure* [p.302], *diaphragm* [p.303], *cervical cap* [p.303], *contraceptive sponge* [p.303], *intrauterine device* (IUD) [p.303], *Mirena* [p.303], *condoms* [p.303], "female condom" [p.303], *birth control pill* [p.303], *Norplant* [p.303], *morning-after pill* [p.303], *Preven* [p.303]

Boldfaced, Page-Referenced Terms

[p.301] coitus _____

_____

[p.301] orgasm _____

_____

[p.301] fertilization _____

_____

Elimination

Scratch through ONE of the three words in each set of parentheses in order to make each statement concerning sexual intercourse correct.

1. Sexual intercourse is technically called (coitus/fertilization/copulation). [p.301]

2. The male sex act involves (lubrication/erection/ejaculation). [p.301]

3. Vasodilation occurs in (the male's penis/the female's genital area/the male's bladder). [p.301]

4. Semen consists of sperm combined with secretions of the (pancreas/prostate gland/seminal vesicles). [p.301]

5. The culmination of the sex act usually involves (ejaculation/orgasm/pregnancy). [p.301]

6. Pregnancy in a female requires (orgasm/sperm/secondary oocyte). [p.301]

7. Fertilization may involve sperm entering the vagina (a few days before ovulation/a few days after ovulation/at any time in the menstrual cycle). [p.301]

Matching

Match each birth control option with its description (capital letters placed in the short blanks). Complete the exercise by selecting the "effectiveness" category and placing that lowercase letter within the parentheses provided with each birth control option.

8. _____ (____) abstinence [p.302]

9. _____ (____) rhythm or sympto-thermal method [p.302]

10. _____ (____) withdrawal [p.302]

11. _____ (____) douching [p.302]

12. _____ (____) spermicidal foams and jellies [pp.302–303]

A. Each vas deferens is severed and tied off
B. Removal of the penis from the vagina before ejaculation
C. A small plastic or metal device that is placed in the uterus and interferes with implantation
D. Rinsing out the vagina with a chemical after intercourse
E. An implant inserted under the skin of a woman's upper arm; contains a hormone that prevents implantation

13. _____ (____) diaphragm plus spermicide [pp.302–303]

14. _____ (____) cervical cap [pp.302–303]

15. _____ (____) contraceptive sponge plus spermicide [pp.302–303]

16. _____ (____) IUD [pp.302–303]

17. _____ (____) condoms (good quality) plus spermicide [pp.302–303]

18. _____ (____) female condom alone [pp.302–303]

19. _____ (____) oral contraceptive (the "pill") [pp.302–303]

20. _____ (____) Depo-Provera [pp.302–303]

21. _____ (____) Norplant [pp.302–303]

22. _____ (____) morning-after pills [pp.302–303]

23. _____ (____) vasectomy [p.302]

24. _____ (____) tubal ligation [p.302]

   a. Extremely effective

   b. Highly effective

   c. Effective

   d. Moderately effective

   e. Unreliable

F. No sexual intercourse

G. The oviducts are cauterized or cut and tied off

H. Injection of progestin to inhibit ovulation

I. Interfere with hormones that control events between ovulation and implantation

J. A soft, disposable disk that contains a spermicide and covers the cervix; inserted up to 24 hours before intercourse

K. Most widely used method of fertility control; contains synthetic estrogens and progesterones; suppresses release of LH and FSH, normally required for eggs to mature

L. Avoiding intercourse during the woman's fertile period; uses daily temperature readings

M. Thin, tight-fitting sheaths of latex or animal skin worn over the penis during intercourse

N. Flexible, dome-shaped device that is inserted into the vagina and positioned over the cervix before intercourse

O. Toxic to sperm; packaged in an applicator and placed in the vagina just before intercourse; not reliable unless used with a diaphragm or condom

P. Variation on the diaphragm but smaller and can be left in place up to three days with a single dose of spermicide

Q. Latex pouch inserted into the vagina

Short Answer

25. List some future options for fertility control. [p.303]

_____

_____

_____

16.8. OPTIONS FOR COPING WITH INFERTILITY [p.304]

16.9. DILEMMAS OF FERTILITY CONTROL [p.305]

Selected Words: *artificial insemination by donor* (AID) [p.304], *in vitro fertilization* (IVF) [p.304], *zygotes* [p.304], *IVF with embryo transfer* [p.305], "surrogate mother" [p.305], GIFT [p.305], ZIFT [p.305]

Choice

For questions 1–13, choose from the following:

a. *in vitro* fertilization   b. intrafallopian transfers   c. artificial insemination

1. _____ AID [p.304]

2. _____ Literally means "fertilization in glass" [p.304]

3. _____ ZIFT [p.305]

4. _____ A technique that can produce several viable embryos at one time; those not used in a given procedure can be frozen and stored for long periods [p.304]

5. _____ GIFT [p.305]

6. _____ The fate of unused embryos has prompted ethical debates. [p.304]

7. _____ A sperm bank provides sperm from an anonymous donor. [p.304]

8. _____ A couple's sperm and oocytes are collected and then placed into an oviduct (fallopian tube); fertilization rate is about 20 percent. [p.305]

9. _____ Zygotes are transferred to a solution that will support further development. [p.304]

10. _____ A fertile female volunteer is inseminated with sperm from a male whose female partner is infertile. [p.298]

11. _____ A single sperm is injected into an egg using a tiny glass needle. [p.305]

12. _____ Oocytes and sperm are brought together in a laboratory dish, where fertilization can give rise to a zygote; the zygote is then placed in one of the woman's oviducts. [p.305]

13. _____ If the donor becomes pregnant, the developing embryo is transferred to the infertile woman's uterus. [p.305]

Short Answer

14. After what stage of pregnancy are abortions particularly controversial? [p.305]

_____

_____

15. How has increasing population influenced attitudes toward abortions in some countries? [p.305]

_____

_____

Self-Quiz

Are you ready for the exam? Test yourself on key concepts by taking the additional tests linked with your BiologyNow CD-ROM.

Choice

For questions 1–2, choose from the following answers:

a. cervix   b. oviduct   c. urethra   d. uterus   e. vulva

1. The _____ is the lower narrowed portion of the uterus. [p.296]

2. The _____ is a pathway from the ovary to the uterus. [p.296]

For questions 3–6, choose from the following answers:

a. Leydig (or interstitial) cells   b. seminiferous tubules   c. vas deferens
d. epididymis   e. Sertoli cells

3. Sperm mature and are stored in the _____. [p.292]

4. The _____ connects a structure on the surface of the testis with the ejaculatory duct. [p.292]

5. Testosterone is produced by the _____. [p.295]

6. Male gametes are produced by meiosis in the _____. [p.294]

Multiple Choice

_____ 7. Male reproductive functions are controlled
by the hormones _____. [p.295]
    a. LH, FSH, and progesterone
    b. estrogen, FSH, and LH
    c. testosterone, LH, and FSH
    d. testosterone and FSH

Choice

For questions 8–12, choose from the following answers:

a. corpus luteum   b. developing early embryo   c. follicle   d. hypothalamus   e. pituitary

8. _____ Provides a mid-cycle surge of LH to trigger ovulation [p.299]

9. _____ Secretes follicle-stimulating hormone (FSH) and luteinizing hormone (LH) [p.298]

10. _____ Cells outside the zona pellucida secrete estrogens; estrogen-containing fluid starts to increase [p.298]

11. _____ Secretes GnRH, which stimulates the pituitary to begin secreting LH and FSH [p.298]

12. _____ Secretes some estrogen and progesterone [p.299]

For questions 13–17, choose from the following answers:

a. contraceptive sponge   b. sympto-thermal or rhythm   c. diaphragm   d. IUD
e. spermicidal foam and spermicidal jelly

_____ 13. The _____ is a small plastic or metal device that is placed into the uterus and interferes with implantation. [p.303]

_____ 14. A soft, disposable disk that contains a spermicide and covers the cervix is a(n) _____. [p.303]

_____ 15. A _____ is a flexible, dome-shaped device that is inserted into the vagina just before intercourse. [p.303]

_____ 16. _____ are toxic to sperm, packaged in an applicator, and placed in the vagina just before intercourse. [p.303]

_____ 17. _____ is the avoidance of intercourse during the woman's fertile period. [p.302]

For questions 18–20, choose from the following answers:

a. *in vitro* fertilization   b. intrafallopian transfers   c. artificial insemination

_____ 18. _____ is the placing of semen into the vagina or uterus by artificial means, usually a syringe, around the time of ovulation. [p.304]

_____ 19. GIFT and ZIFT are means of _____. [p.305]

_____ 20. With _____, conception can occur externally. [p.304]

Chapter Objectives/Review Questions

This section lists general and detailed chapter objectives that can be used as review questions. You can make maximum use of these items by writing answers on a separate sheet of paper. Fill in answers where blanks are provided. To check for accuracy, compare your answers with information given in the chapter or glossary.

1. The primary reproductive organs are sperm-producing _____ [p.291] in males and egg-producing _____ [p.291] in females.
2. In humans the primary reproductive organs also produce _____ _____ [p.291], which influence reproductive functions and _____ [p.291] sexual traits.
3. Where are sperm made? Where do they mature? [p.292]
4. Follow the path of a mature sperm from the seminiferous tubules to the urethral exit. List every structure encountered along the path and state its contribution to the nurture of the sperm. [pp.292–293]
5. List, in order, the stages of spermatogenesis. [pp.294–295]
6. Name the three hormones that directly control sperm formation and that form part of feedback loops among the hypothalamus, anterior pituitary, and testes. [p.295]
7. Diagram the structure of a sperm, label its components, and state the function of each. [pp.294–295]
8. Ovulation is the release of a primary _____ from the ovary. [p.296]
9. Name the event that brings about ovulation, and name the other hormonal events that bring about the onset and finish of menstruation. [pp.298–299]
10. Describe the origin and functions of the corpus luteum. [pp.299]
11. The four hormones that control egg maturation and release as well as changes in the endometrium are _____, _____, _____, and _____; they are part of feedback loops involving the hypothalamus, anterior pituitary, and ovaries. [pp.298–299]
12. In both males and females, _____ from the hypothalamus stimulates the anterior pituitary to release LH and FSH. [pp.295,298]

13. List the physiological events that bring about erection of the penis during sexual stimulation, and explain the process of ejaculation. [p.301]
14. Trace the path of a sperm from the urethral exit to the place where fertilization normally occurs. Mention, in correct sequence, all major structures of the female reproductive tract that are passed along the way, and state the principal function of each structure. [p.301]
15. Two different types of surgical birth control are _____ _____ and _____. [p.302]
16. The _____ is a method of birth control that also helps prevent the spread of sexually transmitted diseases. [p.303]
17. Generally define *artificial insemination* and *AID*. [p.304]
18. Describe *in vitro* fertilization as a method of overcoming infertility. [p.304]
19. Distinguish among ZIFT and GIFT as methods of intrafallopian transfers. [p.305]
20. During what portion of a pregnancy is abortion legal in the United States? [p.305]

Media Menu Review Questions

Questions 1–8 are drawn from the following InfoTrac College Edition article: "Could This Be the End of the Monthly Period?" N. Seppa. *Science News*, August 18, 2001.

1. Birth control pills provide steady doses of progesterone and estrogen, thus preventing the brain from triggering _____.
2. The experimental substances ZK 137 316 and ZK 230 211 stop menstrual bleeding in female monkeys by blocking _____.
3. In both macaques and human females, progesterone induces the buildup of the uterine lining correctly called the _____.
4. While being treated with the experimental substances, the rhesus macaques displayed little or no _____ _____.
5. When taken off the experimental substances, the monkeys returned to having _____ menstrual cycles.
6. The experimental substances could benefit women with heavy menstrual bleeding, _____, and very bad PMS.
7. Using the experimental substance that allows ovulation might be _____, as it would allow the normal production of estrogen.
8. Both substances work by binding to progesterone _____ on the surfaces of cells.

Questions 9–10 are drawn from the following InfoTrac College Edition article: "First Chewable Oral Contraceptive." *FDA Consumer*, March/April 2004.

9. A new oral contraceptive, approved by the FDA in November 2003, is a _____-flavored tablet.
10. The only requirement for insuring the tablet's success is that a woman drink a glass of _____ after taking the pill.

Integrating and Applying Key Concepts

What percentage of humans on the planet today do you think resulted from unplanned pregnancies? As technology allows increasingly better control of fertility, what effect do you think this might have on size of families or age of parents having families? What effect will it have on societies that currently do not have easily available birth control? Assume that technology will continue to be available to some groups or populations before others; what effect do you think this will have on increases or decreases in the relative sizes of populations?

17

DEVELOPMENT AND AGING

Interactive Exercises

Impacts, Issues: Fertility Factors and Mind-Boggling Births [p.309]

17.1. THE SIX STAGES OF EARLY DEVELOPMENT: AN OVERVIEW [pp.310–311]

17.2. THE BEGINNINGS OF YOU—EARLY STEPS IN DEVELOPMENT [pp.312–313]

17.3. VITAL MEMBRANES OUTSIDE THE EMBRYO [pp.314–315]

For additional practice, use the interactive vocabulary exercises linked with your BiologyNow CD-ROM.

Selected Words: morula [p.310], blastomere [p.310], growth and tissue specialization [p.310], capacitation [p.312], trophoblast [p.312], identical twins [p.313], fraternal twins [p.313], ectopic (tubal) pregnancy [p.313], chorionic villi [p.313]

Boldfaced, Page-Referenced Terms

[p.310] gametes _____

_____

[p.310] fertilization _____

_____

[p.310] zygote _____

_____

[p.310] cleavage _____

_____

[p.310] gastrulation _____

_____

[p.310] germ layers _____

_____

[p.310] endoderm _____

_____

[p.310] mesoderm _____

_____

[p.310] ectoderm _____

_____

[p.310] organogenesis _____

_____

[p.310] cell determination _____

[p.310] cell differentiation _____

[p.310] morphogenesis _____

[p.312] ovum (plural: ova) _____

[p.312] blastocyst _____

[p.312] inner cell mass _____

[p.313] implantation _____

[p.314] embryonic disk _____

[p.314] extraembryonic membranes _____

[p.314] yolk sac _____

[p.314] amnion _____

[p.314] allantois _____

[p.314] umbilical cord _____

[p.314] chorion _____

[p.314] placenta _____

Sequence

Arrange the following events in correct chronological sequence. Write the letter of the first step next to 1, the letter of the second step next to 2, and so on.

1. _____   A. Gastrulation [p.310]

2. _____   B. Fertilization [p.310]

3. _____   C. Cleavage [p.310]

4. _____   D. Growth and tissue specialization [p.310]

5. _____   E. Organogenesis (organ formation) [p.310]

6. _____   F. Gamete formation [p.310]

Complete the Table

7. Complete the following table by entering the correct embryonic germ layer (ectoderm, mesoderm, or endoderm) that forms the tissues and organs listed. [p.310]

Germ Layer	Tissues/Organs
a.	Muscle
b.	Nervous tissue
c.	Epithelial lining of the GI tract and lungs
d.	Cardiovascular system (blood vessels, heart)
e.	Epidermis (skin)
f.	Reproductive and excretory organs
g.	Parts of tonsils, thyroid and parathyroid glands
h.	Cartilage and bone
i.	Mammary glands, pituitary gland, subcutaneous glands

Fill-in-the-Blanks

When the three germ layers have formed, they separate into subgroups of cells. This signals the beginning of a phase called (8) _____ [p.310], or organ formation. During this phase, different sets of cells get their basic biological identities, and they give rise to different tissues and organs. The final stage of development is known as growth and tissue (9) _____ [p.310]. There are three key processes of development. The first is cell (10) _____ [p.310]. It establishes the eventual fate of the cell and its descendants. The second process is known as cell (11) _____ [p.311]. This is a gene-guided process by which cells in different locations in the embryo become specialized. The third process is known as (12) _____ [p.311]. This process produces the shape and structure of particular body regions. It also involves localized cell division and growth, and (13) _____ [p.311] of cells and entire tissues from one site to another. Controlled death and folding of sheetlike tissues can also occur in this process.

As an example, morphogenesis at the ends of limb buds first produced (14) _____ -shaped hands [p.311] at the ends of your arms; then skin cells between lobes in the paddles (15) _____ [p.311] on cue, leaving separate fingers. Keep in mind that as a developing organism comes to the end of each developmental stage, the embryo has become more (16) _____ [p.311] than it was in the previous stage.

Matching

Choose the most appropriate description for each term.

17. _____ capacitation [p.312]

18. _____ zona pellucida [p.312]

19. _____ ovum [p.312]

20. _____ identical twins [pp.312–313]

21. _____ fraternal twins [p.313]

22. _____ morula [p.312]

23. _____ blastocyst [p.312]

24. _____ trophoblast [p.312]

25. _____ inner cell mass [p.312]

26. _____ implantation [p.313]

27. _____ ectopic pregnancy [p.313]

A. The solid ball of cells reaching the uterus three or four days after fertilization
B. Produced by separation of the two cells formed by the first cleavage of the zygote; two independent embryos develop
C. About one week after fertilization, epithelial cells of the blastocyst become embedded in the endometrium
D. Acrosome enzymes clear a path through this outer layer of the egg
E. Caused by implantation of a fertilized egg in an oviduct or in the abdominal wall
F. Produced when two different eggs are fertilized at roughly the same time by two different sperm
G. Chemical process that weakens the acrosome membrane of a sperm
H. Surface layer of epithelial cells on a blastocyst
I. The clump of cells located inside the blastocyst that will develop into an embryo
J. Produced along with a polar body by meiotic cell division of the secondary oocyte
K. A ball of cells that consists of a surface epithelium and an inner clump of cells to one side of the ball

Matching

Match each numbered structure on the accompanying illustrations with its correct letter. Some structures are indicated on more than one drawing for clarity. [pp.312–313]

28. _____   A. blastocyst

29. _____   B. endometrium

30. _____   C. fertilization

31. _____   D. four-cell stage

32. _____   E. implantation

33. _____   F. inner cell mass

34. _____   G. morula

35. _____   H. ovary

36. _____   I. oviduct (fallopian tube)

37. _____   J. ovulation

38. _____   K. opening of cervix

39. _____   L. trophoblast

40. _____   M. two-cell stage

41. _____   N. uterus

42. _____   O. zygote

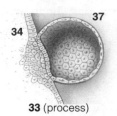

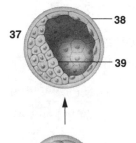

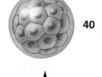

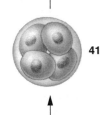

Find these numbered illustrations below, and arrange them in correct sequential order: 29, 30, 31, 33, 40 [pp.312–313]

43. _____ would occur first,

44. _____ second,

45. _____ third,

46. _____ fourth, and

47. _____ last in the sequence.

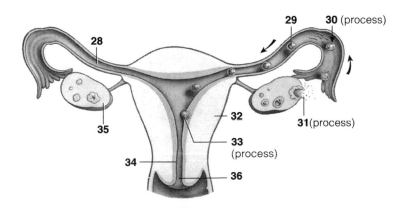

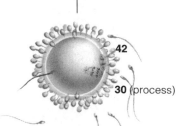

Choice

For questions 48–60, choose from the following:

> a. yolk sac   b. amnion   c. allantois   d. umbilical cord   e. chorion   f. placenta

48. _____ A close association of the embryonic chorion and the superficial cells of the mother's endometrial lining [p.314]

49. _____ A protective membrane around the embryo and other membranes [p.314]

50. _____ Membrane that develops from the trophoblast and continues the secretion of HCG that began when the blastocyst implanted [p.314]

51. _____ Links the embryo with the placenta [p.314]

52. _____ Source of early blood cells and of germ cells that will become the gametes [p.314]

53. _____ Membrane that develops into a fluid-filled sac that surrounds the embryo (later, the fetus) [p.314]

54. _____ Parts of this membrane give rise to the embryo's digestive tube. [p.314]

55. _____ Fluid within this membrane keeps the embryo from drying out, absorbs shocks, and acts as insulation. [p.314]

56. _____ Produces blood vessels that invade the umbilical cord [p.314]

57. _____ Through this tissue, the embryo receives nutrients and oxygen from the mother and sends out wastes to the mother's bloodstream in return. [p.315]

58. _____ The "maternal side" of this structure consists of a layer of endometrial tissue containing arterioles and venules. [p.314]

59. _____ Small projections of this membrane extend into spaces filled with maternal blood. [p.315]

60. _____ In addition to oxygen and nutrients, many other substances taken in by the mother—including alcohol, caffeine, drugs, pesticide residues, the AIDS virus, and toxins in cigarette smoke—can cross this structure. [p.315]

17.4. HOW THE EARLY EMBRYO TAKES SHAPE [pp.316–317]

17.5. THE FIRST EIGHT WEEKS—HUMAN FEATURES EMERGE [pp.318–319]

17.6. DEVELOPMENT OF THE FETUS [pp.320–321]

Selected Words: primitive streak [p.316], *notochord* [p.316], *coelom* [p.317], *spina bifida* [p.317], "vertebrate" [p.318], *miscarriage* [p.319], *vernix caseosa* [p.320], *respiratory distress syndrome* [p.320], *foramen ovale* [p.321], *ductus arteriosus* [p.321], *ductus venosus* [p.321]

Boldfaced, Page-Referenced Terms

[p.316] neural tube _____

_____

[p.316] somites _____

_____

[p.319] fetus _____

_____

Matching

Choose the most appropriate answer for each term.

1. _____ embryonic period [p.316]
2. _____ apoptosis [p.317]
3. _____ gastrulation [p.316]
4. _____ neural tube [p.316]
5. _____ somites [pp.316–317]
6. _____ coelom [p.317]
7. _____ neurulation [p.317]
8. _____ spina bifida [p.317]
9. _____ gonad development [p.319]
10. _____ fetus [p.319]
11. _____ miscarriage [p.319]
12. _____ primitive streak [p.316]
13. _____ morphogenesis [p.317]

A. Paired blocks of mesoderm that give rise to most bones and skeletal muscles of the neck and trunk plus their dermal coverings
B. Begins to develop in both sexes by the second half of the first trimester
C. Appears around day 15 along midline of embryonic disk; first indication of bilateral symmetry
D. Designation for an embryo after eight weeks; organ systems have formed
E. Body-forming process involving folding of cell sheets and migration of cells
F. Begins shortly after fertilization and lasts for eight weeks
G. Important developmental stage initiated by the time a woman has missed her first period
H. Spontaneous expulsion of the uterine contents; occurs in 20 percent of all conceptions
I. First stage in the development of the nervous system from ectodermal cells
J. Enzymatic destruction of cells to help sculpt body parts
K. Its forerunner is the primitive streak; gives rise to the brain and spinal cord
L. Birth defect that occurs when the neural tube fails to develop properly; a portion of the spine may be exposed
M. Formed by spaces that open up in the mesoderm and then coalesce to form a larger cavity

Choice

In the blank beside each structure, write the letter of the embryonic feature that gives rise to it.

        a. neural tube   b. allantois   c. pharyngeal gill arches   d. notochord

14. _____ face, mouth, and neck [p.317]
15. _____ umbilical cord blood vessels [p.316]
16. _____ brain and spinal cord [p.316]
17. _____ vertebral column [p.316]

Fill-in-the-Blanks

When the fetus is three months old, soft, fuzzy hair, the (18) _____ [p.320], covers the fetal body. The skin is wrinkled and protected by a thick, cheesy coating called the (19) _____ _____ [p.320]. The (20) _____ [p.320] trimester of human development extends from the start of the fourth month to the end of the sixth. Facial muscles move and, near the end of the trimester, the mother feels arm and leg movements. Eyelids and eyelashes form.

The (21) _____ [p.320] trimester extends from the seventh month until birth. Not until the middle of the third trimester can the baby survive on its own. Babies born before seven months' gestation often suffer from respiratory (22) _____ [p.320] syndrome. The circulatory system takes a detour on its way to independence. Because the fetus exchanges gases and receives nutrients via the mother's bloodstream prior to birth, the fetal circulatory system develops temporary vessels that bypass the lungs and liver. At birth, normal circulatory routes begin functioning. Two (23) _____ [p.320] arteries within the umbilical cord transport deoxygenated blood and metabolic wastes from the fetus to the placenta. Oxygenated blood, enriched with nutrients, returns from the placenta to the fetus in the (24) _____ [p.321] vein.

Other temporary vessels divert blood past the (25) _____ [p.321] and (26) _____ [p.321]. These organs do not develop as rapidly as some others, because (by way of the placenta) the mother's body can perform their functions. Fetal lungs are (27) _____ [p.321] and do not become functional for gas exchange until the newborn takes its first breaths outside the womb. A little of the blood entering the heart's right (28) _____ [p.321] flows into the right ventricle and moves on to the lungs, but most of it travels through a gap in the interior heart wall called the (29) _____ _____ [p.321] or into an arterial duct, the ductus arteriosus, that entirely bypasses the nonfunctioning lungs.

Likewise, most blood bypasses the fetal liver because the mother's liver performs most liver functions until (30) _____ [p.321]. Nutrient-laden blood from the placenta travels through a venous duct past the liver and on to the (31) _____ [p.321], which pumps it to body tissues. At birth, blood pressure in the heart's left atrium (32) _____ [p.321]. This causes a valvelike flap of tissue to close off the (33) _____ _____ [p.321], which gradually seals. The closure separates the (34) _____ [p.321] and (35) _____ [p.321] circuits of blood flow, and the arterial duct collapses. The (36) _____ [p.321] duct gradually closes during the first few weeks after birth.

Labeling

Identify each indicated part of the following illustrations.

37. _____ _____ [p.316]   44. _____ [p.316]

38. _____ _____ [p.316]   45. _____ _____ [p.316]

39. _____ _____ [p.316]   46. _____ _____ [p.318]

40. _____ _____ [p.316]   47. _____ _____ [p.318]

41. _____ _____ [p.316]   48. _____ [p.318]

42. _____ _____ [p.316]   49. _____ _____ [p.318]

43. _____ _____ [p.316]   50. _____ _____ [p.318]

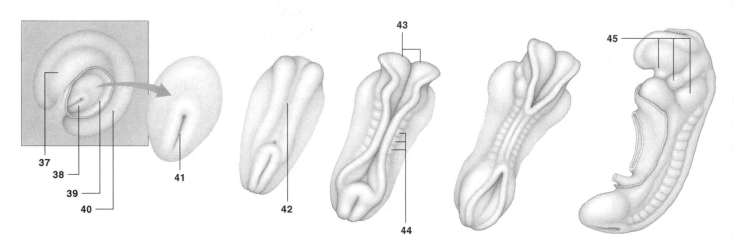

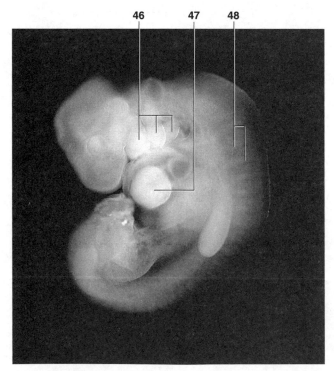

A human embryo at (**4**) weeks after conception.

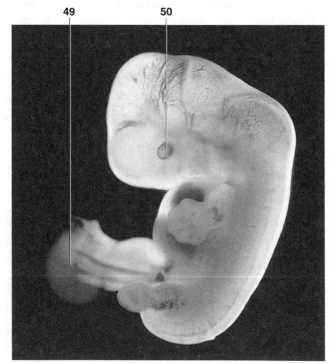

A human embryo at (**5–6**) weeks after conception.

Short Answer

51. Briefly describe the role of hormones in determining whether an embryo develops male or female gonads. [p.319]

_____

_____

_____

17.7. HAS THE AGE OF CLONING ARRIVED? [pp.322–323]

17.8. BIRTH AND BEYOND [p.322]

17.9. MOTHER AS PROVIDER, PROTECTOR, AND POTENTIAL THREAT [pp.324–325]

17.10. PRENATAL DIAGNOSIS: DETECTING BIRTH DEFECTS [p.326]

17.11. THE PATH FROM BIRTH TO ADULTHOOD [p.327]

Selected Words: "cell line" [p.322], *therapeutic cloning* [p.322], "due date" [p.322], "labor" [p.322], *breech position* [p.322], *rubella* [p.325], *fetal alcohol syndrome (FAS)* [p.325], *amniocentesis* [p.326], *chorionic villus sampling (CVS)* [p.326], *preimplantation diagnosis* [p.326], *fetoscopy* [p.326], *neonate* [p.327], *puberty* [p.327], *senescence* [p.327]

Boldfaced, Page-Referenced Term

[p.323] lactation _____

_____

Short Answer

1. Explain the theoretical value of therapeutic cloning. [p.322]

_____

_____

_____

2. How long do most scientists think it will be before babies are cloned? Why? [p.322]

_____

_____

_____

Fill-in-the-Blanks

(3) _____ [p.322], or birth, takes place about thirty-nine weeks after fertilization. The birth process

or (4) "_____" [p.322], begins when smooth muscle in the uterus starts to contract. This process

is divided into three stages. In the first, uterine contractions push the fetus against the mother's

(5) _____ [p.322], which gradually dilates to a diameter of about 10 centimeters, or 4 inches.

Usually, the (6) _____ [p.322] sac ruptures during the first contraction stage. The second stage is the actual (7) _____ [p.322] of the fetus. This stage is usually brief—under (8) _____ [p.322] hours. The baby is expelled through the cervix, usually head first. Complications can develop if the baby begins to emerge in a "bottom first," or (9) _____ [p.322], position, and the attending physician may use hands or forceps to aid the delivery. Uterine contractions of the third labor stage force fluid, blood, and the placenta, or (10) _____ [p.322], from the mother's body. The (11) _____ _____ [p.322] is now severed. Without the placenta to remove wastes, (12) _____ _____ [p.322] builds up in the baby's blood. These and other factors, that include handling by medical personnel, stimulate control centers in the brain that respond by triggering (13) _____ [p.322]—the newborn's crucial first breath. The reminder of this last stage is the scar we call the (14) _____ [p.323], the site where the umbilical cord was once attached. Babies born prematurely can suffer complications because their (15) _____ [p.323] are not developed to the point that they can function independently.

Under the influence of estrogen and progesterone, mammary glands and ducts grew within the mother's breasts. Only the colorless fluid (16) _____ [p.323] is produced for a few days. This fluid is low in fat but rich in proteins, antibodies, minerals, and vitamin A. Then prolactin secreted by the pituitary gland stimulates milk production, or (17) _____ [p.323]. (18) _____ [p.323] released from the pituitary acts to force milk into mammary ducts.

Short Answer

19. Compose a few statements about the critical importance of maternal lifestyle during pregnancy. [pp.318–319]

_____

_____

_____

_____

_____

Dichotomous Choice

Circle one of two possible answers given between parentheses in each statement.

20. (Amniocentesis/CVS) uses tissue from the chorionic villi of the placenta for prenatal diagnosis of genetic defects. [p.326]

21. (Amniocentesis/CVS) samples fluid from within the amnion that contains some fetal cells and chemicals used for prenatal diagnosis of genetic defects. [p.326]

22. Using methods of (CVS/preimplantation diagnosis), an embryo "conceived" by *in vitro* fertilization is analyzed for genetic defects using recombinant DNA technology. [p.326]

23. [Ultrasound/Fetoscopy] uses a fiber-optic endoscope that uses sound waves to diagnose blood disorders in a fetus. [p.326]

24. In the period following pubescence, known as [adulthood/adolescence], physical and mental maturation progress. [p.327]

25. An individual from infancy to about age 12 or 13 years is in the (childhood/pubescent) period. [p.327]

26. During the first two weeks after birth, a newborn is referred to as a (neonate/infant). [p.327]

27. Human bone formation and growth is completed when an individual is an (adolescent/adult). [p.327]

28. An individual is (pubescent/adolescent) when secondary sexual traits develop. [p.327]

29. The progressive cellular and bodily deterioration is built into the life cycle of all organisms; the process is called (senescence/adulthood). [p.327]

17.12. TIME'S TOLL: EVERYBODY AGES [p.328]

17.13. AGING SKIN, MUSCLE, BONES, AND REPRODUCTIVE SYSTEMS [p.329]

17.14. AGE-RELATED CHANGES IN SOME OTHER BODY SYSTEMS [pp.330–331]

Selected Words: telomeres [p.328], *neurofibrillary tangles* [p.330], *beta amyloid* [p.330], *Alzheimer's disease* [p.330], *apolipoprotein E* [p.330], *urinary incontinence* [p.331]

Matching

Choose the most appropriate description for each term.

1. _____ aging or senescence [p.328]

2. _____ cumulative damage to DNA [p.328]

3. _____ aging skin, muscles, and the skeleton [p.329]

4. _____ aging in the cardiovascular and respiratory systems [p.331]

5. _____ aging of the nervous system and senses [pp.330–331]

6. _____ aging of the reproductive system and changes in sexuality [p.329]

7. _____ aging of the immune system [p.331]

8. _____ aging of the digestive system [p.331]

9. _____ aging of the urinary system [p.331]

A. Menopause, hot flashes, and HRT; after about age 50, men begin to take longer to achieve erection and may experience prostate enlargement

B. Muscles of bladder and urethra weaken; urinary incontinence; loss of nephrons

C. Number of T cells falls; B cells become less active; ability to recognize self markers on body cells declines; more prone to autoimmune diseases

D. Free radical damage of proteins, DNA, and other biological molecules; cells lose the capacity for DNA self-repair; genes accumulate; enzymes in short supply

E. Walls of alveoli break down; enlarged heart; smaller and weaker heart muscles; plaques; high blood pressure

F. Number of fibroblasts in dermis decreases; elastic fibers are replaced with more rigid collagen; skin becomes thinner and less elastic; bones become weaker, more porous and brittle; joint breakdown and osteoarthritis

G. Begins around age 40; physical and physiological changes continue until we die

H. Neurofibrillary tangles, beta amyloid protein fragments, and Alzheimer's disease (AD)

I. Glands in mucous membranes of stomach and large intestine gradually deteriorate; fewer digestive enzymes are secreted

Short Answer

10. Explain the effects of cross-linking on collagen, enzymes, and possibly DNA. [pp.328]

_____

_____

_____

Self-Quiz

Are you ready for the exam? Test yourself on key concepts by taking the additional tests linked with your BiologyNow CD-ROM.

Multiple Choice

_____ 1. As it travels down the oviduct, the zygote is subdivided into a multicellular embryo through a process known as _____. [p.312]
   a. meiosis
   b. parthenogenesis
   c. embryonic induction
   d. cleavage
   e. invagination

_____ 2. The morula is transformed into a(n) _____. [p.312]
   a. zygote
   b. blastocyst
   c. gastrula
   d. third germ layer
   e. organ

_____ 3. Muscles differentiate from _____ tissue. [p.310]
   a. ectoderm
   b. mesoderm
   c. endoderm
   d. parthenogenetic
   e. yolk

_____ 4. The nervous system differentiates from _____ tissue. [p.310]
   a. ectoderm
   b. mesoderm
   c. endoderm
   d. parthenogenetic
   e. yolk

_____ 5. The pathway from the ovary to the uterus is the _____. [p.312]
   a. vas deferens
   b. epididymis
   c. coelom
   d. oviduct
   e. blastocyst

_____ 6. The innermost embryonic membrane is the _____. [p.314]
   a. chorion
   b. endoderm
   c. yolk sac
   d. amnion
   e. allantois

Matching

7. _____ HCG [p.313]

8. _____ second trimester [p.320]

9. _____ umbilical cord [p.314]

10. _____ first trimester [pp.318–319]

11. _____ ovum [p.312]

12. _____ fetus [p.319]

A. Outermost extraembryonic membrane; secretes HCG to maintain the uterine lining for three months
B. Period of development extending from the start of the fourth month to the end of the sixth
C. Blastocyst adheres to the uterine lining; cells invade maternal tissues
D. Fluid-filled sac immediately surrounding the embryo
E. Period of development extending from the seventh month until birth

13. _____ amnion [p.314]

14. _____ placenta [p.314]

15. _____ chorion [p.314]

16. _____ lactation [p.323]

17. _____ third trimester [p.320]

18. _____ implantation [p.313]

19. _____ FAS [p.325]

20. _____ birth [p.322]

F. Hormone secreted by the blastocyst; stimulates corpus luteum to continue estrogen and progesterone secretion to maintain uterine lining

G. Results from alcohol use during pregnancy; third most common cause of mental retardation in the United States.

H. Sperm penetration of the oocyte's cytoplasm stimulates its formation in meiosis II

I. Period of development from fertilization to the end of the third month

J. Blood vessels of this structure are used to transport oxygen and nutrients for the embryo

K. Occurs about thirty-nine weeks after fertilization; involves uterine contractions, cervical dilation, amnion rupture, fetal expulsion, and severed umbilical cord

L. Period beginning with completed organ system formation, beginning about the ninth week

M. By way of this tissue, the embryo receives nutrients and oxygen from the mother and sends out wastes to her bloodstream

N. Period during which hormone-primed glands produce milk

Chapter Objectives/Review Questions

This section lists general and detailed chapter objectives that can be used as review questions. You can make maximum use of these items by writing answers on a separate sheet of paper. Fill in answers where blanks are provided. To check for accuracy, compare your answers with information given in the chapter or glossary.

1. When the three germ layers have formed and they split into subpopulations of cells, it marks the onset of _____. [p.310]

2. Name each of the three germ layers, and generally list the organs formed from each. [p.310]

3. The gene-guided process by which cells in different locations in the embryo become specialized is called _____ _____. [p.311]

4. Describe the processes and products of morphogenesis. [p.311]

5. Describe early human embryonic development, and distinguish among the following: zygote, fertilization, cleavage, morula, gastrulation, morphogenesis, and organogenesis. [pp.310–311]

6. What process begins during gastrulation that did not happen during cleavage? [pp.310–311]

7. _____ is the process in which the region of cell membrane covering the sperm cell's acrosome becomes structurally unstable. [p.312]

8. A(n) _____ consists of a surface layer of cells called the *trophoblast* and a clump of cells to one side called the *inner cell mass*. [p.312]

9. How much time passes between fertilization and implantation? [p.313]

10. Describe the process of implantation. [p.313]

11. Name the four membranes that form around the early embryo, and give the major functions of each. [p.314]

12. The embryo and mother exchange substances through the _____. [p.314]

13. The primitive streak is the forerunner of a _____ _____, which will give rise to the brain and spinal cord. [p.316]

14. At what point in the development process does the embryo begin to be referred to as a fetus? [p.319]

15. Generally describe fetal development until birth. [pp.320–321]

16. Describe events occurring during the process of human birth. [pp.322–323]

17. Explain why the mother must be particularly careful of her diet, health habits, and lifestyle during the first trimester after fertilization. [pp.324–325]
18. List and characterize the stages of human development, beginning with a zygote and completing the list with old age. [p.327]
19. List factors affecting various body systems that may contribute to aging, or _____. [pp.329–331]

Media Menu Review Questions

Questions 1–4 are drawn from the following InfoTrac College Edition article: "Embryos Stick to Mothers with Sugar." *United Press International*, January 16, 2003.

1. Embryo implantation appears to involve a natural, occasional _____ coating of the uterus.
2. Understanding how embryos form a connection to their sweetened mothers could help boost the success of _____ treatments.
3. Failure of an embryo to implant causes about _____ of lost pregnancies.
4. Scientists noticed a similarity in attraction to sugars between embryos and white blood cells called _____.

Questions 5–10 are drawn from the following InfoTrac College Edition article: "Inside the Womb: What Scientists Have Learned About Those Amazing First Nine Months—and What it Means for Mothers." J. Madeleine Nash. *Time*, November 11, 2002.

5. Ultrasound imaging has led to our understanding that the most important developmental steps occur before the end of the first _____ months.
6. In the 1980s, researchers found remarkable similarities in the molecular toolkits of organisms throughout the _____ kingdom.
7. Embryonic cells known as _____ cells harbor untapped therapeutic potential.
8. Scientific evidence suggests that maladies such as atherosclerosis and diabetes may trace back to detrimental _____ conditions.
9. In organisms from a fruit fly to a human, the homeotic homeobox, or _____, genes are in charge of establishing the body axis.
10. Of all the long-term health threats during fetal development, maternal _____ may top the list.

Integrating and Applying Key Concepts

If cell differentiation did not occur, how would the human body appear? If controlled cell death did not happen in a human embryo, how would its hands appear? If development did not take place, what would a thirty-year-old human look like?

18

LIFE AT RISK: INFECTIOUS DISEASE

Interactive Exercises

Impacts, Issues: The Face of Aids [p.335]

18.1. VIRUSES AND INFECTIOUS PROTEINS [pp.336–337]

18.2. BACTERIA—THE UNSEEN MULTITUDES [pp.338–339]

For additional practice, use the interactive vocabulary exercises linked with your BiologyNow CD-ROM.

Selected Words: "enveloped" viruses [p.336], *latent* [p.336], *infectious mononucleosis* [p.337], *reverse transcriptase* [p.337], *provirus* [p.337], *Creutzfeldt-Jakob disease* (CJD) [p.337], *bovine spongiform encephalitis* [p.337], *spirochetes* [p.338], *bacterial flagellum* [p.338], *pili* (singular: pilus) [p.338], *prokaryotic fission* [p.338], "fertility" plasmid [p.338], *Salmonella* [p.338], *Streptococcus* [p.338], *strep throat* [p.338], *Borrelia burgdorferi* [p.338], *antiviral* drugs [p.339], "super bugs" [p.339], *Staphylococcus aureus* [p.339]

Boldfaced, Page-Referenced Terms

[p.336] virus _____

_____

[p.337] retrovirus _____

_____

[p.337] prions _____

_____

[p.339] antibiotic _____

_____

Short Answer

1. A virus has been described as "bad news wrapped in protein." What is the "bad news" component? [p.336]

_____

_____

_____

True/False

If the statement is true, write a "T" in the blank. If the statement is false, make it correct by writing the word(s) in the blank that should take the place of the underlined word(s).

_____ 2. A virus <u>is</u> a cell. [p.336]

_____ 3. A virus invades a host cell in order to <u>feed</u>. [p.336]

_____ 4. Viruses that infect bacteria are <u>bacteriophages</u>. [p.336]

_____ 5. The protein coat immediately surrounding the nucleic acid of a virus is called the <u>envelope</u>. [p.336]

_____ 6. The genetic material of a virus is DNA <u>or</u> RNA. [p.336]

_____ 7. Proteins extending from the viral coat bind to receptors on <u>other viruses</u>. [p.330]

Sequencing

Show the correct order of the events listed by placing their letters in the blanks beside the numbers. [p.336]

8. _____
9. _____
10. _____
11. _____
12. _____
13. _____

A. Assembly of viral nucleic acids and proteins into new virus particles
B. Entry of the virus or its genetic material into the host cell's cytoplasm
C. Synthesis of enzymes and other proteins, using host cell machinery
D. Release of new virus particles from the cell
E. Attachment to the host cell
F. Replication of viral nucleic acid

Dichotomous Choice

Circle one of two possible answers given between parentheses in each statement.

14. A given virus can often attack only one type of cell because of differences in (receptors/lipids) at the cell surface. [p.336]

15. Usually, a virus (kills its host quickly/enters a period of latency). [p.336]

16. Latent viruses are (inactivated/reactivated) by stressors like sunburn, illness, and emotional upsets. [p.336]

17. Type 1 Herpes simplex can remain latent inside clusters of (nerve/epithelial) cells. [p.336]

18. Infectious mononucleosis is caused by the (type 2 Herpes simplex/Epstein-Barr) virus. [p.337]

19. Viruses that use reverse transcriptase to copy DNA from viral RNA are called (retroviruses/latent viruses). [p.337]

20. Creutzfeldt-Jakob disease and mad cow disease (bovine spongiform encephalitis) are caused by (retroviruses/prions). [p.337]

21. A prion is a misfolded (lipid/protein). [p.337]

Fill-in-the-Blanks

Bacteria are (22) _____ [p.338] with no nucleus or other membrane-bound organelles. In most, there is a cell wall made of (23) _____ [p.338]. A ball-shaped bacterium is a(n) (24) _____ [p.338], a rod-shaped bacterium is a(n) (25) _____ [p.338], and a twisted bacterium is a(n) (26) _____ [p.338]. Some bacteria have a protein filament called a(n) (27) _____ [p.338] for locomotion. (28) _____ [p.338] are filaments used to stick to surfaces. Bacteria reproduce by (29) _____ _____ [p.338]. Bacterial cells have (30) _____ [p.338] circular chromosome, making cell division simple. Some bacteria can divide as often as every (31) _____ [p.338] minutes. Bacterial cells may have small circles of DNA containing just a few genes. These are called (32) _____ [p.338]. One type, a (33) "_____" [p.338] plasmid, carries genes that allow the bacterium to transfer plasmid DNA to other bacterial cells.

Short Answer

34. Name some ways in which humans use bacteria for our benefit. [p.338]

_____

_____

_____

35. List some factors that have brought about antibiotic-resistant "super-bugs." [p.339]

_____

_____

_____

Dichotomous Choice

Circle one of two possible answers given between parentheses in each statement.

36. Antibiotics were discovered in the (1920s/1940s). [p.339]

37. Most antibiotics are produced by (humans/bacteria and fungi). [p.339]

38. Antibiotics may be effective against (bacteria/viruses). [p.339]

39. The body's defenses against viruses include proteins called (enzymes/interferons). [p.339]

40. Antibiotics wipe out (susceptible/resistant) bacteria. [p.339]

41. Resistance conferred on bacteria by chance mutations are very serious because bacteria reproduce rapidly and spread genes via (sexual reproduction/plasmids). [p.339]

42. Staph A is one of several highly pathogenic bacteria that may soon be (susceptible/resistant) to all available antibiotics. [p.339]

18.3. INFECTIOUS PROTOZOA AND WORMS [p.340]

18.4. INFECTIOUS FOES NEW AND OLD [p.341]

18.5. CHARACTERISTICS AND PATTERNS OF INFECTIOUS DISEASES [pp.342–343]

Selected Words: Entamoeba histolytica [p.340], amoebic dysentery [p.340], Giardia lamblia [p.340], giardiasis [p.340], Trypanosoma brucei [p.340], African trypanosomiasis [p.340], sleeping sickness [p.340], cryptosporidiosis [p.340], Cryptosporidium parvum [p.340], Ascaris [p.340], malaria [p.341], Plasmodium [p.341], Anopheles [p.341], sickle cell anemia [p.341], tuberculosis (TB) [p.341], Mycobacterium tuberculosis [p.341], intermediate host [p.342], ringworm [p.342], gonorrhea [p.343], emerging disease [p.343], re-emerging diseases [p.343]

Boldfaced, Page-Referenced Terms

[p.342] disease vector _____

_____

[p.342] nosocomial infection _____

_____

[p.342] sporadic disease _____

_____

[p.342] endemic disease _____

_____

[p.342] epidemic _____

_____

[p.342] pandemic _____

_____

[p.343] virulence _____

_____

Matching

1. _____ giardiasis
2. _____ pinworm
3. _____ African sleeping sickness
4. _____ amoebic dysentery
5. _____ cryptosporidiosis

A. *Entamoeba histolytica*, common in areas without sewage treatment [p.340]
B. *Giardia lamblia*, widespread cause of explosive diarrhea and "rotten egg" belches [p.340]
C. *Trypanosoma brucei*, spread by tsetse fly; invades central nervous system and is fatal if untreated [p.340]
D. Most common worm infection in developed countries; infection route is oral–fecal [p.340]
E. *Cryptosporidium parvum*, watery diarrhea, due to fecal contamination of food or water [p.340]

Elimination

Scratch through the incorrect answer within each set of parentheses.

6. (*Cryptosporidium parvum/Giardia lamblia/Trypanosoma brucei*) produce symptoms including severe diarrhea. [p.340]

7. Factors involved in the spread of infections by protozoa and worms include (poor sanitation/sexual intercourse/eating contaminated meat). [p.340]

8. (*Cryptosporidium/Ascaris*/tapeworms) are examples of human worm parasites. [p.340]

Choice

Place the letter of the relevant disease beside the information given.

      a. malaria   b. SARS   c. West Nile virus   d. tuberculosis

9. _____ Severe Acute Respiratory Syndrome [p.341]

10. _____ caused by the protozoan *Plasmodium* [p.341]

11. _____ caused by a bacillus bacterium [p.341]

12. _____ antibiotics taken diligently for at least a year usually provides a cure [p.341]

13. _____ parasite is carried by *Anopheles* mosquito [p.341]

14. _____ extreme cases lead to paralysis and death [p.341]

15. _____ caused by a type of coronavirus [p.341]

16. _____ possessing one copy of sickle-cell anemia gene provides some protection [p.341]

17. _____ no serious illness in over 90 percent of those infected [p.341]

18. _____ tropical Africa is stronghold, but disease is spreading [p.341]

19. _____ more of a risk for those living in crowded, unsanitary conditions [p.341]

20. _____ symptoms include shaking and chills due to very high fever [p.341]

21. _____ disease appeared first in China in 2002 [p.341]

22. _____ spread by mosquitoes that have fed on infected birds [p.341]

23. _____ first transmission to humans was probably from civets [p.341]

24. _____ re-emerged as a major health threat in 1993 [p.341]

25. _____ kills about 3 million people a year [p.341]

26. _____ effects evident in ancient mummies [p.341]

27. _____ IV drug use and compromised immune systems (such as with HIV) create risk [p.341]

28. _____ first appeared in the United States in 2002; 284 out of 4,000 infections led to death [p.341]

29. _____ causative organism is a bacillus bacterium, genus *Mycobacterium*

Matching

30. _____ virulence

31. _____ sporadic disease

32. _____ contact with a vector

33. _____ nosocomial infection

34. _____ endemic disease

35. _____ inhaling pathogens

36. _____ pandemic

37. _____ direct contact

38. _____ epidemic

39. _____ indirect contact

A. Insects carry a pathogen from an infected organism to a new host [p.342]
B. Acquired in a hospital [p.342]
C. Ability of a pathogen to cause serious disease; depends on rate of invasion, severity of damage, and targeted tissues [p.343]
D. Epidemic breaks out in several countries in a given time span; AIDS is an example [p.342]
E. Transfer of pathogen by a handshake, kiss, or other contact [p.342]
F. Transfer of a pathogen by a cough or sneeze, such as with influenza [p.342]
G. Breaks out irregularly and affects relatively few people, such as whooping cough [p.342]
H. Disease rate increases to a level above what was predicted, such as cholera in Peru in 1991 [p.342]
I. Transfer of a pathogen by touching objects previously in contact with an infected person, or contaminated food or water [p.342]
J. Disease that occurs more or less continuously such as the common cold or ringworm [p.342]

Short Answer

40. Explain how the immune system's response to gonorrhea is an example of a response gone awry. [p.343]

_____

_____

_____

41. How do most disease-causing bacteria harm their host? [p.343]

_____

_____

_____

42. List three factors that contribute to the spread of new and re-emerging infectious diseases. [p.343]

_____

_____

_____

Complete the Table

43. Complete the following table by entering the correct disease category, description of category, or example. [p.343]

Disease Category	Description	Example
Specific to one organism	one specific animal, such as humans, are reservoir	a.
b.	c.	SARS, rabies
d.	diseases that are becoming threatening for the first time	e.
f.	diseases that were controlled but are becoming threatening again	g.

18.6. THE HUMAN IMMUNODEFICIENCY VIRUS AND AIDS [pp.344–345]

18.7. TREATING AND PREVENTING HIV INFECTION AND AIDS [p.346]

Selected Words: "indicator diseases" [p.344], *transcriptase* [p.344], *transcription* [p.344], drug "cocktail" [p.346], "entry inhibitors" [p.346]

Boldfaced, Page-Referenced Terms

[p.344] human immunodeficiency virus (HIV) _____

_____

True/False

If the statement is true, write a "T" in the blank. If the statement is false, make it correct by writing the word(s) in the blank that should take the place of the underlined word(s).

_____ 1. AIDS is <u>one</u> disease caused by infection with HIV. [p.344]

_____ 2. HIV destroys T cells, crippling the <u>cardiovascular</u> system. [p.344]

_____ 3. Certain types of pneumonia, cancer, yeast infections, and drug-resistant tuberculosis are "<u>indicator diseases</u>" of AIDS. [p.344]

_____ 4. The HIV virus can enter the body from infected body fluids that come into contact with <u>cuts or abrasions</u>. [p.344]

_____ 5. The most common mode of transmission of HIV in the United States is <u>intravenous drug use</u>. [p.344]

_____ 6. HIV <u>is</u> easily transmitted through food, air, water, and insect bites. [p.344]

_____ 7. Infected mothers <u>cannot</u> transfer HIV to their babies during pregnancy, birth, and breast-feeding. [p.344]

_____ 8. On a worldwide level, the overwhelming majority of HIV-positive individuals <u>are</u> homosexuals. [p.344]

_____ 9. In recent years, more young adults in the United States have died from <u>hepatitis C</u> than from any other cause. [p.344]

_____ 10. HIV is a retrovirus with <u>RNA</u> as its genetic material. [p.344]

_____ 11. Reverse transcriptase makes DNA that <u>inserts into</u> the host cell chromosome. [p.344]

_____ 12. HIV can infect macrophages and <u>bone marrow</u> cells. [p.344]

_____ 13. About 5 percent of HIV-infected people don't develop symptoms for 10 years or more after diagnosis, possibly due to infection with a less virulent strain. [p.345]

_____ 14. Individuals born with a mutation resulting in the absence of the CD4 receptor on some cells <u>are</u> immune to AIDS and HIV. [p.345]

Sequencing

Show the correct order of the events listed by placing their letters in the blanks beside the numbers. [p.345]

15. _____

16. _____

17. _____

18. _____

19. _____

a. Copies of HIV produced by replication inside infected cells circulate in bloodstream, producing flulike symptoms.

b. Huge reservoirs of infected T cells accumulate in lymph nodes.

c. The body mounts an immune response, but infection rate overwhelms it.

d. HIV infects a person and docks with a CD4 receptor on helper T cell or macrophage.

e. The number of helper T cells drops and symptoms appear such as weight loss, fatigue, nausea, night sweats, enlarged lymph nodes, and eventually an "indicator disease."

True/False

If the statement is true, write a "T" in the blank. If the statement is false, make it correct by writing the word(s) in the blank that should take the place of the underlined word(s).

_____ 20. Current drugs cannot cure AIDS because HIV genes are inserted in the host cell <u>DNA</u>. [p.346]

_____ 21. Protease inhibitors target steps required for <u>replication</u> of new virus particles. [p.346]

_____ 22. Present treatments for AIDS consist of an <u>antibiotic</u> and two anti-HIV drugs. [p.346]

_____ 23. Drugs called "entry inhibitors" prevent HIV from entering the <u>bloodstream</u>. [p.346]

_____ 24. The biggest problem in developing a vaccine against HIV is that it mutates extremely <u>rapidly</u>. [p.346]

_____ 25. An easy and inexpensive test for HIV requires a sample of <u>saliva</u>. [p.346]

18.8. PROTECTING YOURSELF—AND OTHERS—FROM STDs [p.347]

18.9. COMMON STDs CAUSED BY BACTERIA [pp.348–349]

18.10. A ROGUE'S GALLERY OF VIRAL STDs AND OTHERS [pp.350–351]

Selected Words: *Chlamydia trachomatis* [p.348], *chlamydia* [p.348], *pelvic inflammatory disease* (PID) [p.348], *Neisseria gonorrhea* [p.348], *syphilis* [p.349], *Treponema pallidum* [p.349], *primary stage* of syphilis [p.349], *secondary stage* of syphilis [p.349], *tertiary stage* of syphilis [p.349], *genital warts* [p.350], *Pap smear* [p.350], *hepatitis C* [p.351], *Haemophilus ducreyi* [p.351], *pubic lice* [p.351], "crabs" [p.351], "nits" [p.351], *Candida albicans* [p.351], candidiasis [p.351], *Trichomonas vaginalis* [p.351], trichomoniasis [p.351]

Boldfaced, Page-Referenced Terms

[p.350] genital herpes _____

_____

[p.350] human papillomavirus (HPV) _____

_____

[p.351] hepatitis B virus (HBV) _____

_____

[p.351] chancroid _____

_____

Choice

For questions 1–9, write the letter of the causative organism in the blank beside the name of each disease.

        a. bacterium   b. virus   c. animal   d. protozoan or fungus

1. _____ chlamydial infection [p.348]
2. _____ gonorrhea [p.348]
3. _____ syphilis [p.349]
4. _____ genital herpes [p.350]
5. _____ genital warts [p.350]
6. _____ hepatitis B or C [p.351]
7. _____ chancroid [p.351]
8. _____ pubic lice [p.351]
9. _____ vaginitis [p.351]

For questions 10–18, choose from the following treatments for each disease:

a. antibiotics
b. interferon
c. surgery or by freezing or burning the affected area
d. the only treatment is rest

e. antiparasitic drugs
f. antifungal drugs
g. no cure exists

10. _____ chlamydial infection [p.348]
11. _____ gonorrhea [pp.348–349]
12. _____ syphilis [p.349]
13. _____ genital herpes [p.350]
14. _____ genital warts [p.350]
15. _____ hepatitis B or C [p.351]
16. _____ chancroid [p.351]
17. _____ pubic lice [p.351]
18. _____ vaginitis [p.351]

Short Answer

19. What does "PID" stand for? Name two STDs that can lead to this condition. What contributory behavior do these two diseases have in common in female infections? [pp.348–349]

_____

_____

_____

20. Briefly describe the three stages of syphilis, and when each develops. [p.349]

_____

_____

Self-Quiz

Are you ready for the exam? Test yourself on key concepts by taking the additional tests linked with your BiologyNow CD-ROM.

For questions 1–14, choose the letter that applies from the following:

a. AIDS
b. gonorrhea
c. syphilis
d. chlamydial infection
e. hepatitis
f. pelvic inflammatory disease
g. genital herpes
h. genital warts
i. vaginitis

1. _____ The virus slowly cripples the immune system and opens the door to "opportunistic" infections. [p.344]
2. _____ The pathogen causing this disease enters a host by contact between infected body fluids and a cut or abrasion. [p.344]
3. _____ Serious complication of some STDs; severe abdominal pain, scarred oviducts leading to abnormal pregnancies and sterility. [p.348]
4. _____ Caused by a parasitic bacterium that invades cells of the genital and urinary tracts; results in swollen lymph nodes, and often leads to PID. [p.348]
5. _____ A bacterium, *Neisseria gonorrhea,* enters mucous membranes and causes more noticeable symptoms in males; prompt treatment quickly cures this disease. [p.348]
6. _____ Being cured of infection does not confer on the person immunity to the bacterium that causes the disease; the use of condoms can prevent infection. [p.349]
7. _____ Caused by a motile, corkscrew-shaped bacterium, *Treponema pallidum.* [p.349]
8. _____ Produces a chancre (localized ulcer) one to eight weeks following infection. [p.349]
9. _____ The tertiary stage of this disease begins when pathogens enter the brain, with the ultimate result being insanity and death. [p.349]
10. _____ Between flare-ups, the virus remains latent within the nervous system. [p.350]
11. _____ Can lead to lesions in the eyes that cause blindness in babies born to mothers with the disease. [p.350]
12. _____ Benign, painless clustered growths on the penis, cervix, or around the anus; probable cause of cervical cancer. [p.350]
13. _____ This infection damages the liver and is far more contagious than HIV. [p.351]
14. _____ Yeast infections and trichomoniasis are examples of this type of infection found in females. [p.351]

Chapter Objectives/Review Questions

This section lists general and detailed chapter objectives that can be used as review questions. You can make maximum use of these items by writing answers on a separate sheet of paper. To check for accuracy, compare your answers with information given in the chapter or glossary.

1. Describe the principal body structures of bacteria and viruses (inside and outside). [pp.336–338]
2. Describe the reproductive cycle of viruses. [p.336]
3. Explain how some bacteria have become antiobiotic resistant. [p.339]
4. Describe four ways that a person can be infected with HIV. [p.344]
5. Describe in detail which tissues HIV attacks, the viral behaviors that allow HIV to remain undetected in the body, the general progress of attack, and the symptoms ultimately presented. [pp.344–345]
6. List the names and symptoms of ten diseases that can be sexually transmitted by humans to their partners. [pp.344,348–351]
7. For each of ten STDs, state how you can avoid being infected, or, if in spite of all your precautions, you did become infected, what would be the likely course of treatment. [pp.344–351]

Media Menu Review Questions

Questions 1–6 are drawn from the following InfoTrac College Edition article: "SARS Virus' Genome Hints at Independent Evolution." B. Harder. *Science News*, April 26, 2003.

1. SARS emerged from _____ in February 2003 and has killed over _____ people thus far.
2. European scientists determined that the SARS virus is a type of _____ never detected before the current outbreak.
3. By April 2003, the _____ of SARS had been completely sequenced.
4. Sequencing studies in Great Britain, the United States, and China indicate that the virus _____ rapidly.
5. Comparing the DNA sequence of SARS to other coronaviruses indicates that it is _____ to other families of coronaviruses.
6. The proposed explanation for this data is that the virus has had the capacity to infect people but only recently encountered conditions that _____ its spread; or that the virus only recently "jumped" from a wild _____ by mutation.

Questions 7–10 are drawn from the following InfoTrac College Edition article: "Genital Herpes Linked to Cervical Cancer." N. Seppa. *Science News*, November 9, 2002.

7. Nearly every woman who contracts cancer of the cervix is also infected with the human _____ (HPV).
8. Not all women with HPV get cervical cancer; it may be that a factor in who does contract this form of cancer is whether that woman also has _____ _____.
9. A study by Jennifer Smith in France showed that women with cervical cancer were _____ percent more likely to have herpes than were those without the cancer.
10. It may be that the tissue damage and/or inflammation caused by genital herpes has the effect of accelerating the growth or spread of _____.

Integrating and Applying Key Concepts

STDs can be at the least annoying and at worst, deadly. Adult humans can have a happy and fulfilling sex life without ever contracting any of the pathogens discussed in this chapter. Make a written outline of the approach you would like to use with a child of your own to make certain he or she understands not only the pleasures that can be achieved, but also the possible dangers of sexual activity.

19

CELL REPRODUCTION

Impacts, Issues: Henrietta's Immortal Cells

DIVIDING CELLS BRIDGE GENERATIONS
    Division of the "parent" nucleus sorts the DNA
        into nuclei for daughter cells
    Chromosomes are DNA "packages" in the cell
        nucleus
    Two sets of chromosomes = a diploid cell

THE CELL CYCLE

A CLOSER LOOK AT CHROMOSOMES
    In a chromosome, DNA interacts with proteins
    Spindles attach to chromosomes and move them

THE FOUR STAGES OF MITOSIS
    Mitosis begins with prophase
    Next comes metaphase
    Anaphase, then telophase, follow

HOW THE CYTOPLASM DIVIDES

CONCERNS AND CONTROVERSIES OVER IRRADIATION

MEIOSIS—THE BEGINNINGS OF EGGS AND SPERM
    Meiosis: Two divisions, not one
    Meiosis is the first step in the formation of
        gametes

A VISUAL TOUR OF THE STAGES OF MEIOSIS

THE SECOND STAGE OF MEIOSIS—NEW COMBINATIONS OF PARENTS' TRAITS
    In prophase I, genes may be rearranged
    In metaphase I, maternal and paternal
        chromosomes are shuffled

MEIOSIS AND MITOSIS COMPARED

Interactive Exercises

Impacts, Issues: Henrietta's Immortal Cells [p.355]

19.1. DIVIDING CELLS BRIDGE GENERATIONS [pp.356–357]

For additional practice, use the interactive vocabulary exercises linked with your BiologyNow CD-ROM.

Selected Words: "immortal" cell lineage [p.355], *chromatin* [p.357], *karyotype* [p.357], *homologues* [p.357]

Boldfaced, Page-Referenced Terms

[p.356] reproduction _____

_____

[p.356] life cycle _____

_____

[p.356] mitosis _____

_____

[p.356] meiosis _____

_____

[p.356] germ cells _____

_____

[p.357] chromosome _____

_____

[p.357] chromosome number _____

_____

[p.357] diploid cell (2*n*) _____

_____

[p.357] homologous chromosomes _____

_____

Short Answer

1. Define and describe the value of *HeLa cells* to biologists. [p.355]

_____

_____

_____

_____

Matching

Choose the most appropriate description for each term.

2. _____ karyotype [p.357]

3. _____ diploid cell [p.357]

4. _____ meiosis [p.356]

5. _____ homologous chromosomes [p.357]

6. _____ chromosome [p.357]

7. _____ germ cells [p.356]

8. _____ life cycle [p.356]

9. _____ somatic [p.356]

10. _____ chromosome number [p.357]

11. _____ mitosis [p.356]

12. _____ reproduction [p.356]

A. Mechanism for dividing nuclei of germ cells; first stage in sexual reproduction
B. Cells in which meiosis occurs—oogonia and spermatogonia
C. When a parent cell produces a new generation of cells, or when parents produce a new individual
D. Indicates the sum of all chromosomes normally present in a cell
E. Photographic arrangement of chromosomes shown with their homologous partners
F. Nuclear division process involved in growth, replacement, and repair
G. A pair of physically similar chromosomes (one from each parent) with genetic instructions for the same traits
H. Single DNA molecule combined with protein
I. Refers to body cells
J. A recurring series of events in which individuals grow, develop, maintain themselves, and reproduce according to instructions encoded in DNA
K. Refers to any cell possessing two of each type of chromosome

19.2. THE CELL CYCLE [p.358]

19.3. A CLOSER LOOK AT CHROMOSOMES [p.359]

Selected Words: histones [p.359], nucleosome [p.359]

Boldfaced, Page-Referenced Terms

[p.358] cell cycle _____

_____

[p.358] interphase _____

_____

[p.359] sister chromatids _____

_____

[p.359] centromere _____

_____

[p.359] spindle _____

_____

Labeling

Identify the stage in the cell cycle indicated by each number. [p.358]

1. _____
2. _____
3. _____
4. _____
5. _____
6. _____
7. _____
8. _____
9. _____
10. _____

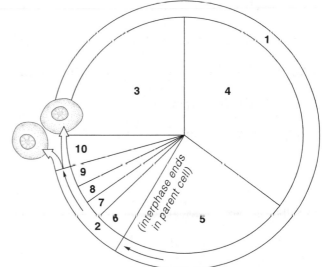

Matching

Link each of the following time spans to the parts of the cell cycle by placing numbers from the diagram above into the blanks below. Answers may be used more than once.

11. _____ Period after DNA replication; cell prepares to divide [p.356]

12. _____ Chromosomes are sorted into two sets, followed by cytoplasmic division [p.356]

13. _____ DNA replication occurs [p.356]

14. _____ Period of cell growth before DNA replication [p.356]

15. _____ Longest phase of the cell cycle [p.356]

16. _____ Period of cytoplasmic division [p.356]

17. _____ Period that includes G_1, S, and G_2 [p.356]

Dichotomous Choice

Circle one of the two possible answers given between parentheses in each statement.

18. A chromosome consists of DNA attached to proteins called (spindles/histones). [p.359]

19. A unit of DNA complexed with a histone is a (nucleosome/centromere). [p.359]

20. When DNA instructions need to be read, the nucleosome packing in that region (loosens/tightens). [p.359]

21. As a cell nucleus prepares for division, each chromosome (condenses/elongates). [p.359]

22. Each replicated chromosome consists of sister (centromeres/chromatids). [p.359]

23. The "pinched in" region of a replicated chromosome is called a (chromatid/centromere) and is where (microtubules/histones) attach. [p.359]

24. The (centromere/spindle) consists of two sets of microtubules extending from the two poles of the cell. [p.359]

25. Colchicine is a plant-derived poison that affects (microtubules/chromosomes); it is used in research to stop mitosis in dividing cells. [p.359]

19.4. THE FOUR STAGES OF MITOSIS [pp.360–361]

19.5. HOW THE CYTOPLASM DIVIDES [p.362]

19.6. CONCERNS AND CONTROVERSIES OVER IRRADIATION [p.363]

Selected Words: "prometaphase" [p.360]

Boldfaced, Page-Referenced Terms

[p.360] prophase _____

_____

[p.360] metaphase _____

_____

[p.360] anaphase _____

_____

[p.360] telophase _____

_____

[p.362] cytokinesis _____

_____

[p.352] cleavage furrow _____

_____

Labeling-Matching

Identify each of the following mitotic stages by entering the correct stage in the blank beneath the sketch. Select from *late prophase, transition to metaphase, interphase—parent cell, metaphase, early prophase, telophase, interphase—daughter cells,* and *anaphase.* Complete the exercise by matching and entering the letter of the correct phase description in the parentheses after each label. [pp.360–361]

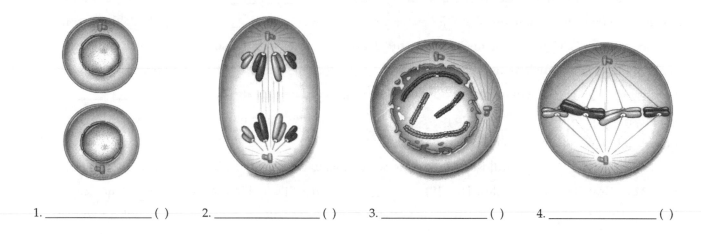

1. _____ ( )   2. _____ ( )   3. _____ ( )   4. _____ ( )

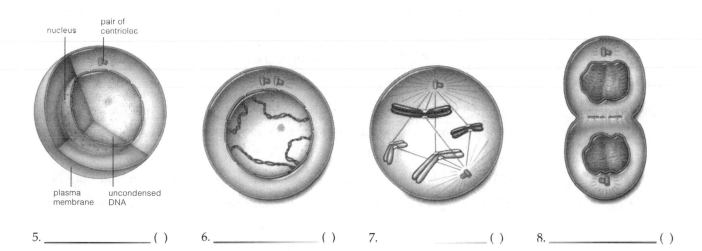

5. _____ ( )   6. _____ ( )   7. _____ ( )   8. _____ ( )

A. Attachment between two sister chromatids of each chromosome breaks; the two are now chromosomes in their own right; they move to opposite spindles.

B. Microtubules that form the spindle apparatus enter the nuclear region; microtubules become attached to the sister chromatids of each chromosome.

C. The DNA and its associated proteins start to condense into the threadlike chromosome form.

D. Chromosomes are now fully condensed and lined up at the equator of the spindle.

E. DNA is duplicated and the cell prepares for division.

F. Two daughter cells have formed, each diploid and with two of each type of chromosome.

G. Chromosomes continue to condense; new microtubules are assembled, and they move one of two centrioles toward the opposite end of the cell: the nuclear envelope begins to break up.

H. New patches of membrane join to form nuclear envelopes around the decondensing chromosomes.

Fill-in-the-Blanks

(9) _____ [p.360], the first stage of mitosis, is evident when chromosomes become visible in the light microscope as threadlike forms. Each was duplicated earlier, during (10) _____ [p.360]. Each already consists of two (11) _____ _____ [p.360] joined together at the (12) _____ [p.360]. Early in prophase, the sister chromatids of each (13) _____ [p.360] twist and fold into a more compact form. By late prophase, all the chromosomes will be (14) _____ [p.360] into thick, rod-shaped forms. Cytoplasmic microtubules break apart into their tubulin subunits. The subunits reassemble near the nucleus, as new (15) _____ [p.360]. Many cells have two barrel-shaped (16) _____ [p.360] that were duplicated during interphase. By the time (17) _____ [p.360] is under way, the cell has two pairs of them. Microtubules begin moving one pair to the opposite pole of the newly forming spindle.

(18) _____ [p.360] is a time during which the nuclear envelope breaks up completely, into numerous tiny, flattened vesicles. The (19) _____ [p.360] begin interacting with the microtubules. When each chromosome is harnessed by microtubules from both (20) _____ [p.360], a two-way pull orients the two sister (21) _____ [p.360] of a chromosome toward opposite poles. When all the duplicated chromosomes are aligned midway between the poles of a completed spindle, the cell is in (22) _____ [p.361].

During (23) _____ [p.361], the two sister chromatids of each chromosome separate and move to opposite poles by two mechanisms: Microtubules attached to the centromere regions shorten, pulling the chromosomes to the poles, and the spindle elongates when overlapping microtubules ratchet past each other and push the spindle poles farther apart. Once separated from its sister, each (24) _____ [p.361] now becomes a chromosome in its own right.

(25) _____ [p.361] begins when the two clusters of chromosomes arrive at opposite spindle poles. Chromosomes are no longer harnessed to microtubules, and they return to a more threadlike form. Vesicles of the old (26) _____ _____ [p.361] fuse to form, patch by patch, a new nuclear envelope that separates each cluster of chromosomes from the cytoplasm. With mitosis, each new nucleus has the same chromosome (27) _____ [p.361] as the parent nucleus. Once the two nuclei form, (28) _____ [p.361] is over—and so is (29) _____ [p.361].

Matching

Match the descriptions of cytokinesis with the four diagrams. [p.362]

30. _____

31. _____

32. _____

33. _____

A. Shrinkage of the microfilament ring pulls the cell surface inward.
B. Mitosis is complete and the spindle is disassembling.
C. Contractions continue; the cell is pinched in two.
D. Just beneath the plasma membrane, microfilament rings at the former spindle equator begin to contract.

Short Answer

34. List some common natural sources of ionizing radiation. [p.363]

_____

_____

_____

35. Summarize the effects of ionizing radiation on cells. [p.363]

_____

_____

_____

36. Name some uses of irradiation in medicine and the food industry. [363]

_____

_____

_____

19.7. MEIOSIS—THE BEGINNINGS OF EGGS AND SPERM [pp.364–365]

19.8. A VISUAL TOUR OF THE STAGES OF MEIOSIS [pp.366–367]

Selected Words: spermatogonia [p.364], oogonia [p.364]

Boldfaced, Page-Referenced Terms

[p.364] reductional division _____

_____

[p.364] haploid number (*n*) _____

_____

[p.364] spermatogenesis _____

_____

[p.365] oogenesis _____

_____

True/False

If the statement is true, write a "T" in the blank. If the statement is false, make it correct by changing the underlined word(s) and writing the correct word(s) in the answer blank.

_____ 1. Meiosis occurs in dividing germ cells in ovaries or testes. [p.364]

_____ 2. Sperm and eggs are haploid cells referred to as germ cells. [p.364]

_____ 3. Meiosis involves a reductional division that halves the diploid chromosome number. [p.364]

_____ 4. Haploid cells possess pairs of homologous chromosomes. [p.364]

_____ 5. Gametes formed by meiosis have one member of each pair of homologous chromosomes possessed by the species. [p.364]

Comparison

Fill in the chart below in order to compare and contrast mitosis and meiosis. Assume that both processes begin with diploid (2*n*) cells. [p.364]

	Mitosis	*Meiosis*
6. # of times DNA replication occurs	1	
7. # of times cell division occurs	1	
8. # of resulting cells	2	
9. ploidy (*n* or 2*n*) of final daughter cells	2*n*	
10. purpose in human life cycle	growth, repair	

Matching

To review the major events of meiosis I and II, match the following written descriptions with the appropriate sketch. To simplify the process, assume that the cell in this model initially has only one pair of homologous chromosomes (one from a paternal source and one from a maternal source) and that crossing over does not occur. Complete the exercise by indicating diploid (2n) or haploid (n) chromosome number of the cell chosen in the parentheses following each blank.

11. _____ (____) A pair of homologous chromosomes prior to S of interphase in a diploid germ cell [p.364]

12. _____ (____) While the germ cell is in S of interphase, chromosomes are duplicated through DNA replication; the two sister chromatids are attached at the centromere [p.364]

13. _____ (____) During meiosis I, each duplicated chromosome lines up with its partner, homologue to homologue [p.364]

14. _____ (____) Also during meiosis I, the chromosome partners separate from each other in anaphase I; cytokinesis occurs, and each chromosome goes to a different cell [p.364]

15. _____ (____) During meiosis II (in two cells), the sister chromatids of each chromosome are separated from each other; four haploid nuclei form; cytokinesis results in four cells (potential gametes) [p.364]

A.   B.   C.   D.   E.

Sequence

Arrange the following entities in correct order of development, entering a 1 by the stage that appears first and a 5 by the stage that completes the process of spermatogenesis. Complete the exercise by indicating if each cell is n or 2n in the parentheses following each blank. Refer to Figure 19.10 in the text. [pp.364–365]

16. _____ (____) primary spermatocyte

17. _____ (____) sperm

18. _____ (____) spermatid

19. _____ (____) spermatogonium

20. _____ (____) secondary spermatocyte

Short Answer

21. List the major differences between oogenesis and spermatogenesis. [p.365]

_____

_____

_____

Labeling-Matching

This section will help you review the details of meiosis I and II. Identify each of the following meiotic stages by entering the correct stage of either meiosis I or meiosis II in the blank beneath the sketch. Choose from *prophase I, metaphase I, anaphase I, telophase I, prophase II, metaphase II, anaphase II,* and *telophase II.* Complete the exercise by matching and entering the letter of the correct phase description in the parentheses following each label. [pp.366–367]

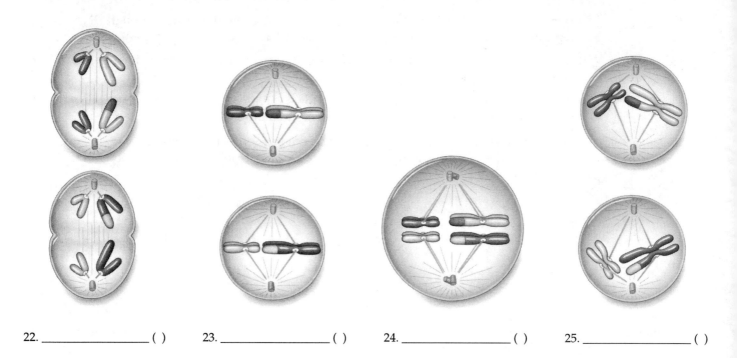

22. _____ ( )   23. _____ ( )   24. _____ ( )   25. _____ ( )

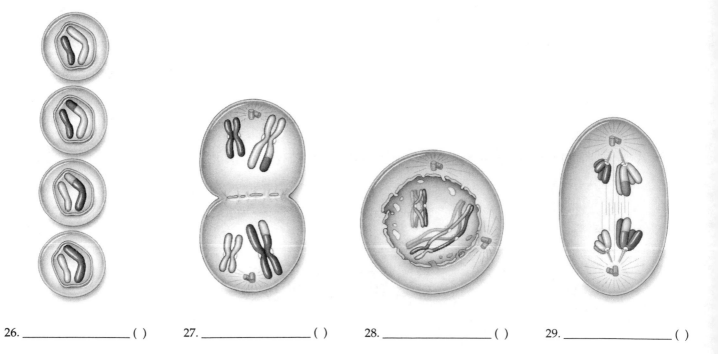

26. _____ ( )   27. _____ ( )   28. _____ ( )   29. _____ ( )

A. The spindle is now fully formed; all chromosomes are positioned midway between the poles of one cell.

B. Centrioles have moved to opposite poles in each of two cells; a spindle forms; and microtubules attach the duplicated chromosomes to the spindle and begin moving them.

C. Four daughter nuclei form; when the cytoplasm divides, each new cell has a haploid number of chromosomes, all in the unduplicated state; one or all cells may develop into gametes.

D. In one cell, each duplicated chromosome is pulled away from its homologue; the two are moved to opposite spindle poles.

E. Each chromosome is drawn up close to its homologue; crossing over and swapping of segments (genetic recombination) occurs.

F. All the chromosomes are now aligned at the spindle's equator; this is occurring in two haploid cells.

G. Two haploid cells form, but chromosomes are still in the duplicated state.

H. Chromatids of each chromosome separate; former "sister chromatids" are now chromosomes in their own right and are moved to opposite poles.

19.9. THE SECOND STAGE OF MEIOSIS—NEW COMBINATIONS OF PARENTS' TRAITS [pp.368–369]

19.10. MEIOSIS AND MITOSIS COMPARED [pp.370–371]

Selected Words: nonsister chromatids [p.369], *maternal* chromosomes [p.369], *paternal* chromosomes [p.369], *8,388,608 combinations* [p.369]

Boldfaced, Page-Referenced Terms

[p.368] crossing over _____

_____

[p.369] genetic recombination _____

_____

[p.369] alleles _____

_____

[p.369] disjunction _____

_____

Matching

Choose the most appropriate description for each term.

1. _____ crossing over [pp.368–369]
2. _____ nonsister chromatids [p.369]
3. _____ alleles [p.369]
4. _____ genetic recombination [p.369]
5. _____ maternal chromosomes [p.369]
6. _____ paternal chromosomes [p.369]
7. _____ disjunction [p.369]

A. Chromosomes inherited from mother
B. Interaction between two nonsister chromatids of a pair of homologues during prophase I
C. The separation of each homologue from its partner during anaphase I
D. One chromatid from each of a homologous pair of chromosomes
E. Alternative forms of genes
F. The result of chromosomes exchanging pieces between nonsister chromatids during prophase I
G. Chromosomes inherited from father

Complete the Table

8. Complete the following table by entering the word *mitosis* or *meiosis* in the blank adjacent to the statement describing one of the two processes. [pp.370–371]

Type of Cell Division

Description

a.	Involves one division
b.	Functions in growth, repair, and maintenance
c.	Daughter cells are haploid
d.	Occurs in germ cells
e.	Involves two divisions
f.	Daughter cells have one chromosome from each homologous pair
g.	Daughter cells have the diploid chromosome number
h.	Completed when four daughter cells are formed
i.	Functions to form gametes

Matching

The cell model used in this exercise has two pairs of homologous chromosomes, one long pair and one short pair. Match the descriptions to the letters of the sketches below. [pp.370–371]

9. _____ One cell at the beginning of meiosis II

10. _____ A daughter cell at the end of meiosis II

11. _____ Metaphase I of meiosis

12. _____ Metaphase of mitosis

13. _____ G₁ in a daughter cell following mitosis

14. _____ Prophase of mitosis

A

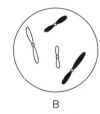
B

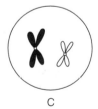
C

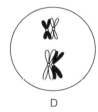
D

E

F

The following questions refer to the preceding sketches. (Write the number in the blank.) [pp.370–371]

15. _____ How many chromosomes are present in cell E?

16. _____ How many chromatids are present in cell E?

17. _____ How many chromatids are present in cell C?

18. _____ How many chromatids are present in cell D?

19. _____ How many chromosomes are present in cell F?

Self-Quiz

Are you ready for the exam? Test yourself on key concepts by taking the additional tests linked with your BiologyNow CD-ROM.

_____ 1. The DNA copying process occurs _____. [p.358]
   a. during the S phase of interphase
   b. during the G_2 phase of interphase
   c. during prophase of mitosis
   d. during the G_1 phase of interphase

_____ 2. In eukaryotic cells, which of the following occurs during mitosis? [p.358]
   a. cytoplasmic division
   b. replication of DNA
   c. a long growth period
   d. sorting of chromosomes into two nuclei

_____ 3. Being "diploid" means having _____. [p.357]
   a. two chromosomes of each type in somatic cells
   b. twice the parental chromosome number
   c. half the parental chromosome number
   d. one chromosome of each type in somatic cells

_____ 4. Somatic cells are _____ cells; germ cells are _____ cells. [p.356]
   a. meiotic; body
   b. body; body
   c. meiotic; meiotic
   d. body; meiotic

_____ 5. If a parent cell has sixteen chromosomes and undergoes mitosis, the resulting cells will have _____ chromosomes. [pp.360–361]
   a. sixty-four
   b. thirty-two
   c. sixteen
   d. eight
   e. four

_____ 6. The correct order of the stages of mitosis is _____. [pp.360–361]
   a. prophase, metaphase, telophase, anaphase
   b. telophase, anaphase, metaphase, prophase
   c. telophase, prophase, metaphase, anaphase
   d. anaphase, prophase, telophase, metaphase
   e. prophase, metaphase, anaphase, telophase

_____ 7. Which of the following does NOT occur in prophase I of meiosis? [pp.366,368]
   a. a cytoplasmic division
   b. a cluster of four chromatids (two chromosomes)
   c. homologues pairing tightly
   d. crossing over

_____ 8. Crossing over is one of the most important events in meiosis because _____. [pp.368–369]
   a. it produces new combinations of alleles on chromosomes
   b. homologous chromosomes must be separated into different daughter cells
   c. the number of chromosomes allotted to each daughter cell must be halved
   d. homologous chromatids must be separated into different daughter cells

___ 9. If a parent cell has sixteen chromosomes and undergoes meiosis, the resulting cells will have _____ chromosomes. [pp.366–367]
   a. sixty-four
   b. thirty-two
   c. sixteen
   d. eight
   e. four

___10. Which of the following does NOT increase genetic variation? [pp.368–369]
   a. crossing over
   b. random fertilization
   c. prophase of mitosis
   d. random homologue alignments at metaphase I

___11. Which of the following is the most correct sequence of events in human life cycles? [p.364]
   a. meiosis → fertilization → gametes → diploid organism
   b. diploid organism → meiosis → gametes → fertilization
   c. fertilization → gametes → diploid organism → meiosis
   d. diploid organism → fertilization → meiosis → gametes

___12. In sexually reproducing organisms, the zygote is _____. [p.365]
   a. an exact genetic copy of the female parent
   b. an exact genetic copy of the male parent
   c. a genetic mixture of traits from both male parent and female parent
   d. unlike either parent genetically

___13. The cell in the diagram is a diploid that has three pairs of chromosomes. From the number and pattern of chromosomes, the cell could be in _____. [p.361]
   a. the first division of meiosis
   b. the second division of meiosis
   c. mitosis
   d. either b or c

Chapter Objectives/Review Questions

This section lists general and detailed chapter objectives that can be used as review questions. You can make maximum use of these by writing answers on a separate sheet of paper. Fill in answers where blanks are provided. To check for accuracy, compare your answers with information given in the chapter or glossary.

1. What substance contains the instructions for making proteins? [p.356]
2. Distinguish between somatic cells and germ cells as to their location and function. [p.356]
3. *Mitosis* and *meiosis* refer to the sorting and packaging of the cell's _____. [p.357]
4. Define the term *karyotype*. [p.357]
5. Any cell having two of each type of chromosome is said to be a _____ cell. [p.357]
6. State the criteria that two chromosomes must meet to be called "homologous chromosomes." [p.357]
7. List, in order, the stages and the various activities occurring during the eukaryotic cell cycle. [p.358]
8. The eukaryotic chromosome is composed of _____ and _____. [p.359]
9. The two attached threads of a duplicated chromosome are known as sister _____. [p.359]
10. Describe the function of the portion of a chromosome known as a *centromere*. [p.359]
11. While diploid germ cells are in the S phase of interphase, each chromosome is _____; each chromosome is then composed of two sister _____. [pp.358–359]
12. Briefly explain reductional division using the concepts of homologous chromosomes, diploid, and haploid. [p.364]

13. If the diploid chromosome number for humans is 46, the haploid number is _____. [p.364]
14. What are the two different divisions of meiosis called? [p.365]
15. Describe spermatogenesis in human males. [pp.364–365]
16. Describe oogenesis in human females. [p.365]
17. In prophase of meiosis I, homologous chromosomes exchange corresponding genes in a process known as _____. How is this process beneficial? [pp.368–369]
18. What occurs during metaphase I that also produces genetic variation? [p.369]
19. In comparing mitosis and meiosis, explain why it may be said that meiosis I is the reductional division, and meiosis II is exactly like a mitotic division. [pp.370–371]

Media Menu Review Questions

Questions 1–10 are drawn from the following InfoTrac College Edition article: "Telomere Tales." Ricki Lewis. *BioScience*, December 1998.

1. Research at two facilities confirms that providing normal somatic (body) cells with the enzyme _____ extends the length of their chromosome tips (_____) and renders them immortal, yet healthy.
2. Telomeres at the ends of all chromosomes are made up of repeats of the _____ sequence TTAGGG.
3. As telomeres lose terminal bases with each cell division, they eventually shorten to the point that signals division to _____.
4. Although a permanent end to division is normal for many body cells, telomerase in highly proliferative tissues, such as bone marrow and some epithelium, works to continually _____ chromosome tips.
5. Some hint at the importance of telomeres was noted in the early 1960s, with a significant discovery in 1961 showing that cells in culture divide a _____ number of times, usually 40–60.
6. In 1990, it was discovered that though telomeres of human somatic cells shorten with each cell division, those of _____ cells do not.
7. Also in the 1990s, the link between telomeres and _____ was established by the finding that telomeres are lost at an accelerated rate in people with rapid-aging disorders.
8. The connection between telomerase and cancer is complex, as illustrated by an experiment in which mice lacking telomerase still could get cancer—confirming that cancer development often involves the participation of several _____.
9. Adding telomerase to cell cultures allows them to become "immortal" but they do not become _____.
10. Telomerase research has both research and clinical applications such as diagnosing cancer, making bone marrow _____ safer, treating AIDS, and reducing or delaying signs of _____.

Integrating and Applying Key Concepts

Runaway cell division is characteristic of cancer. Imagine the various points of the mitotic process that might be sabotaged in cancerous cells in order to halt their multiplication. Then try to imagine how one might discriminate between cancerous and normal cells in order to guide those methods of sabotage most effective in combating cancer.

20

OBSERVABLE PATTERNS OF INHERITANCE

Interactive Exercises

Impacts, Issues: Designer Genes? [p.375]

20.1. BASIC CONCEPTS OF HEREDITY [p.376]

20.2. ONE CHROMOSOME, ONE COPY OF A GENE [p.377]

20.3. FIGURING GENETIC PROBABILITIES [pp.378–379]

For additional practice, use the interactive vocabulary exercises linked with your BiologyNow CD-ROM.

Selected Words: eugenic engineering [p.375], *"natural history"* [p.376], Mendel's *"factors"* [p.376], *pair of homologous chromosomes* [p.376], *gene locus* [p.376], *pair of alleles* [p.376], *pair of genes* [p.376], *first* (and second) *filial* generation [p.378]

Boldfaced, Page-Referenced Terms

[p.376] genes _____

_____

[p.376] allele _____

_____

[p.376] homozygous condition _____

_____

[p.376] heterozygous condition _____

_____

[p.376] dominant _____

_____

[p.376] recessive _____

_____

[p.376] genotype _____

_____

[p.376] phenotype _____

_____

[p.377] monohybrid cross _____

_____

[p.377] segregation _____

_____

[p.378] P (generation) _____

_____

[p.378] F$_1$ (generation) _____

_____

[p.378] F$_2$ (generation) _____

_____

[p.378] Punnett square _____

_____

[p.378] probability _____

_____

Matching

Choose the most appropriate description for each term.

1. _____ eugenic engineering [p.375]
2. _____ alleles [p.376]
3. _____ heterozygous [p.376]
4. _____ dominant allele [p.376]
5. _____ phenotype [p.376]
6. _____ genes [p.376]
7. _____ recessive allele [p.376]
8. _____ homozygous [p.376]
9. _____ diploid cell [p.376]
10. _____ gene locus [p.376]
11. _____ genotype [p.376]

A. The different versions of a gene
B. Location of a specific gene on a specific chromosome
C. Describes an individual having a pair of different alleles
D. Gene whose effect is masked by its partner allele
E. Refers to an individual's actual traits
F. Refers to the alleles inherited by an individual
G. Manipulation of parental genes so that a child would have traits desirable to parents
H. Describes an individual with two identical alleles
I. Units of information about specific traits; passed from parents to offspring
J. Has two copies of each gene, on pairs of homologous chromosomes
K. Gene whose effect masks the effect of its partner

Dichotomous Choice

Circle one of two possible answers given between parentheses in each statement.

12. A genetic cross involving only one trait is a (homologous/monohybrid) cross. [p.377]

13. When eggs or sperm are produced, each receives one allele for each trait because each has one copy of each (chromosome/gamete). [p.377]

14. The separation of homologous chromosomes into different gametes is called (segregation/differentiation). [p.377]

15. Gametes are always (haploid/diploid). [p.377]

16. The offspring of a cross between two organisms represents the (P/F_1) generation. [p.378]

17. If a P generation cross is $CC \times cc$, half of the F_2 generation will have the genotype (CC/Cc). [p.378]

18. A (Probability/Punnett) square is a convenient way of determining the probable outcome of a cross. [p.378]

19. The number between zero and one that expresses the likelihood of an event is the (ratio/probability). [p.378]

20. The outcomes predicted by probability (have to/don't have to) turn up in a given family. [p.378]

21. If a ¼ probability for the inheritance of a given phenotype is calculated for the first child in a family, this probability (does/doesn't) change for the second child. [p.378]

Sequence

Place the steps for completing a Punnett square problem in the correct sequence by placing the letter of the first step in the first blank, and so on. [p.378]

22. _____
23. _____
24. _____
25. _____
26. _____

A. Determine the possible gametes that each parent can make.
B. Draw a square with each possible female gamete at the top of a column, and each possible male gamete beside a row.
C. Find the answer to the problem by presenting the genotypes in the Punnett square as a ratio, and/or interpreting the resulting phenotypic ratio.
D. Determine the genotypes of the parents.
E. Fill in the Punnett square by indicating genotypes of possible offspring at the intersection of each column and row.

Problems

27. In humans, chin fissure (represented by C) is dominant over smooth chin (represented by c). A woman who is heterozygous for this trait marries a man with a smooth chin. [pp.377–379]

    a. What two types of gametes would the woman make? _____

    _____

    b. What one type of gamete would the man make? _____

    _____

    c. Place these on the appropriate sides of the Punnett square on the following page.

    d. Fill in the four squares of the Punnett square. What ratio of genotypes would you predict in the

       offspring? _____

    _____

e. What ratio of phenotypes would you predict for the offspring? _____

_____

f. If the couple has one child with a smooth chin, what is the probability of their second child having a chin fissure? _____

_____

28. The previous problem can also be solved by using simple probability equations. The likelihood of the woman producing either *C* or *c* gametes is ½. The probability of the man producing *c* gametes is 1. [p.379]

a. Show the equation and solution for predicting the probability of this couple having a *Cc* child.

_____

_____

b. Show the equation and solution for predicting the probability of this couple having a *cc* child.

_____

_____

c. What fraction of the offspring is predicted to have a chin fissure? What fraction is predicted to have a smooth chin? _____

_____

20.4. THE TESTCROSS: A TOOL FOR DISCOVERING GENOTYPES [p.380]

20.5. HOW GENES FOR DIFFERENT TRAITS ARE SORTED INTO GAMETES [pp.380–381]

20.6. MORE GENE-SORTING POSSIBILITIES [p.382]

Selected Words: "simple dominance" [p.380]

Boldfaced, Page-Referenced Terms

[p.380] testcross _____

_____

[p.380] independent assortment _____

_____

[p.380] dihybrid cross_____

_____

Matching

Choose the most appropriate description for each term.

1. _____ independent assortment [p.380]
2. _____ testcross [p.380]
3. _____ dihybrid cross [p.380]
4. _____ simple dominance [p.380]

A. Follows the inheritance of two traits
B. Method used to determine the genotype of a parent with a dominant phenotype
C. Alleles for different traits are sorted into gametes independently of one another.
D. With two contrasting alleles, one is dominant and the other is recessive.

Problems

Researchers use a simple technique called a testcross to determine an unknown genotype. In the testcross, one individual's genotype is known to be homozygous recessive for the trait being studied. The other individual exhibits the dominant trait that could result if the individual were homozygous dominant or heterozygous dominant. Although the testcross method is reserved for nonhuman organisms, the principles behind it can be applied to human matings.

5. In humans, the gene for the ability to taste phenyl-thio-carbamine (PTC), *A*, is dominant; its allele for the inability to taste PTC is *a*. Two parents, one a taster and one a nontaster, have two children. One child is a taster, and one child is a nontaster. With the testcross in mind, determine the genotype of the parent who is a taster. State the reasons for your answer. [p.380]

_____

_____

_____

6. In humans, a dominant gene *C* results in chin fissure, and its allele *c* codes for smooth chin. A man who lacks chin fissure had a mother who exhibited chin fissure and a father who lacked chin fissure. The man is married to a woman with chin fissure whose mother lacked chin fissure but whose father had chin fissure. The man and woman are expecting their first child. State the probability that the child will have a smooth chin or a chin fissure. [p.380]

_____

_____

_____

7. Albinos cannot form the pigments that normally produce skin, hair, and eye color, so albinos have white hair and pink eyes and skin (because the blood shows through). To be an albino, one must be homozygous recessive (*aa*) for the pair of genes that codes for the key enzyme in pigment production. Suppose a woman of normal pigmentation (*A*) with an albino mother marries an albino man. State the kinds of pigmentation possible for this couple's children, and specify the ratio of each kind of child the couple is likely to have. Show the genotype(s) and state the phenotype(s). [p.380]

_____

_____

_____

Problems

When working genetics problems dealing with two gene pairs, one can visualize the independent assortment of gene pairs located on nonhomologous chromosomes into gametes by use of a fork-line device. Assume that in humans, pigmented eyes (*B*) are dominant (an eye color other than blue) over blue (*b*), and right-handedness (*R*) is dominant over left-handedness (*r*). To learn to solve a problem, cross the parents *BbRr* × *BbRr*. A sixteen-block Punnett square is required, with gametes from each parent arrayed on two sides of the Punnett square (see Figure 20.10 in the text). The gametes receive genes through independent assortment using a fork-line method. [pp.380–382]

8. Array the gametes at the right on two sides of the Punnett square; combine these haploid gametes to form diploid zygotes within the squares. After filling in the square, enter the probability ratios derived within the Punnett square for the phenotypes listed beside the blanks. [pp.376–378]

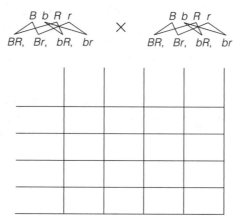

a. _____ pigmented eyes, right-handed

b. _____ pigmented eyes, left-handed

c. _____ blue-eyed, right-handed

d. _____ blue-eyed, left-handed

9. The problem above can also be worked using probability equations. To do this, the two monohybrid situations within the dihybrid problem are separated. Then, multiplication is used to predict phenotypes for the next generation. [p.382]

a. _____ For the trait of eye pigmentation, what is the probability of the cross *Bb* × *Bb* producing offspring with the phenotype of pigmented eyes?

b. _____ For the trait of eye pigmentation, what is the probability of the cross *Bb* × *Bb* producing offspring with the phenotype of blue eyes?

c. _____ For the trait of handedness, what is the probability of the cross *Bb* × *Bb* producing offspring with the right-handed phenotype?

d. _____ For the trait of handedness, what is the probability of the cross *Bb* × *Bb* producing offspring with the left-handed phenotype?

e. Using multiplication, determine the probability of the offspring having pigmented eyes and being right-handed. Show the equation.

f. What is the probability of the offspring having pigmented eyes and being left-handed? Show the equation.

_____ × _____ = _____

g. What is the probability of the offspring having blue eyes and being right-handed? Show the equation.

_____ × _____ = _____

h. What is the probability of the offspring having blue eyes and being left-handed? Show the equation.

_____ × _____ = _____

20.7. SINGLE GENES, VARYING EFFECTS [p.383]

20.8. OTHER GENE IMPACTS AND INTERACTIONS [pp.384–385]

20.9. CUSTOM CURES? [p.385]

Selected Words: "expressed" [p.383], *Marfan's syndrome* [p.383], *sickle cell anemia* [p.383], *sickle cell trait* [p.383], "dormant" genes [p.383], *polydactyly* [p.384], *campodactyly* [p.384], "variable" expression [p.384], *pharmacogenetics* [p.385]

Boldfaced, Page-Referenced Terms

[p.383] pleiotropy _____

_____

[p.383] codominance _____

_____

[p.383] multiple allele system _____

_____

[p.384] penetrance _____

_____

[p.385] polygenic traits _____

_____

[p.385] continuous variation _____

_____

Complete the Table

1. Complete the following table by supplying the principle of inheritance illustrated by each example. Choose from *pleiotropy, polygenic inheritance, multiple alleles,* and *codominance.*

Type of Inheritance	Example
a.	Marfan's syndrome exhibits a phenotype of several seemingly unrelated effects. [p.383]
b.	Blood type involves a gene system of three alleles, I^A, I^B, and i. [p.383]
c.	I^A and I^B alleles for blood type are both expressed in an individual. [p.383]
d.	Human eye and skin color show continuous variation within a population. [pp.384–385]

Short Answer

2. Why would doctors suggest the addition of the food additive butyrate to the diet of an infant who tests positive for sickle cell anemia?

_____

_____

_____

Choice

For questions 3–12, choose from the following concepts concerned with variable gene expression:

a. codominance
b. pleiotropy
c. penetrance
d. incomplete penetrance
e. variable expressivity
f. polygenic traits
g. pharmacogenetics

3. _____ Continuous variation is seen when there is a range of differences for a trait in a population. [p.385]

4. _____ Some people who inherit the dominant allele for polydactyly have the normal number of digits, and others have more. [p.384]

5. _____ This form of inheritance is an explanation for traits such as human eye color and skin color. [p.385]

6. _____ Drugs may be developed that are designed for a particular mix of alleles. [p.385]

7. _____ One gene has several seemingly unrelated effects. [p.383]

8. _____ Both of a pair of contrasting alleles that specify different phenotypes are expressed in a heterozygous individual. [p.383]

9. _____ One hundred percent of persons who are homozygous recessive for the cystic fibrosis gene have cystic fibrosis disease. [p.384]

10. _____ The first step of identifying genes that control common reactions to drugs is being done now. [p.385]

11. _____ Some persons with campodactyly have immobile, bent fingers on both hands. In others, the trait shows up on one hand only. [p.384]

12. _____ $I^A I^B$ [p.383]

Self-Quiz

Are you ready for the exam? Test yourself on key concepts by taking the additional tests linked with your BiologyNow CD-ROM.

_____ 1. The best statement of Mendel's principle of independent assortment is that _____. [p.380]
   a. one allele is always dominant to another
   b. hereditary units from the male and female parents are blended in the offspring
   c. the two hereditary units that influence a certain trait separate during gamete formation
   d. each hereditary unit is inherited separately from other hereditary units

_____ 2. One of two or more alternative forms of a gene for a single trait is a(n) _____. [p.376]
   a. chiasma
   b. allele
   c. autosome
   d. locus

_____ 3. In the F$_2$ generation of a monohybrid cross involving complete dominance, the expected phenotypic ratio is _____. [p.378]
   a. 3:1
   b. 1:1:1:1
   c. 1:2:1
   d. 1:1

_____ 4. Ethical concerns about "designer babies" stem from possible future technology called _____. [p.385]
   a. the human genome project
   b. gene therapy
   c. selective breeding
   d. eugenic engineering

_____ 5. In a testcross, individuals with a dominant phenotype are crossed to an individual known to be _____ for the trait. [p.380]
   a. heterozygous
   b. homozygous dominant
   c. homozygous
   d. homozygous recessive

_____ 6. A man with type A blood could be the father of any of the following except _____. [p.383]
   a. a child with type A blood
   b. a child with type B blood
   c. a child with type O blood
   d. a child with type AB blood

_____ 7. A single gene that affects several seemingly unrelated aspects of an individual's phenotype is said to be _____. [p.383]
   a. pleiotropic
   b. epistatic
   c. mosaic
   d. continuous

_____ 8. Suppose two individuals, each heterozygous for the same characteristic, are crossed. The characteristic involves complete dominance. The expected genotypic ratio of their progeny is _____. [p.378]
   a. 1:2:1
   b. 1:1
   c. 100 percent of one genotype
   d. 3:1

_____ 9. If a homozygous dominant individual is mated to a homozygous recessive individual, the predicted genotypic ratio of the offspring will be _____. [p.378]
   a. 1:1
   b. 1:2:1
   c. 100 percent of one genotype
   d. 3:1

_____ 10. The trait of skin color in humans exhibits _____. [p.385]
   a. pleiotropy
   b. epistasis
   c. mosaic
   d. continuous variation

Chapter Objectives/Review Questions

This section lists general and detailed chapter objectives that can be used as review questions. You can make maximum use of these items by writing answers on a separate sheet of paper. Fill in answers where blanks are provided. To check for accuracy, compare your answers with information given in the chapter or glossary.

1. Define the term *eugenic engineering*. Discuss possible benefits and risks related to this field. [p.375]
2. _____ _____ was the university-educated monk who experimented with garden peas and first discovered basic laws of heredity. [p.376]
3. _____ are units of information about specific traits; they are passed from parents to offspring. [p.376]
4. Define *allele*; how many alleles are present in the genotype *Tt? tt? TT?* [p.376]
5. Having two of the same allele for a trait is a _____ condition; if the two alleles are different, this is a _____ condition. [p.376]
6. Distinguish a dominant allele from a recessive allele. [p.376]
7. _____ refers to the genes present in an individual; _____ refers to an individual's observable traits. [p.376]
8. In a _____ cross, the inheritance of a single trait is followed. [p.377]
9. Mendel's theory of _____ states that during meiosis, the two genes of each pair separate from each other and end up in different gametes. [p.377]
10. Explain why probability is useful to genetics. [p.378]
11. Define the *testcross* and cite an example. [p.380]
12. Mendel's theory of _____ _____ states that alleles for different traits are sorted into gametes independently of each other. [p.380]
13. Explain how the steps used to predict offspring of a dihybrid cross are different from those used for a monohybrid cross. [pp.380–382]
14. Explain *codominance*. [p.383]
15. Explain why sickle cell trait and type AB blood are good examples of codominance. [p.383]
16. Marfan's syndrome and sickle cell anemia are both examples of _____. [p.383]
17. Why is polydactyly said to be an incompletely penetrant genotype? [p.380]
18. Campodactyly is said to show _____ expressivity. [p.384]
19. State the criterion that qualifies a trait to be polygenic. [p.385]
20. Define *pharmacogenetics* and explain the possible benefits to patients being treated for a particular disease. [p.385]

Media Menu Review Questions

Questions 1–2 are drawn from the following InfoTrac College Edition article: "The New World of Tailored Treatments." Amy Tsao. *Business Online Weekly*, November 7, 2003.

1. _____ is the use of genomic knowledge to tailor treatments that best suit the individual patient's needs.
2. Because drugs produced this way would benefit smaller groups of patients, drug and biotech companies might face initial monetary _____—but it would greatly reduce the research _____ needed to bring a drug to market.

Questions 3–9 are drawn from the following InfoTrac College Edition article: "Discovering the Genetics of Autism." Margaret A. Pericak-Vance. *USA Today*, January 2003.

3. Autism is caused by a combination of genetic and environmental factors, not poor _____.
4. Those with autism have difficulty interpreting the variety of social cues people use to infer how another person _____ or _____.
5. Data allowing scientists to distinguish clearly autistic patients is an important first step in discovering the specific _____ involved.

6. Genes called _____ genes make an individual more susceptible to a disease or disorder, but they do not by themselves cause the problem.
7. _____ may be able to find more effective drugs for autism by customizing medicines to an individual's genetic makeup.
8. Scientists look for genes by looking for patterns of DNA called _____ that are linked to the mutated gene in question.
9. Chromosomal regions suspected of containing genes for autism have been found on chromosomes _____ and _____, with a possible "risk gene" on chromosome _____.

Question 10 is drawn from the following InfoTrac College Edition article: "Origins: A Gene for Speech." Cornelia Stolze. *Time*, August 26, 2002.

10. Researchers studying a London family with a rare disorder that prevents normal _____ now know that the gene involved is present in all primates, but that slight mutations over time have possibly resulted in humans being able to _____ the mouth, lips, and tongue in ways that allow speech.

Integrating and Applying Key Concepts

Solve the following genetics problem: In humans, hypotrichosis (sparse body hair) is caused by gene *h* that is recessive to normal, *H*. Pituitary dwarfism is caused by a deficiency of a growth hormone controlled by a recessive allele, *p*; normal growth is caused by dominant gene *P*. A hypotrichotic male who is homozygous for normal stature marries a woman who is homozygous for normal body hair but is a pituitary dwarf. Predict the appearance of their children. If a mating were to occur between two genotypes like those of their children, what prediction would you make concerning the appearance and the proportions of the characteristics those children might have?

21

CHROMOSOMES AND HUMAN GENETICS

Interactive Exercises

Impacts, Issues: Menacing Mucus [p.389]

21.1. CHROMOSOMES AND INHERITANCE [p.390]

21.2. PICTURING CHROMOSOMES WITH KARYOTYPES [p.391]

For additional practice, use the interactive vocabulary exercises linked with your BiologyNow CD-ROM.

Selected Words: *locus* [p.390], *karyotype* [p.391], "fixed" [p.391]

Boldfaced, Page-Referenced Terms

[p.390] genes _____

_____

[p.390] homologous chromosomes _____

_____

[p.390] alleles _____

_____

[p.390] crossing over _____

_____

[p.390] genetic recombination _____

_____

[p.390] sex chromosomes _____

_____

[p.390] autosomes _____

_____

Matching

1. _____ crossing over [p.390]
2. _____ XX [p.390]
3. _____ karyotype [p.391]
4. _____ alleles [p.390]
5. _____ genes [p.390]
6. _____ genetic recombination [p.390]
7. _____ autosomes [p.390]
8. _____ linked [p.390]
9. _____ XY [p.390]
10. _____ 30,000 [p.390]
11. _____ diploid (2*n*) [p.390]
12. _____ homologous [p.390]
13. _____ sex chromosomes [p.390]
14. _____ locus [p.390]

A. X and Y chromosomes
B. Display of photographed chromosomes from a single body cell
C. Inherited units of information located on chromosomes
D. Male sex chromosomes
E. Female sex chromosomes
F. Different forms of a gene
G. New combinations of alleles not present in a parent
H. Location of a particular gene on a chromosome
I. Chromosomes other than the sex chromosomes
J. Exchange of segments between homologous chromosomes
K. Refers to genes physically located on the same chromosome
L. Cell with a pair of each chromosome
M. Approximate number of human genes
N. Refers to a pair of chromosomes alike in length, shape, and genes

Sequence

In the blanks beside the numbers, write the letters of the steps taken to prepare a karyotype in the correct order. [p.391]

15. _____   A. Colchicine is added to arrest mitosis at metaphase.

16. _____   B. The cells are placed on a slide, fixed, and stained.

17. _____   C. The cell culture is centrifuged.

18. _____   D. Cells are transferred to a saline solution, causing them to swell.

19. _____   E. The chromosomes are cut out and arranged by size with the sex chromosomes last.

20. _____   F. Cells are cultured in a growth medium.

21. _____   G. The chromosomes are photographed through a microscope.

21.3. HOW SEX IS DETERMINED [pp.392–393]

21.4. LINKED GENES [p.393]

21.5. HUMAN GENETIC ANALYSIS [pp.394–395]

Selected Words: *SRY* [p.392], "switched off" [p.393], Barr body [p.393], *anhidrotic ectodermal dysplasia* [p.393], "linked" [p.393], *carrier* [p.394], *abnormality* [p.395], *disorder* [p.395], *syndrome* [p.395]

Boldfaced, Page-Referenced Terms

[p.392] X chromosome _____

_____

[p.392] Y chromosome _____

_____

[p.392] X-linked genes _____

_____

[p.392] Y-linked genes _____

_____

[p.393] X inactivation _____

_____

[p.394] pedigree chart_____

_____

Complete the Table

1. Complete the Punnett square table, which brings Y-bearing and X-bearing sperm together randomly in fertilization. [p.392]

Dichotomous Choice

Answer the following questions related to the completed Punnett square.

2. Human males transmit their Y chromosome only to their (sons/daughters). [p.392]

3. Human males receive their X chromosome only from their (mothers/fathers). [p.392]

4. Human mothers and fathers each provide an X chromosome for their (sons/daughters). [p.392]

5. While the Y chromosome is responsible for masculinizing an embryo, most of the genes on the X chromosome deal with (sexual/nonsexual) traits. [p.392]

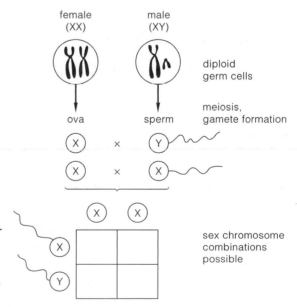

Matching

Choose the most appropriate description for each term.

6. _____ X inactivation [p.393]

7. _____ Barr body [p.393]

8. _____ linked [p.393]

9. _____ genetic counseling [p.395]

10. _____ pedigree [p.394]

11. _____ carrier [p.394]

12. _____ abnormality [p.395]

13. _____ genetic disorder [p.395]

14. _____ syndrome [p.395]

15. _____ *SRY* gene [p.392]

A. Designates a person who is heterozygous for a recessive trait that causes a genetic disease

B. An inherited condition causing mild to severe medical problems

C. Refers to genes located on the same chromosome

D. Early development mechanism that randomly switches off all or most of the genes on one of a female's X chromosomes

E. Characterized by a set of symptoms that usually occur together

F. A chart that is constructed to show the genetic connections among individuals

G. Y chromosome master gene that causes a new individual to develop testes

H. Deviation from the average, such as having six toes on each foot rather than five

I. Condensed, inactivated X chromosome that may be seen under a light microscope

J. Involves determining parents' genotypes and providing information needed to help them decide whether to conceive

Short Answer

16. Explain the relationship between distance between linked genes and recombination due to crossing over. [p.393]

_____

_____

_____

21.6. INHERITANCE OF GENES ON AUTOSOMES [pp.396–397]

21.7. INHERITANCE OF GENES ON THE X CHROMOSOME [pp.398–399]

21.8. SEX-INFLUENCED INHERITANCE [p.400]

Selected Words: *phenylketonuria* (PKU) [p.396], *Tay-Sachs disease* [p.396], *achondroplasia* [p.397], *familial hypercholesterolemia* [p.397], *hemophilia* [p.398], "muscular dystrophy" [p.399], *Duchenne muscular dystrophy* (DMD) [p.399], *dystrophin* [p.399], *red-green color blindness* [p.399], *faulty enamel trait* [p.399], *testicular feminizing syndrome* [p.399], "androgen insensitivity" [p.399], *sex-influenced* traits [p.400]

Choice

For questions 1–15, choose from the following patterns of inheritance; some items may require more than one letter.

a. autosomal recessive
b. autosomal dominant
c. X-linked recessive
d. X-linked dominant
e. sex-influenced inheritance

1. _____ The trait is expressed in heterozygotes (of either sex). [p.396]

2. _____ A genetic disease cannot be detected in a heterozygote (of either sex). [p.396]

3. _____ The recessive phenotype shows up far more often in males than in females. [p.398]

4. _____ Both parents of a person with a genetic disease may be heterozygous normal. [p.396]

5. _____ If one parent is heterozygous for a genetic disease and the other is homozygous recessive, there is a 50 percent chance that any child of theirs will inherit the disease. [p.396]

6. _____ The allele is expressed in homozygotes or heterozygotes. [p.396]

7. _____ Heterozygous normal parents can expect that one-fourth of their children will be affected by a disorder inherited in this manner. [p.396]

8. _____ These traits appear more frequently in one sex, or the phenotype differs depending on whether the individual is male or female. [p.400]

9. _____ A son cannot inherit the recessive allele from his father, but a daughter can. [p.398]

10. _____ Females can mask this gene, males cannot. [p.398]

11. _____ Individuals displaying this type of disorder will always be homozygous for the trait. [p.396]

12. _____ The allele behaves as a dominant in one sex and a recessive in the opposite sex; genes for these traits are carried on autosomes. [p.400]

13. _____ Heterozygous females will transmit the recessive allele to half their sons and half their daughters. Males can only transmit such traits to their daughters. [p.398]

14. _____ Heterozygous females will display the trait. Males are not homozygous or heterozygous for the trait. [p.399]

15. _____ All of the daughters of a man with a genetic disease will have the disease, but he cannot pass the allele to any of his sons. [p.399]

Problems

16. The autosomal allele that causes albinism, *a*, is recessive to the allele for normal pigmentation, *A*. A normally pigmented woman whose father is an albino marries an albino man whose parents are normal. They have three children, two normal and one albino. Give the genotypes for each person listed. [p.396]

_____

_____

_____

_____

17. Huntington disorder is a rare form of autosomal dominant inheritance, *II*; the normal gene is *h*. The disease causes progressive degeneration of the nervous system with onset exhibited near middle age. An apparently normal man in his early twenties learns that his father has recently been diagnosed as having Huntington disorder. What are the chances that the son will develop Huntington disorder? [pp.396–397]

_____

_____

18. A color-blind man and a woman with normal vision whose father was color blind have a son. Color blindness, in this case, is caused by an X-linked recessive gene. If only the male offspring are considered, what is the probability that this couple's son is color blind? [pp.398–399]

_____

_____

19. Hemophilia A is caused by an X-linked recessive gene. A woman who is seemingly normal but whose father was a hemophiliac marries a normal man. What proportion of their sons will have hemophilia? What proportion of their daughters will have hemophilia? What proportion of their daughters will be carriers? [pp.398–399]

20. The following pedigree shows the pattern of inheritance of color blindness in a family (people with the trait are indicated by blackened circles). What is the chance that the third-generation female indicated by the arrow (below) will have a color-blind son if she marries a normal male? A color-blind male? [pp.398–399]

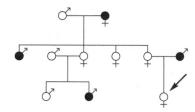

21. A nonbald man marries a nonbald woman whose mother was bald. What is the probability that their first child will be bald? [p.400]

Complete the Table

22. Complete the following table by indicating whether the genetic disorder listed is caused by inheritance that is autosomal recessive, autosomal dominant, X-linked recessive, X-linked dominant, or sex influenced.

Genetic Disorder	Inheritance Pattern
a. Achondroplasia [p.397]	
b. Hemophilia [p.398]	
c. Pattern baldness [p.400]	
d. Huntington disease [pp.396–397]	
e. Duchenne muscular dystrophy [pp.398–399]	
f. Faulty enamel trait [p.399]	
g. Red-green color blindness [pp.398–399]	
h. Phenylketonuria [p.396]	
i. Cystic fibrosis [p.396]	
j. Tay-Sachs disease [p.396]	
k. Familial hypercholesterolemia [p.397]	

21.9. HOW A CHROMOSOME'S STRUCTURE CAN CHANGE [pp.400–401]

21.10. CHANGES IN CHROMOSOME NUMBER [pp.402–403]

Selected Words: deletion [p.400], *cri-du-chat* [p.400], *duplications* [p.401], *translocation* [p.401], *aneuploidy* [p.402], *polyploidy* [p.402], Down syndrome [p.402], *Turner syndrome* [p.402], XXX females [p.402], Klinefelter syndrome [p.402], XYY condition [p.402]

Boldfaced, Page-Referenced Terms

[p.401] mutation _____

_____

[p.402] nondisjunction _____

_____

[p.402] trisomy_____

_____

[p.402] monosomy_____

_____

Labeling-Matching

On rare occasions, chromosome structure becomes abnormally rearranged. Such changes may have profound effects on the phenotype of an organism. Label the following diagrams of abnormal chromosome structure as a deletion, a duplication, an inversion, or a translocation. Complete the exercise by matching and entering the letter of the proper description in the parentheses following each label. [pp.400–401]

A. The loss of a chromosome segment; an example is cri-du-chat disorder.

B. A gene sequence repeated several to many times on a chromosome.

C. The transfer of part of one chromosome to a nonhomologous chromosome; an example is when chromosome 14 ends up with a segment of chromosome 8.

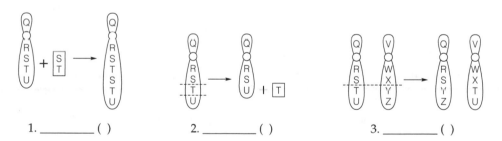

1. _____ ( )   2. _____ ( )   3. _____ ( )

Short Answer

4. Define the term mutation. Then list the three possible consequences of a mutation. [p.401]

_____

_____

_____

Complete the Table

5. Complete the following table of important terms associated with chromosome number change in organisms. Choose from *monosomy, nondisjunction, aneuploidy, trisomy,* and *polyploidy.*

Category of Change	Description
a.	New individual does not have an exact multiple of the normal haploid set of 23 chromosomes [p.402]
b.	New individuals have three or more of each type of chromosome; a lethal condition in humans [p.402]
c.	One or more pair of chromosomes fail to separate during mitosis or meiosis [p.402]
d.	A gamete with an extra chromosome ($n + 1$) unites with a normal gamete at fertilization; the new individual will have three of one type of chromosome ($2n + 1$) [p.402]
e.	If a gamete is missing a chromosome, then the new individual resulting from fertilization will have only one of one type of chromosome ($2n - 1$) [p.402]

Short Answer

6. How does aneuploidy occur? [p.402]

_____

_____

_____

Choice

For questions 7–15, choose from the following:

    a. Down syndrome   b. Turner syndrome   c. Klinefelter syndrome   d. XYY condition

7. _____ XXY male [p.403]

8. _____ Ovaries nonfunctional and secondary sexual traits fail to develop at puberty [pp.402–403]

9. _____ Testes smaller than normal, sparse body hair, and some breast enlargement [p.403]

10. _____ Could only be caused by a nondisjunction in males [p.403]

11. _____ Moderate to severe mental retardation; about 40 percent of those affected develop heart defects [p.402]

12. _____ XO female; often miscarried early in pregnancy; abnormal female phenotype [pp.402–403]

13. _____ Males that tend to be taller than average but most are phenotypically normal [p.403]

14. _____ Injections of testosterone reverse feminized traits but not the mental retardation or sterility [p.403]

15. _____ Trisomy 21; abnormal skeleton development and muscles weaker than normal [p.402]

Self-Quiz

Are you ready for the exam? Test yourself on key concepts by taking the additional tests linked with your BiologyNow CD-ROM.

___ 1. The evidence that human females have only one functional X chromosome is provided by _____. [p.393]
   a. the Y chromosome
   b. the *SRY* gene
   c. sex determination
   d. the Barr body

___ 2. Chromosomes other than those involved in sex determination are known as _____. [p.390]
   a. nucleosomes
   b. heterosomes
   c. alleles
   d. autosomes

___ 3. The farther apart two genes are on a chromosome, _____. [p.393]
   a. the less likely that crossing over and recombination will occur between them
   b. the greater will be the frequency of crossing over and recombination between them
   c. the more likely they are to be in two different linkage groups
   d. the more likely they are to be deleted or duplicated

___ 4. Karyotype analysis is _____. [p.391]
   a. a means of detecting and reducing mutagenic agents
   b. a surgical technique that separates chromosomes that have failed to segregate properly during meiosis II
   c. used to detect chromosomal mutations
   d. a process that substitutes defective alleles with normal ones

___ 5. Which of the following is NOT the result of an autosomal dominant gene? [pp.396–397]
   a. Huntington disease
   b. Cystic fibrosis
   c. Achondroplasia
   d. Familial hypercholesterolemia

___ 6. Red-green color blindness is an X-linked recessive trait in humans. A color-blind woman and a man with normal vision have a son. What are the chances that the son is color blind? If the parents ever have a daughter, what is the probability that the daughter will be color blind? (Consider only female offspring.) [pp.398–399]
   a. 100 percent; 0 percent
   b. 50 percent; 0 percent
   c. 100 percent; 100 percent
   d. 50 percent; 100 percent

___ 7. Suppose that a hemophilic male (X-linked recessive allele) and a female carrier for the hemophilic trait have a nonhemophilic daughter with Turner syndrome. Nondisjunction could have occurred in _____. [pp.398,402–403]
   a. both parents
   b. neither parent
   c. the father only
   d. the mother only

___ 8. Nondisjunction involving the X chromosome occurs during oogenesis and produces two kinds of eggs, XX and O (no X chromosome). If normal Y sperm fertilize the two types, which genotypes are possible? [pp.402–403]
   a. XX and XY
   b. XXY and YO
   c. XYY and XO
   d. XYY and YO

___ 9. A person with 2*n* + 1 chromosomes is _____. [p.402]
   a. monosomic
   b. aneuploid
   c. polyploid
   d. tetraploid

_____ 10. Through a nondisjunction, a female has a
juvenile phenotype, 45 chromosomes,
nonfunctional ovaries, and a webbed neck.
This fits the description of _____.
[pp.402–403]
a. Klinefelter syndrome
b. XYY condition
c. Turner syndrome
d. XXX condition

Chapter Objectives/Review Questions

This section lists general and detailed chapter objectives that can be used as review questions. You can make maximum use of these items by writing answers on a separate sheet of paper. Fill in answers where blanks are provided. To check for accuracy, compare your answers with information given in the chapter or glossary.

1. The units of information about heritable traits are known as _____. [p.390]
2. Diploid (2*n*) cells have pairs of _____ chromosomes. [p.390]
3. _____ are different forms of the same gene that arise through mutation. [p.390]
4. State the circumstances required for crossing over, and describe the significance of the results. [p.390]
5. Name and describe the sex chromosomes in human males and females. [p.390]
6. Distinguish between chromosomes and autosomes. [p.390]
7. Define *karyotype*; briefly describe its preparation and value. [p.391]
8. Distinguish between an X-linked gene and a Y-linked gene. [p.392]
9. How are the sex chromosomes involved in sex determination? [p.392]
10. Explain meiotic segregation of sex chromosomes to gametes and the subsequent random fertilization that determines sex in many organisms. [p.392]
11. A newly identified region of the Y chromosome called _____ appears to be the master gene for sex determination. [p.392]
12. Explain the process of X inactivation and the Barr body. [p.393]
13. The term _____ refers to genes being located on the same chromosome. [p.393]
14. State the relationship between crossover frequency and the location of genes on a chromosome. [p.393]
15. A(n) _____ chart or diagram is used to study genetic connections between individuals. [p.394]
16. With respect to genetic traits, _____ simply means deviation from the average, and a genetic _____ is an inherited condition that causes mild to severe medical problems. [p.395]
17. A(n) _____-_____ trait appears more frequently in one sex than the other; pattern baldness is an example. [p.400]
18. When gametes or cells of an affected individual end up with one extra or one less than the correct number of chromosomes (not an exact multiple of the normal haploid set), it is known as _____; relate this to monosomy and trisomy. [p.402]
19. Having three or more sets of chromosomes is called _____. [p.402]
20. _____ is the failure of the chromosome pairs to separate during either mitosis or meiosis (most significant during gamete formation). [p.402]
21. Trisomy 21 is known as _____ syndrome; Turner syndrome has the chromosome constitution, _____; most _____ females develop normally; XXY chromosome constitution is _____ syndrome; taller-than-average males with normal male phenotypes have the _____ condition. [pp.402–403]
22. Describe the cause and characteristics of Turner syndrome, XXX condition, Klinefelter syndrome, and the XYY condition. [pp.402–403]

Media Menu Review Questions

Questions 1–3 are drawn from the following InfoTrac College Edition article: "The Sexes: New Insights in the X and Y Chromosomes." Bob Beale. *The Scientist,* July 23, 2001.

1. Of the approximate 31,000 human genes, men and women differ only in the sex chromosomes, and only a few _____ genes seem to be involved.
2. Genes for _____ production exist on both the Y and X chromosomes.
3. Research in 1990 showed that the Y chromosome gene _____ causes the development of testes and maleness, although other genes may complicate the results.

Questions 4–9 are drawn from the following InfoTrac College Edition article: "What Makes You Who You Are: Which is Stronger—Nature or Nurture?" Matt Ridley. *Time,* June 2, 2003.

4. Just _____ genes would be enough to make every human being unique.
5. It may be that _____ consists of nothing more than switching genes on and off.
6. The common human fear of snakes may not be inherited as that, but as a _____ to learn a fear of snakes.
7. Francis Galton coined the phrase _____ vs. _____, a concept that led him to espouse _____ adopted by the Nazis to justify their treatment of Jews.
8. To make grand changes in the body plan of animals, all you need to do is switch the same genes on and off in different _____.
9. Early scientific discoveries, such as Pavlovian conditioning, had hinted at the importance of interplay between heredity and the _____.

Questions 10–15 are drawn from the following InfoTrac College Edition article: "Delving Into Huntington's Disease." Virginia Goolkasian. *USA Today,* September 2001.

10. The gene for Huntington's Disease (HD) was discovered in _____, followed by the development of a _____ test for the gene.
11. A child of a person with HD has a _____ chance of inheriting the disease.
12. HD involves a gradual deterioration of mind and body due to cell death in the _____.
13. While some people at risk for HD opt not to be tested, many choose testing when faced with life decisions such as marriage and _____.
14. Research has shown that, though there is no _____ for HD, an enriched environment can minimize the impact of symptoms.
15. Since the discovery of the HD gene, one of the biggest breakthroughs has been the development of _____ mice that are used to study possible medications for the disease.

Integrating and Applying Key Concepts

1. The parents of a young boy bring him to their doctor. They explain that the boy does not seem to be going through the same vocal developmental stages as his older brother. The doctor orders a common cytogenetics test to be done, and it reveals that the young boy's cells contain two X chromosomes and one Y chromosome. Describe the test that the doctor ordered, and explain how and when such a genetic result, XXY, most logically occurred. What treatment would the doctor most likely prescribe?
2. Solve the following genetics problem. Show rationale, genotypes, and phenotypes. A husband sues his wife for divorce, arguing that she has been unfaithful. His wife gave birth to a girl with a fissure in the iris of her eye, an X-linked recessive trait. Both parents have normal eye structure. Can the genetic facts be used to argue for the husband's suit? Explain your answer.

22

DNA, GENES, AND BIOTECHNOLOGY

Interactive Exercises

Impacts, Issues: Ricin and Your Ribosomes [p.407]

22.1. DNA: A DOUBLE HELIX [pp.408–409]

22.2. PASSING ON GENETIC INSTRUCTIONS [pp.410–411]

For additional practice, use the interactive vocabulary exercises linked with your BiologyNow CD-ROM.

Selected Words: *thymine dimers* [p.410], *expansion mutations* [p.411], *fragile X syndrome* [p.411], *Huntington disorder* [p.411], *neurofibromatosis* [p.411]

Boldfaced, Page-Referenced Terms

[p.408] nucleotides _____

_____

[p.408] base pairs _____

_____

[p.409] gene _____

_____

[p.410] DNA replication _____

_____

[p.410] semiconservative replication _____

_____

[p.410] DNA polymerases _____

_____

[p.411] gene mutations _____

_____

[p.411] base-pair substitution _____

_____

Short Answer

1. What is the source of ricin? How does this toxin work? Is there an antidote for ricin poisoning? [p.407]

    _____

    _____

2. List the three parts of a nucleotide. [p.408]

    _____

    _____

    _____

Complete the Table

3. Complete the table below showing important characteristics of the four nucleotides found in DNA. [p.408]

Name of Nucleotide	Abbreviation	Single or Double Ring	Base It Pairs With
a.	A	e.	i.
b.	G	f.	j.
c.	T	g.	k.
d.	C	h.	l.

True/False

If the statement is true, write a "T" in the blank. If false, explain why by changing one or more of the underlined words.

_____ 4. DNA is composed of <u>two</u> different types of nucleotides. [p.408]

_____ 5. In the DNA of every species, the amount of <u>adenine</u> present always equals the amount of thymine, and the amount of <u>cytosine</u> always equals the amount of guanine (A = T and C = G). [p.408]

_____ 6. In a nucleotide, the phosphate group is attached to the <u>nitrogen-containing base</u>. [p.408]

_____ 7. <u>Watson and Crick</u> are given credit for determining the structure of DNA. [p.408]

_____ 8. Nucleotides are linked together by <u>covalent</u> bonds to form strands of DNA. [p.408]

_____ 9. The shape of the DNA molecule is often described as a double <u>ladder</u>. [p.408]

_____ 10. The two DNA strands run in <u>opposite</u> directions. [p.408]

_____ 11. The two strands of DNA are held together by <u>covalent</u> bonds between pairs of bases. [p.408]

Labeling

Identify each indicated part of the following DNA illustration. Choose from these answers: *phosphate group, double-ring nitrogen base, single-ring nitrogen base, nucleotide, deoxyribose, hydrogen bond, covalent bond.*

12. _____ [p.408]

13. _____ [p.408]

14. _____ -ring nitrogen base [p.408]

15. _____ -ring nitrogen base [p.408]

16. _____ -ring nitrogen base [p.408]

17. _____ -ring nitrogen base [p.408]

18. A complete _____ [p.408]

19. _____ bond [p.408]

20. _____ bond [p.408]

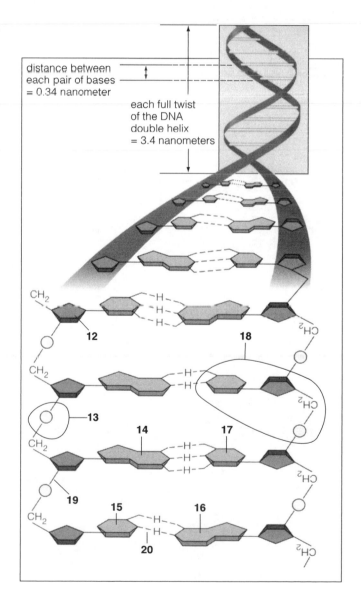

distance between each pair of bases = 0.34 nanometer

each full twist of the DNA double helix = 3.4 nanometers

Short Answer

21. Define the word *gene.* [p.409]

_____

_____

_____

Labeling

22. The term *semiconservative replication* refers to the fact that each new DNA molecule resulting from the replication process is "half old, half new." In the accompanying illustration, complete the replication required in the middle of the molecule by adding the required letters representing the missing nucleotide bases. Recall that ATP energy and the appropriate enzymes are required in order to complete this process. [p.410]

T- _____   _____ -A

G- _____   _____ -C

A- _____   _____ -T

C- _____   _____ -G

C- _____   _____ -G

C- _____   _____ -G

old new   new old

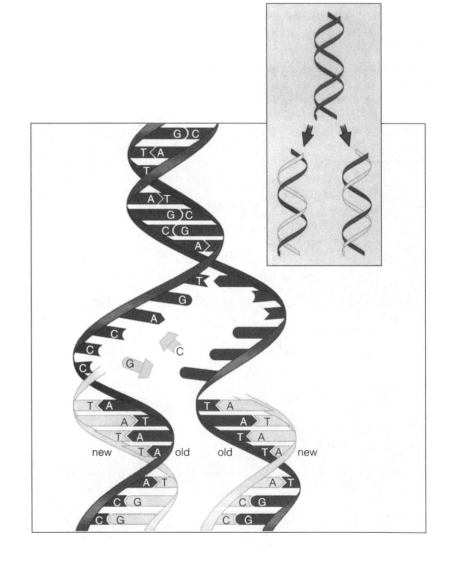

Fill-in-the-Blanks

DNA replication occurs as the double helix unwinds and hydrogen bonds between strands are broken, exposing stretches of its nucleotide (23) _____ [p.410]. Each parent strand remains intact as a new companion strand is assembled on it, one (24) _____ [p.410] at a time. DNA replication is said to be (25) _____ [p.410] because one strand of each completed DNA molecule is "new" and the other is from the starting molecule.

Individual nucleotides are assembled into the new portion of the double helix by enzymes called (26) _____ _____ [p.410]. These, along with DNA ligases and other enzymes, also function in DNA (27) _____ [p.410]. It has been estimated that a human cell must repair breaks in a single strand of DNA up to (28) _____ _____ [p.410] times every hour. When errors that occur during DNA replication are not detected and corrected, the result is a(n) (29) _____ [p.410].

DNA is also vulnerable to damage from chemicals, ionizing radiation, and (30) _____ _____ [p.410]. UV light can cause a type of damage known as a (31) _____ _____ [p.410] in which adjacent thymine bases become linked.

Small-scale changes in the nucleotide sequence are called (32) _____ _____ [p.411]. In a (33) _____-_____ _____ [p.411], one base is wrongly paired with another. Sickle cell anemia is caused by this type of mutation. Other mutations involve an extra base (34) _____ [p.411] into a gene, or a base (35) _____ [p.411] from a gene. When a particular nucleotide sequence is repeated over and over, this is called a(n) (36) _____ [p.411] mutation, such as causes fragile X syndrome. Bits of DNA that move from one location to another are called (37) _____ _____ [p.411]. They cause mutations, such as the one responsible for neurofibromatosis.

Mutations usually give rise to (38) _____ [p.411] that don't function properly. Whereas mutations can occur in any cell, they are only inherited when they take place in the cells of (39) _____ [p.411] that produce gametes. Mutations may be harmful, neutral, or (40) _____ [p.411].

22.3. DNA INTO RNA—THE FIRST STEP IN MAKING PROTEINS [pp.412–413]

22.4. READING THE GENETIC CODE [p.414–415]

22.5. TRANSLATING THE GENETIC CODE INTO PROTEIN [pp.416–417]

Selected Words: nucleotide "cap" [p.412], "pre-mRNA" [p.413], "base triplets" [p.414], "three-bases-at-a-time" [p.414], "start" signal [p.414], "hook" site of tRNA [p.414], "wobble effect" [p.415], *initiation* [p.416], *elongation* [p.416], *termination* [p.416], *polysomes* [p.417]

Boldfaced, Page-Referenced Terms

[p.412] RNA _____

_____

[p.412] uracil _____

_____

[p.412] transcription _____

_____

[p.412] translation _____

_____

[p.412] ribosomal RNA (rRNA) _____

_____

[p.412] messenger RNA (mRNA) _____

_____

[p.412] transfer RNA (tRNA) _____

_____

[p.412] RNA polymerases _____

_____

[p.412] promoter _____

_____

[p.413] introns _____

_____

[p.413] exons _____

_____

[p.413] regulatory proteins _____

_____

[p.414] codons _____

_____

[p.414] genetic code _____

_____

[p.414] anticodon _____

_____

Complete the Table

1. Complete the following table to summarize the molecular differences between DNA and RNA. [p.412]

DNA Sugar	DNA Bases	RNA Sugar	RNA Bases
a.	b.	c.	d.

Fill-in-the-Blanks

The two steps from genes to proteins are called (2) _____ [p.412] and (3) _____ [p.412]. In (4) _____ [p.412], single-stranded molecules of RNA are assembled on DNA templates in the nucleus. In (5) _____ [p.412], the RNA molecules are shipped from the nucleus into the cytoplasm, where they are used as templates for assembling (6) _____ [p.412] chains. Following translation, one or more of these chains are folded into (7) _____ [p.412] molecules.

Complete the Table

8. Three types of RNA are translated from DNA in the nucleus (from genes that code only for RNA). Complete the following table, which summarizes information about these RNA molecules. [p.412]

RNA Molecule	Abbreviation	Description/Function
Ribosomal RNA	a.	
Messenger RNA	b.	
Transfer RNA	c.	

Short Answer

9. Cite the key differences between DNA replication and transcription (both processes involve DNA and occur in the nucleus). [p.412]

_____

_____

_____

_____

_____

Sequence

Arrange the steps of transcription in hierarchical order, with the earliest step first and the latest step last.

10. _____   A. A termination base sequence serves as a signal to release the RNA transcript. [p.413]

11. _____   B. Proteins help position an RNA polymerase on the DNA so that it binds with promoter. [p.412]

12. _____   C. Newly formed pre-RNA is modified; introns are removed and exons are spliced together. [p.413]

13. _____   D. Mature mRNA leaves the nucleus.

14. _____   E. A promoter is a base sequence that signals the start of a gene. [p.412]

15. _____   F. RNA polymerase moves along the DNA, joining RNA nucleotides together. [p.413]

Completion

16. Suppose the following line represents the DNA strand that will act as a template for the production of mRNA through the process of transcription. Complete the blanks below the DNA strand with the sequence of complementary bases that will represent the message carried by mRNA from DNA to the ribosome in the cytoplasm. [pp.412–413]

TAC — AAG — ATA — ACA — TTA — TTT — CCT — ACC — GTC — ATC

____ – ____ – ____ – ____ – ____ – ____ – ____ – ____ – ____ – ____

(transcribed single strand of mRNA)

Labeling-Matching

Newly transcribed mRNA contains more genetic information than is necessary to code for a chain of amino acids. Before the mRNA leaves the nucleus for its ribosome destination, an editing process occurs as certain portions of nonessential information are snipped out. Identify each indicated part of the following illustration; use abbreviations for the nucleic acids. Choose from *tail, introns, DNA, mature RNA transcript, cap,* and *exons.* Complete the exercise by matching and entering the letter of the description in the parentheses after each label. [p.413]

17. _____ ( )

18. _____ ( )

19. _____ ( )

20. _____ ( )

21. _____ ( )

22. _____ _____ _____ ( )

A. Regions that are translated into proteins [p.413]
B. Base sequences that do not get translated into an amino acid sequence [p.413]
C. mRNA ready to leave the nucleus [p.413]
D. Trailing end of a pre-mRNA transcript [p.412]
E. The region of the DNA template strand to be copied [p.412]
F. A nucleotide added to mRNA as transcription starts [p.412]

Short Answer

23. Describe the function of regulatory proteins. [p.413]

_____

_____

_____

Matching

24. _____ codon [p.414]

25. _____ sixty-four [p.414]

26. _____ genetic code [p.414]

27. _____ ribosome [p.415]

28. _____ wobble effect [p.415]

29. _____ anticodon [p.414]

30. _____ the "stop" codons [p.414]

31. _____ three-bases-at-a-time [p.414]

A. Composed of two subunits made in the nucleus and sent to the cytoplasm
B. Reading frame of the nucleotide bases in mRNA
C. UAA, UAG, UGA (three of sixty-four codons)
D. A sequence of three nucleotide bases on tRNA that can pair with a specific mRNA codon
E. Flexibility in codon–anticodon pairing at the third base
F. Name for each base triplet in mRNA
G. The number of codons in the genetic code
H. Provides cells with basic instructions for synthesizing proteins

Complete the Table

32. Complete the following table, which distinguishes the stages of translation. [pp.416–417]

Translation Stage	Description
a.	An initiator tRNA binds with the small ribosomal subunit and attaches to one end of the mRNA; this unit moves along the mRNA until it encounters the start codon (AUG) of the mRNA transcript; after this, a large ribosomal unit binds with a small one.
b.	A polypeptide chain forms as the mRNA strand passes between the ribosomal subunits; some proteins in the ribosome function as enzymes, joining amino acids together in the sequence dictated by mRNA codons; they catalyze the formation of a peptide bond between adjacent amino acids.
c.	A stop codon is reached and there is no corresponding anticodon; now the ribosome interacts with certain release factor proteins; this causes the ribosome as well as the polypeptide chain to detach from the mRNA; the detached chain may join the cytoplasmic pool of free proteins or enter the cytomembrane system.

33. Find your answer to question 16, and enter it on the line provided. Deduce the composition of the tRNA anticodons that would pair with the specific mRNA codons as these tRNAs deliver the amino acids that are identified here to the binding sites of the small ribosomal subunit. [p.416]

mRNA _____ _____ _____ _____ – _____ – _____ – _____ – _____ – _____ – _____

tRNA _____ – _____ – _____ – _____ – _____ – _____ – _____ – _____ – _____ – _____

34. From the mRNA transcript in exercise 33, use Figure 22.10 in the text to identify the composition of the amino acids in the polypeptide sequence.

_____ – _____ – _____ – _____ – _____ – _____ – _____ – _____ — _____ – _____ [p.414]

(amino acids)

Labeling-Matching

A summary of the flow of genetic information in protein synthesis is useful as an overview. Identify the indicated parts of the following illustration by filling in the blanks with the names of the appropriate <u>structures</u> or <u>functions</u>. Choose from the following: *mRNA, tRNA, polypeptide, ribosome subunits, ribosome, translation,* and *transcription.* Complete the exercise by matching and entering the letter of the description (A–G) in the parentheses after each label.

35. _____ ( )

36. _____ _____ ( )

37. _____ _____ ( )

38. _____ _____ ( )

39. _____ ( )

40. _____ ( )

41. _____ ( )

A. Carries the genetic code from the nucleus to the cytoplasm where it will be translated into a protein product [p.414]

B. Includes three stages: initiation, chain elongation, and chain termination [p.416]

C. Once released, it may function in cytoplasm or enter rough ER for transport [p.417]

D. Join together at initiation of translation [p.415]

E. RNA molecules are produced on DNA templates in the nucleus [p.412]

F. Place where translation occurs [p.415]

G. Will pick up specific amino acid for delivery to ribosome for translation [p.414]

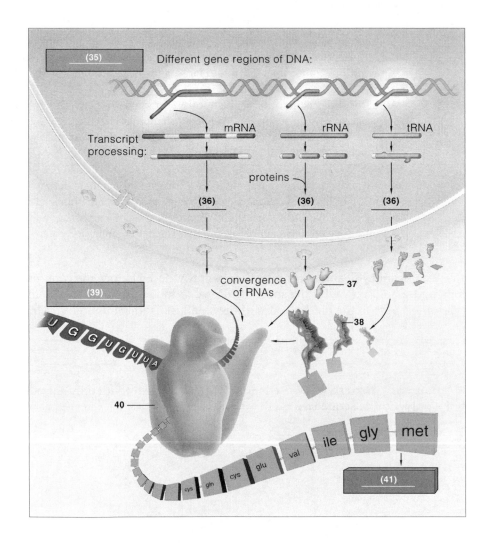

22.6. TOOLS FOR "ENGINEERING" GENES [pp.418–419]

22.7. "SEQUENCING" DNA [p.420]

Selected Words: *restriction enzymes* [p.418], "sticky" [p.418], "cloned" [p.418], *cloning vector* [p.418], *probe* [p.420]

Boldfaced, Page-Referenced Terms

[p.418] recombinant DNA technology _____

_____

[p.418] genetic engineering _____

_____

[p.418] genome _____

_____

[p.419] polymerase chain reaction (PCR) _____

_____

[p.420] automated DNA sequencing _____

_____

[p.420] gene library _____

_____

Short Answer

1. Describe and distinguish between the bacterial chromosome and plasmids present in a bacterial cell. [p.418]

_____

_____

_____

Fill-in-the-Blanks

Nature has conducted countless genetic experiments through (2) _____ [p.418] and other events.
Humans now are creating genetic change by using (3) _____ _____ [p.418] technology
in which researchers cut and splice together gene regions from different (4) _____ [p.418]. The
modified molecules are then inserted into bacteria or other cells that can rapidly (5) _____ [p.418]
genetic material and then divide. This new technology also is the basis for (6) _____
_____ [p.418], in which genes are isolated, modified, and inserted back into the same organism
or a different one.

Bacterial (7) _____ [p.418] are circular DNA molecules containing only a few genes. A(n)
(8) _____ _____ [p.418] can cut apart specific sequences of DNA. Fragments of DNA
with the same (9) "_____" [p.418] ends will combine and form a recombinant DNA molecule.

Foreign DNA inserted into a plasmid is called a DNA (10) _____ [p.418] because a bacterium will replicate the engineered plasmid, producing many "clones." As bacteria with engineered plasmids reproduce, they make much more of, or (11) _____ [p.419], the foreign DNA.

Choice

a. DNA clones
b. plasmids
c. cloning vectors
d. recombinant DNA technology

e. restriction enzymes
f. DNA amplification
g. genome
h. polymerase chain reaction

12. _____ Specialized knowledge that forms the basis for genetic engineering [p.418]

13. _____ All the DNA in a haploid set of a species' chromosome [p.418]

14. _____ Rapid way to copy DNA fragments [p.419]

15. _____ Molecules that cut DNA molecules [p.418]

16. _____ Rapid division of DNA clones in a cloning library [p.419]

17. _____ Small, circular DNA molecules that carry only a few genes [p.418]

18. _____ Multiple, identical copies of DNA fragments that have been inserted into bacterial plasmids [p.418]

19. _____ Plasmids that serve to deliver foreign DNA into a host cell [p.418]

Short Answer

20. How does PCR differ from the use of cloning vectors for DNA amplification? [p.419]

_____

_____

_____

21. What is the function of primers used in PCR? [p.419]

_____

_____

_____

Blanks

_____ 22. Determining the order of nucleotides in a gene or other piece of DNA [p.420]

_____ 23. Short stretch of radioactively labeled DNA used to find a gene of interest [p.420]

_____ 24. A mixture of DNA fragments containing genes of interest [p.420]

22.8. MAPPING THE HUMAN GENOME [pp.420–421]

22.9. SOME APPLICATIONS OF BIOTECHNOLOGY [pp.422–423]

22.10. ISSUES FOR A BIOTECHNOLOGICAL SOCIETY [p.424]

22.11. ENGINEERING BACTERIA, ANIMALS, AND PLANTS [p.425]

22.12. MR. JEFFERSON'S GENES [p.426]

Selected Words: "junk DNA" [p.420], SNPs ("snips") [p.421], *amyotrophic lateral sclerosis* [p.421], *transfection* [p.422], *severe combined immune deficiency*—SCID-XI [p.422], "bubble babies" [p.422], "suicide tags" [p.423], "lipoplexes" [p.423], *restriction fragment length polymorphisms* (RFLPs) [p.423], "superweeds" [p.424], "GM" or genetically modified [p.424], *bioremediation* [p.424], "micro-injected" [p.426], "supermouse" [p.426]

Boldfaced, Page-Referenced Terms

[p.421] DNA chip _____

_____

[p.422] gene therapy _____

_____

[p.423] DNA fingerprint_____

_____

True/False

If the statement is true, write a "T" in the blank. If false, explain why by changing one or more of the underlined words.

_____ 1. In 2002, the sequence of <u>genes</u> in the human genome was determined. [p.420]

_____ 2. The human genome consists of about 2.9 billion nucleotide pairs with perhaps <u>100,000</u> genes. [p.420]

_____ 3. The coding portions of human DNA make up about <u>80</u> percent of our DNA. [p.420]

_____ 4. There are over 1.4 million SNPs in the human genome, including many that result in different <u>alleles</u>. [p.421]

_____ 5. DNA <u>chips</u> are microarrays of thousands of DNA sequences stamped onto a glass plate, and used to pinpoint gene activity. [p.421]

_____ 6. The findings of the Human Genome Project may allow therapeutic drugs to be <u>customized</u> for individuals and particular situations. [p.421]

_____ 7. The Human Genome Project seeks to create maps of where specific genes are on chromosomes; this <u>has not</u> been successful. [p.421]

Fill-in-the-Blanks

In (8) _____ _____ [p.422], one or more normal genes are inserted into a person's body to replace mutated genes and correct a genetic defect. Smaller genes can be carried into a host cell by (9) _____ [p.422], while larger ones must enter in some other way. In (10) _____ [p.422], DNA is integrated into exposed cells in a laboratory culture, but it has not been very successful. Many gene therapy trials involve inserting a small gene into a (11) _____ [p.422] that then infects a host cell and carries the correct gene in. Research with SCID has used a retrovirus to insert a normal (12) _____ into stem cells of patients [p.422]. In a trial of 11 children, (13) _____ [p.422] were able to leave their isolation tents. In a similar study with cystic fibrosis, only (14) _____ [p.422] percent of affected cells took up the normal gene. Gene therapy has been most successful at treating

(15) _____ [p.422]. Genes for a(n) (16) _____ [p.422] have been introduced into cultured tumor cells and the cells returned to the body. This may stimulate T cells to recognize cancerous cells and (17) _____ [p.423] them. (18) _____ [p.423] are laboratory-made packets in which a plasmid is encased in a lipid that helps it enter a cell. The plasmid carries a gene that can trigger an enormous (19) _____ [p.423] response against cancer cells. Other trials are adding the gene for a powerful (20) _____-_____ [p.423] agent to lymphocytes that home in on tumors.

Dichotomous Choice

Circle one of two possible answers given between parentheses in each statement.

21. The unique set of (DNA fragments/coding genes) in each individual is called a DNA fingerprint. [p.423]

22. The short DNA segments that differ greatly from person to person are called (tandem repeats/riff-lips). [p.423]

23. In DNA fingerprinting, DNA fragments are separated by a procedure called (column chromatography/ gel electrophoresis). [p.423]

24. RFLPs can be detected by electrophoresis after being cut out of DNA by (polymerase enzymes/ restriction enzymes). [p.423]

Short Answer

25. List four hotly debated concerns about human manipulation of DNA. [p.424]

_____

_____

_____

_____

26. Name some benefits of human manipulation of DNA. [p.424]

_____

_____

_____

_____

Fill-in-the-Blanks

Any genetically engineered organism that carries foreign genes is (27) _____ [p.425].

(28) _____ [p.425] in bioengineered bacteria carry many human genes so that useful human (29) _____ [p.425] can be made easily. Growth hormone, (30) _____ [p.425], and interferon are examples of such products. Animal cells can be (31) _____-_____ [p.425] with foreign DNA. Using mice, goats, and other animals to produce compounds used to treat disease eliminates the risk of transmitting (32) _____ [p.425] associated with obtaining these products from humans. Plants may be engineered to improve traits such as (33) _____ [p.425] to a pathogen or herbicide, and to improve crop yields.

Self-Quiz

Are you ready for the exam? Test yourself on key concepts by taking the additional tests linked with your BiologyNow CD-ROM.

____ 1. A DNA molecule is built from four kinds of _____. [p.408]
- a. base pairs
- b. nitrogen-containing bases
- c. phosphates
- d. nucleotides

____ 2. In DNA, base pairing occurs between _____. [p.408]
- a. cytosine and uracil
- b. adenine and guanine
- c. adenine and uracil
- d. adenine and thymine

____ 3. A single strand of DNA with the base-pairing sequence C-G-A-T-T-G is compatible only with the sequence _____. [p.408]
- a. C-G-A-T-T-G
- b. G-C-T-A-A-G
- c. T-A-G-C-C-T
- d. G-C-T-A-A-C

____ 4. During DNA replication, individual nucleotides are assembled onto a parent DNA strand by _____. [p.410]
- a. thymine dimers
- b. DNA polymerases
- c. ribosomes
- d. codons

____ 5. Transcription _____. [p.412]
- a. occurs on the surface of the ribosome
- b. is the final process in the assembly of a protein
- c. occurs during the synthesis of RNA by use of a DNA template
- d. is catalyzed by DNA polymerase

____ 6. _____ carries amino acids to ribosomes, where they are linked into the primary structure of a polypeptide. [p.412]
- a. mRNA
- b. tRNA
- c. An intron
- d. rRNA

____ 7. Transfer RNA differs from other types of RNA because it _____. [p.412]
- a. transfers genetic instructions from cell nucleus to cytoplasm
- b. specifies the amino acid sequence of a particular protein
- c. carries an amino acid at one end
- d. contains codons

____ 8. _____ is an enzyme that dominates the process of transcription. [p.412]
- a. RNA polymerase
- b. DNA polymerase
- c. Phenylketonuriase
- d. Transfer RNA

____ 9. _____ and _____ are found in RNA but not in DNA. [p.412]
- a. Deoxyribose; thymine
- b. Deoxyribose; uracil
- c. Uracil; ribose
- d. Thymine; ribose

____ 10. Each "word" in the DNA and mRNA language consists of _____ "letters." [p.414]
- a. three
- b. four
- c. five
- d. more than five

____ 11. The genetic code is composed of _____ different kinds of codons. [p.414]
- a. three
- b. twenty
- c. sixteen
- d. sixty-four

____ 12. Initiation, elongation, and termination are all stages of _____. [p.416]
- a. replication
- b. translation
- c. transcription
- d. mutagenesis

___13. Small, circular molecules of DNA in bacteria, often used as cloning vectors, are called _____. [p.418]
   a. plasmids
   b. DNA probes
   c. RFLPs
   d. cDNA

___14. _____ enzymes are used to cut genes in recombinant DNA research. [p.418]
   a. Ligase
   b. Restriction
   c. Transcriptase
   d. DNA polymerase

___15. Genetically engineered bacteria may become major weapons in cleaning up environmental pollution, a process called _____. [p.424]
   a. recombinant DNA technology
   b. bioremediation
   c. DNA library construction
   d. DNA sequencing

___16. A DNA probe is a known DNA sequence that is _____. [p.420]
   a. resistant to antibiotics
   b. used to make restriction enzymes
   c. radioactively labeled
   d. found in human noncoding DNA

___17. Amplification results in _____. [p.419]
   a. plasmid integration
   b. bacterial conjugation
   c. cloned DNA
   d. production of DNA ligase

___18. A commonly used method of DNA amplification is _____. [p.419]
   a. polymerase chain reaction
   b. gene expression
   c. genome mapping
   d. RFLPs

___19. Restriction fragment length polymorphisms are valuable because they _____. [p.423]
   a. reduce the risks of genetic engineering
   b. provide an easy way to sequence the human genome
   c. allow fragmenting DNA without enzymes
   d. identify individuals by unique DNA sequences

___20. Transgenic species are those that carry _____. [p.425]
   a. micro-injected DNA
   b. one or more foreign genes
   c. human genes
   d. transfected cells

Chapter Objectives/Review Questions

This section lists general and detailed chapter objectives that can be used as review questions. You can make maximum use of these items by writing answers on a separate sheet of paper. Fill in answers where blanks are provided. To check for accuracy, compare your answers with information given in the chapter or glossary.

1. Aided by clues in X-ray images, _____ and _____ painstakingly devised a correct model of the DNA molecule. [p.408]
2. Explain what is meant by pairing of nitrogen-containing bases (base pairing), and explain the mechanism that causes bases of one DNA strand to join with bases of the other strand. [p.408]
3. Assume that the two parent strands of DNA have been separated and that the base sequence on one parent strand is A-T-T-C-G-C; the base sequence that will complement that parent strand during replication is _____ . [p.408]
4. Explain what is meant by "Each parent strand is conserved in each new DNA molecule." [p.410]
5. Briefly describe the spontaneous DNA mutations known as *base-pair substitutions, insertions, deletions,* and *expansion mutations*. [p.411]
6. Cite an example of a change in one DNA base pair that has profound effects on the human phenotype. [p.411]
7. State how RNA differs from DNA in structure and function, and indicate which features RNA has in common with DNA. [p.412]

8. _____ RNA combines with certain proteins to form a ribosome; _____ RNA carries genetic information for protein construction from the nucleus to the cytoplasm; _____ RNA picks up specific amino acids and moves them to the area of mRNA and ribosome. [p.412]

9. Describe the process of transcription and indicate three ways in which it differs from replication. [p.412]

10. What mRNA code would be formed from the following DNA code: TAC-CAT-GAG-ACC-GCC-ACT? [p.412]

11. Transcription starts at a(n) _____, a specific sequence of bases on one of the two DNA strands that signals the start of a gene. [p.412]

12. Actual coding portions of a newly transcribed mRNA are called _____; _____ are the noncoding portions. [p.413]

13. What is the function of regulatory proteins in the process of transcription? [p.413]

14. Explain the triplet nature of the genetic code. [p.414]

15. Describe the relationships between DNA and mRNA, and between mRNA and tRNA that insure the synthesis of the correct protein. [p.414]

16. Name the three stages of translation and tell what happens in each. [p.416]

17. How has Nature conducted genetic experiments through time? [p.418]

18. Define recombinant DNA technology. [p.418]

19. _____ are small, circular, self-replicating molecules of DNA or RNA within a bacterial cell. [p.418]

20. Some bacteria produce _____ enzymes that cut apart DNA molecules at specific base sequences; such DNA fragments may have "_____ ends" capable of base-pairing with other DNA molecules. [p.418]

21. Describe PCR as a major method of DNA amplification. [p.419]

22. Explain what a DNA probe is, and what a gene library is. [p.420]

23. Discuss the findings of the Human Genome Project. [p.420]

24. Define "gene therapy" and describe techniques used in gene therapy trials. [pp.422–423]

25. Why is a "DNA fingerprint" given that name? [p.423]

26. DNA fragments from a human population show unique restriction _____ length _____. [p.423]

27. Discuss concerns that people have about biotechnology. [p.424]

28. How common are genetically modified foods in the United States? [p.424]

29. What are transgenic organisms, and how are they "created" by humans? [p.425]

30. Has DNA analysis confirmed or refuted circumstantial evidence of a sexual relationship between President Thomas Jefferson and one of his slaves? [p.426]

Media Menu Review Questions

Questions 1–5 are drawn from the following InfoTrac College Edition article: "Bringing the Genome to You: 50 Years of DNA." Carl J. Lauter. *Generations*, Summer 2003.

1. April 2003 was the 50th anniversary of the publication of Watson and Crick's description of _____ and the culmination of the _____ of the human genome.

2. The 108th U.S. Congress passed a resolution designating April as "Human Genome _____" and April 25th as _____ Day.

3. The data from the Human Genome Project is available on the Internet with no _____ on use.

4. In his closing remarks, Francis Collins indicated that the completion of the genome is the first step in determining how genes influence _____ and wellness.

5. Collins predicted that perhaps by 2010, we will have predictive genetic tests, interventions to reduce risks, and _____. By 2020, there will be gene-based _____ drugs for major diseases.

Questions 6–8 are drawn from the following InfoTrac College Edition article: "Frankenfoods v. Luddites: GM Crops." *The Economist*, June 7, 2003.

6. Under European regulations, the government can object to GM crops only if it can prove they pose a _____ to human health or to the _____.
7. The European Parliament has backed a proposal to require the tracing of all _____ ingredients.
8. In Great Britain, the GM crops do not seem as _____ as they did four years ago, with 41 percent of people not bothered by them vs. 29 percent in 2000.

Questions 9–10 are drawn from the following InfoTrac College Edition article: "Cancer Clue: RNA-Destroying Enzyme May Thwart Prostate-Tumor Growth." J. Travis. *Science News*, January 26, 2002.

9. By studying families that include several men with prostate cancer, scientists have identified a _____ - _____ gene on chromosome 1 that may be involved with this type of cancer.
10. A mutation in both RNAse L alleles would allow a cancerous cell to avoid _____ and divide without limits.

Integrating and Applying Key Concepts

Genes code for specific polypeptide sequences. Not every substance in living cells is a polypeptide. Explain how genes might be involved in the production of a storage carbohydrate (such as glycogen) that is constructed from simple sugars.

23

GENES AND DISEASE: CANCER

Interactive Exercises

Impacts, Issues: The Body Betrayed [p.431]

23.1. CANCER: CELL CONTROLS GO AWRY [pp.432–433]

23.2. THE GENETIC TRIGGERS FOR CANCER [pp.434–435]

23.3. ASSESSING THE CANCER RISK FROM ENVIRONMENTAL CHEMICALS [p.436]

For additional practice, use the interactive vocabulary exercises linked with your BiologyNow CD-ROM.

Selected Words: "breast cancer susceptibility genes" [p.431], *neoplasm* [p.432], *benign* tumor [p.432], *angiogenin* [p.433], *retinoblastoma* [p.434], "precarcinogens" [p.435], cancer "promoters" [p.435], "nonself" tags [p.435], "cancer epidemic" [p.436]

Boldfaced, Page-Referenced Terms

[p.432] tumor _____

_____

[p.432] dysplasia _____

_____

[p.433] metastasis _____

_____

[p.434] carcinogenesis _____

_____

[p.434] proto-oncogenes _____

_____

[p.434] oncogene _____

_____

[p.434] tumor suppressor gene _____

_____

[p.435] carcinogens _____

_____

Matching

Choose the most appropriate description for each term.

1. _____ dysplasia [p.432]
2. _____ neoplasm [p.432]
3. _____ malignant [p.433]
4. _____ cancer cell [p.432]
5. _____ tumor [p.432]
6. _____ metastasis [p.433]
7. _____ angiogenin [p.433]
8. _____ benign [p.432]

A. Growth factor secreted by cancer cells; promotes blood vessel growth around cancer cells
B. Cell lacking clear structural specializations; displays abnormally large nuclei, less cytoplasm, and disorganized cytoskeleton
C. Cancer cells breaking away from a tumor and establishing new cancer sites
D. Refers to a noncancerous, slow-growing, and well-differentiated tumor; often enclosed by a capsule
E. Common name for a defined mass of overgrown tissue
F. "Bad form"; abnormal changes in tissue cells
G. Means "new growth"; a defined mass of tissue overgrowth
H. A cancer with the ability to metastasize and cause harm

Dichotomous Choice

Circle one of two possible answers given between parentheses in each statement.

9. The transformation of a normal cell into a cancerous one is called (metastasis/carcinogenesis). [p.434]

10. (Oncogenes/Proto-oncogenes) are normal genes regulating cell growth and development. [p.434]

11. A DNA segment capable of inducing cancer in a normal cell is a(n) (oncogene/proto-oncogene). [p.434]

12. An oncogene (does/does not) respond to controls over cell division. [p.434]

13. An oncogene acting alone (does/does not) cause malignant cancer. [p.434]

14. The onset of cancer requires mutations in several genes, including mutation of at least one (tumor suppressor gene/proto-oncogene). [p.434]

15. Retinoblastoma, or childhood eye cancer, develops when a child inherits (one/two) normal copies of a tumor suppressor gene. [p.434]

16. The p53 protein stops cell (growth/division) when cells are damaged, preventing mutations from being passed on to daughter cells. [p.434]

17. When p53 mutates, the resulting faulty protein seems to (inhibit/promote) the development of cancer, possibly by activating an oncogene. [p.434]

18. The mutation of an oncogene or a related suppressor gene, or a change in chromosome structure that moves an oncogene into a new position, can cause the oncogene to be (activated/inactivated). [p.434]

Labeling

Identify each numbered part of the following illustration that traces the steps of carcinogenesis. [p.434]

19. _____

20. _____ _____

21. _____-_____

22. _____ _____ _____

23. _____ _____

24. _____

25. _____

26. _____

27. _____

28. _____

29. _____

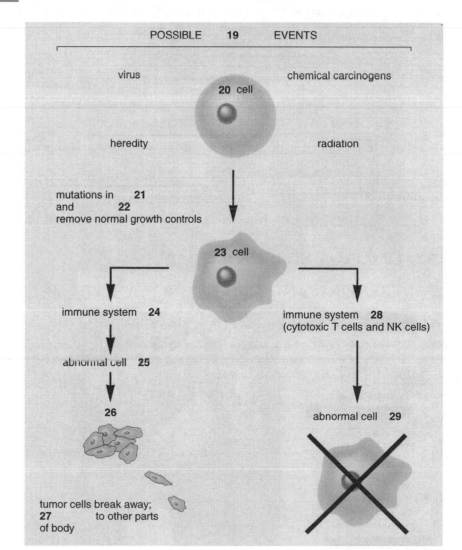

Choice

Choose one of these routes to cancer for each of the statements below. [p.435]

a. inherited susceptibility
b. viruses
c. chemical carcinogens
d. radiation
e. immunity breakdowns

30. _____ Sources include tanning lamps, diagnostic X-rays, radioactive materials

31. _____ May alter a proto-oncogene by inserting into host cell DNA

32. _____ Plays a major role in about 5 percent of cancers, including breast, colorectal, and lung cancer

33. _____ Factor in increase in cancers as we grow older

34. _____ Examples are asbestos, vinyl chloride, and benzene, aflotoxin produced by fungi

35. _____ May carry an oncogene into the host cell DNA

36. _____ Some are precarcinogens, others are promoters

37. _____ Passed in gamete to next generation

38. _____ Antigens that must bind with lymphocytes are not displayed by cancer cells

39. _____ Sun exposure is the greatest risk in this category

Fill-in-the-Blanks

Factors in our environment lead to about (40) _____ [p.436] of all cancers. Government statistics indicate that about (41) _____ [p.436] percent of the food in American supermarkets contains detectable residues of one or more of the active ingredients used in (42) _____ [p.436]. Residues of pesticides can be removed from food by washing, but it is difficult to avoid contact with pesticides used in community (43) _____ [p.436] to control mosquitoes. In addition to agricultural chemicals, (44) _____ [p.436] chemicals have also been linked to cancer.

23.4. DIAGNOSING CANCER [p.437]

23.5. TREATING AND PREVENTING CANCER [p.438]

Selected Words: DNA probe [p.437], "anticancer lifestyle" [p.438], *adjuvant therapy* [p.438], *chronic myelogenous leukemia* [p.438], cancer "vaccines" [p.438]

Boldfaced, Page-Referenced Terms

[p.437] tumor markers _____

_____

[p.437] biopsy_____

_____

[p.438] chemotherapy _____

_____

[p.438] immunotherapy _____

_____

Completion

Complete the following statements of the seven warning signs of cancer. [p.437]

1. Change in _____ or _____ habits and function

2. A _____ that does not heal

3. Unusual _____ or _____ discharge

4. Thickening or _____

5. _____ or difficulty swallowing

6. Obvious change in a _____ or _____

7. Nagging _____ or _____

Matching

Choose the most appropriate description for each term.

8. _____ interleukins [p.438]

9. _____ medical imaging [p.437]

10. _____ monoclonal antibodies [pp.437,438]

11. _____ tumor markers [p.437]

12. _____ adjuvant therapy [p.438]

13. _____ immunotherapy [p.438]

14. _____ cancer screening [p.437]

15. _____ interferon [p.438]

16. _____ biopsy [p.437]

17. _____ chemotherapy [p.438]

18. _____ cancer "vaccines" [p.438]

19. _____ DNA probe [p.437]

A. Treatment with substances that trigger a strong immune response against cancer cells

B. The use of drugs to kill cancer cells

C. Definitive cancer detection tool; a small piece of tissue is removed from the body and analyzed

D. Substances produced by cancer cells or normal cells in response to cancer; detected in blood tests

E. Produced by lymphocytes; potential anti-cancer weapons

F. Diagnosis using MRI, X-rays, ultrasound, and CT

G. A treatment combining surgery and a less toxic dose of chemotherapy

H. Examples are mammograms, Pap test, testicle self-examination, and breast self-examination

I. Pinpoint location of cancer and act as "magic bullets" that deliver lethal doses of radiation or anticancer drugs to tumor cells

J. Stimulate cytotoxic T cells to recognize and destroy cells with an abnormal surface protein

K. Activates cytotoxic T cells and natural killer cells that recognize and kill various types of cancer

L. Segment of radioactively labeled DNA used to locate gene mutations or alleles associated with some types of inherited cancers

23.6. SOME MAJOR TYPES OF CANCER [p.439]

23.7. CANCERS OF THE BREAST AND REPRODUCTIVE SYSTEM [pp.440–441]

23.8. A SURVEY OF OTHER COMMON CANCERS [pp.442–443]

Selected Words: *sarcomas* [p.439], *carcinomas* [p.439], *adenocarcinoma* [p.439], *lymphomas* [p.439], *leukemias* [p.439], *modified radical mastectomy* [p.440], *lumpectomy* [p.440], *Pap smear* [p.441], *squamous cell carcinomas* [p.442], *adenocarcinomas* (or *large-cell carcinomas*) [p.442], *small-cell carcinoma* [p.442], *Wilms' tumor* [p.442], *non-Hodgkin lymphoma* [p.442], *Hodgkin's disease* [p.442], *Burkitt lymphoma* [p.442]

Fill-in-the-Blanks

Cancers of connective tissues (muscle and bone) are (1) _____ [p.439]. Cancers that arise from epithelium, including cells of the skin and epithelial linings of internal organs, are known as (2) _____ [p.439]. (3) _____ [p.439] begin in the body or ducts of a gland. (4) _____ [p.439] are cancers of lymphoid tissues in organs such as lymph nodes. Cancers of blood-forming regions such as stem cells in bone marrow are (5) _____ [p.439].

Choice

For questions 6–36, choose from the following major types of cancer:

a. breast
b. uterine and ovarian
c. testicular and prostate
d. oral and lung

e. digestive system and related organs
f. urinary system
g. blood and lymphatic system
h. skin

6. _____ Risk factors are fair skin and exposure to UV radiation in sunlight [p.443]

7. _____ Hodgkin's disease [p.442]

8. _____ Warning signs include a change in bowel habits, rectal bleeding, and blood in the feces [p.442]

9. _____ Most common among smokers and people who use smokeless tobacco, especially if they are also heavy alcohol drinkers [p.442]

10. _____ About one woman in eight develops this type of cancer [p.440]

11. _____ Kills more people than any other type of cancer [p.442]

12. _____ Non-Hodgkin lymphoma [p.442]

13. _____ These cancers are usually adenocarcinomas of duct cells and often are not detected until they have spread to other organs [p.442]

14. _____ Develop in lymphoid tissues [p.442]

15. _____ Cancers in which stem cells in bone marrow overproduce white blood cells [p.443]

16. _____ Malignant melanoma is the most dangerous kind and is a cancer of melanin-producing cells [p.443]

17. _____ Cancers associated with heavy alcohol consumption and a diet rich in smoked, pickled, and salted foods [p.442]

18. _____ Burkitt lymphoma [p.442]

19. _____ Nonsmokers, particularly children and spouses inhaling secondhand tobacco smoke, have a significant risk for this cancer [p.442]

20. _____ The second leading cause of cancer deaths in men after lung cancer [p.441]

21. _____ Self-examination should occur every month, about a week after the menstrual period [p.440]

22. _____ This cancer is often lethal because symptoms, mainly an enlarged abdomen, do not appear until the cancer is advanced and has metastasized [p.441]

23. _____ A blood test called the PSA test screens for tumor marker associated with this type of cancer [p.441]

24. _____ This cancer has now surpassed breast cancer as the leading cancer killer of women [p.442]

25. _____ Squamous cell carcinomas, adenocarcinomas, large-cell carcinomas, and small-cell carcinomas [p.442]

26. _____ Wilms' tumor, one of the most common of all childhood cancers [p.442]

27. _____ Basal cell carcinomas and squamous cell carcinomas usually are easily treated by minor surgery [p.443]

28. _____ Cigarette smoking is a risk for pancreatic cancer [p.442]

29. _____ Lumpectomy and modified radical mastectomy are treatments [p.440]

30. _____ Risk seems to increase with infection—such as HIV—that impairs immune system function [p.442]

31. _____ Drugs like tamoxifen are increasingly used to shrink these tumors [p.440]

32. _____ Pap smear detects precancerous phases of cervical cancer [p.441]

33. _____ Chemotherapy using two compounds—vincristine and vinblastine—derived from species of periwinkle plants is effective [p.443]

34. _____ Warning signals include a nagging cough, shortness of breath, chest pain, bloody sputum, unexplained weight loss, and frequent respiratory infections or pneumonia [p.442]

35. _____ Risk factors include Down syndrome, exposure to chemicals such as benzene, and radiation exposure [p.443]

36. _____ Some types are the most common childhood cancers [p.443]

Self-Quiz

Are you ready for the exam? Test yourself on key concepts by taking the additional tests linked with your BiologyNow CD-ROM.

Multiple Choice

____ 1. A defined mass of tissue known as a *tumor* can also be called a _____. [p.432]
  a. benign tumor
  b. neoplasm
  c. malignant tumor
  d. capsule

____ 2. Cancer cells that break away from a primary tumor and migrate to establish new cancer sites are said to be undergoing _____. [p.433]
  a. activation
  b. dysplasia
  c. carcinogenesis
  d. metastasis

____ 3. Of the following, which one is NOT characteristic of cancer cells? [pp.432–433]
  a. normal appearance
  b. uncontrolled growth
  c. metastasis
  d. lack strong cell-to-cell adhesion

____ 4. The p53 gene _____. [p.434]
  a. initiates metastasis
  b. mutates and becomes a malignant cell
  c. is a suppressor gene effective in many tissue types
  d. is a type of oncogene

____ 5. A DNA segment capable of inducing cancer in a normal cell is a(n) _____. [p.434]
  a. tumor suppressor gene
  b. oncogene
  c. p53 gene
  d. proto-oncogene

____ 6. Carcinogenesis is _____. [p.434]
  a. the transformation of a normal cell into a cancerous one
  b. the production of carcinogens
  c. the dying process of tumor suppressor genes
  d. a breakdown in immunity that leads to cancer

____ 7. Of the following, which may serve as a trigger for carcinogenesis? [p.435]
   a. viruses
   b. chemicals
   c. radiation
   d. heredity
   e. All of the above can trigger carcinogenesis.

____ 8. _____ is a cancer screening technique. [p.437]
   a. Chemotherapy
   b. MRI
   c. Immune therapy
   d. Digital rectal examination
   e. Stimulation of cytotoxic T cells to recognize and destroy cancer cells

____ 9. Adenocarcinomas are cancers of _____. [p.439]
   a. lymph nodes
   b. connective tissues
   c. blood-forming regions
   d. a gland or its ducts
   e. epithelium

Choice

For questions 10–15, select the correct type of cancer from the following list:

a. Oral and lung cancer
b. Cancer of the skin and epithelial linings of internal organs
c. Cancer of the digestive system and related organs
d. Cancer of the blood and lymphatic system
e. Female breast cancer, cancer of male and female reproductive system

10. PSA is a test for _____. [p.441]
11. A malignant melanoma is a(n) _____. [p.443]
12. Low-dose mammography detects, and lumpectomy may treat _____. [p.440]
13. Hodgkin's disease is a(n) _____. [p.442]
14. Smoking and smokeless tobacco cause _____. [p.442]
15. Leukemia is a(n) _____. [p.443]

Chapter Objectives/Review Questions

This section lists general and detailed chapter objectives that can be used as review questions. You can make maximum use of these items by writing answers on a separate sheet of paper. Fill in answers where blanks are provided. To check for accuracy, compare your answers with information given in the chapter or glossary.

1. If cells overgrow, the result is a defined mass of tissue called a(n) _____. [p.432]
2. _____ is an abnormal change in the sizes, shapes, and organization of cells in a tissue. [p.432]
3. List the characteristics that distinguish benign and malignant tumors. [p.432]
4. Briefly comment on the meaning of the following characteristics of cancer cells: structural abnormalities, uncontrolled growth, and metastasis. [pp.432–433]
5. Write a definition for the term *proto-oncogene*. [p.434]
6. A(n) _____ is a DNA segment that can induce cancer in a normal cell. [p.434]
7. In addition to an oncogene, the onset of malignant cancer requires the mutation of at least one _____ _____ gene. [p.434]
8. Explain the importance of the p53 gene. [p.434]
9. List various factors that may trigger expression of an oncogene. [p.434]
10. List five routes to carcinogenesis other than oncogenes. [p.435]
11. Give brief statements that relate cancer to agricultural and industrial chemicals. [p.436]

12. Define *cancer screening, tumor markers, biopsy,* and *medical imaging.* [p.437]
13. List the recommended cancer screening tests. [p.437]
14. _____ is the use of drugs to kill dividing cells. [p.438]
15. Why is loss of hair cells, stem cells, lymphocytes, and epithelial cells a side effect of chemotherapy? [p.438]
16. List reasons that immunotherapy holds promise for cancer treatment. [p.438]
17. Identify these types of cancers by their source tissue: sarcomas, carcinomas, adenocarcinomas, lymphomas, and leukemias. [p.439]

Media Menu Review Questions

Questions 1–10 are drawn from the following InfoTrac College Edition article: "How You Can Lower Your Cancer Risk." *Harvard Health Letter,* August 2002.

1. Current research supports traditional ideas about avoiding cancer, but puts a new emphasis on staying trim and _____.
2. Staying lean ranks second only to avoiding _____ as the most effective way you can improve your cancer odds.
3. Low fat and nonfat are not the same as low _____, as they often have added sugar.
4. Vigorous exercise may improve the odds against cancer by decreasing the exposure of the breasts to _____, reducing _____ and growth factors that increase cell turnover.
5. Hundreds of studies show that high fruit and vegetable consumption is associated with a lower risk of cancers of the _____, stomach, mouth, pharynx, and _____.
6. Although new studies show no link between fruit and vegetable consumption and reduction in colon cancer, it may be that people who eat fruits and vegetables also have many _____ habits.
7. Red meat is associated with an increased chance of _____ cancer, especially when cooked at _____ temperatures.
8. The common thread connecting sugar, obesity, diabetes, and cancer may be insulin and proteins called insulinlike _____ factors that may trigger cell proliferation.
9. The evidence on _____ and cancer is inconclusive, but it does have many other health benefits.
10. The optimum cancer-preventing diet could vary with a person's metabolism, which can be traced to particular _____.

Integrating and Applying Key Concepts

A large number of chemicals, environmental pollutants, and several types of radiation are recognized as being carcinogenic. Can you think of a common effect these various and different carcinogens might have on a cell to induce cancer?

Why do you think that cancer is more common in older people than in younger people?

24

PRINCIPLES OF EVOLUTION

Interactive Exercises

Impacts, Issues: Measuring Time [p.447]

24.1. A LITTLE EVOLUTIONARY HISTORY [p.448]

24.2. A KEY EVOLUTIONARY IDEA: INDIVIDUALS VARY [p.449]

For additional practice, use the interactive vocabulary exercises linked with your BiologyNow CD-ROM.

Selected Words: *Homo sapiens* [p.447], HMS *Beagle* [p.448], "natural selection" [p.448], *morphological* traits [p.449], *physiological* traits [p.449], *behavioral* traits [p.449]

Boldfaced, Page-Referenced Terms

[p.448] evolution _____

_____

[p.449] microevolution _____

_____

[p.449] macroevolution_____

_____

[p.449] population _____

_____

[p.449] gene pool _____

_____

Complete the Table

1. Several key figures and events in the life of Charles Darwin led him to his conclusions about natural selection and evolution. Summarize these influences by completing the following table. [p.448]

Key Figures/Events	Importance to Synthesis of Evolutionary Theory
a.	Botanist at Cambridge University who perceived Darwin's real interests; arranged for Darwin to become a ship's naturalist
b.	Institution that granted Darwin a degree in theology
c.	British ship that carried Darwin on a five-year voyage (as a naturalist) around the world
d.	Clergyman and economist who proposed that any population tends to outgrow its resources, and its members must compete for what is available
e.	Evolutionary process that Darwin proposed after considering his observations and the ideas of other thinkers

Matching

Choose the most appropriate description for each term.

2. _____ gene pool [p.449]

3. _____ behavioral traits [p.449]

4. _____ evolution [p.448]

5. _____ macroevolution [p.449]

6. _____ microevolution [p.449]

7. _____ morphological traits [p.449]

8. _____ physiological traits [p.449]

9. _____ population [p.449]

A. Cumulative genetic changes that may give rise to new species
B. The collective genes with their alleles present in a population
C. Functions of body structures
D. A group of individuals of the same species occupying a given area
E. Responses to certain basic stimuli
F. The large-scale patterns, trends, and rates of change among groups of species
G. General form and function
H. Genetic changes in lines of descent through successive generations

24.3. MICROEVOLUTION: HOW NEW SPECIES ARISE [pp.450–451]

Selected Words: On the Origin of Species [p.450], "survival of the fittest" [p.450], *founder effect* [p.450], "reproductively isolated" [p.451], *divergence* [p.451], *gradualism* [p.451], *punctuated equilibrium* [p.451]

Boldfaced, Page-Referenced Terms

[p.450] theory of evolution by natural selection _____

_____

[p.450] adaptation _____

_____

[p.450] genetic drift _____

_____

[p.450] gene flow _____

_____

[p.451] species _____

_____

Dichotomous Choice

Circle one of two possible answers given between parentheses in each statement.

1. The variant alleles in species result from heritable changes in DNA called (natural selection/mutations). [p.450]

2. (Gene flow/Natural selection) is the difference in survival and reproduction that has occurred among individuals that differ in one or more traits. [p.450]

3. Charles Darwin presented his ideas in a book called (*On the Origin of Species/The Origin of Man*) [p.450]

4. Evolution and natural selection are considered by most scientists to be (theories/principles). [p.450]

5. Through evolution over time, organisms come to have characteristics that suit them to conditions in a particular environment; this is (reproductive isolation/adaptation). [p.450]

6. The blonde hair and blue eyes of Finns, along with many genetic disorders most common in the Finnish population, can be explained by (natural selection/the founder effect). [p.450]

7. The founder effect is an example of rapid change leading to unique features in a small population, a phenomenon known as (genetic drift/gene flow). [p.450]

8. Allele frequencies can change as individuals leave a population or new individuals enter; this is referred to as (gene flow/genetic drift). [p.450]

9. Reproductive isolation develops when (genetic drift/gene flow) between two populations stops. [p.451]

10. Two species are reproductively isolated when they have enough genetic differences that they cannot produce (any offspring/fertile offspring). [p.451]

11. Different breeding seasons, different mating rituals, and changes in body structures that prevent mating serve as (hybridization/isolating) mechanisms that reduce the likelihood of interbreeding between populations. [p.451]

12. The buildup of differences in allele frequencies among isolated populations is called (convergence/divergence). [p.451]

13. The emergence of new species through many small changes in form over long spans of time is called (punctuated equilibrium/gradualism). [p.451]

14. The statement "Most evolutionary change occurs in bursts" refers to (punctuated equilibrium/gradualism). [p.451]

15. A unit of one or more populations of individuals that can interbreed under natural conditions and produce fertile offspring is a(n) (ecosystem/species). [p.451]

16. Dramatic environmental changes such as an ice age may be the driving force altering the physical environment of adapted populations of organisms; this supports the (gradualism/punctuated equilibrium) model of speciation. [p.451]

Complete the Table

17. Complete the following table to review the major components of Darwin's Theory of Evolution. [p.450]

Concept	Description
a.	Differences among individuals that can be passed to subsequent generations
b.	Phrase meaning that some traits provide survival advantages to some individuals
c.	Refers to longer survival and more reproduction of those with the best adaptations for an environment
d.	Occurs when some forms of a trait become more or less common as a result of selection
e.	Effect of shifts in gene pools on life on Earth

Short Answer

18. The following illustration depicts the divergence of one species into two as time passes. Answer the questions below the illustration. [p.451]

    a. In what areas is there distinctly one species? _____

    b. Between what letters does the divergence begin? _____

    c. What letter represents the time when divergence is probably complete? _____

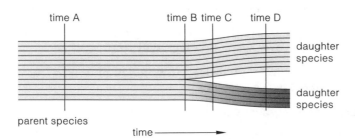

19. In what way do biologists disagree about microevolution? Is it likely that one view will prove the other wrong? Explain. [p.451]

_____

_____

_____

24.4. LOOKING AT FOSSILS AND BIOGEOGRAPHY [pp.452–453]

24.5. COMPARING THE FORM AND DEVELOPMENT OF BODY PARTS [pp.454–455]

24.6. COMPARING BIOCHEMISTRY [p.456]

Selected Words: plate tectonics [p.453], "supercontinent" [p.453], *vestigial* [p.455], "molecular clock" [p.456]

Boldfaced, Page-Referenced Terms

[p.452] fossil _____

_____

[p.452] fossilization _____

_____

[p.453] radiometric dating _____

_____

[p.453] biogeography _____

_____

[p.454] comparative morphology _____

_____

[p.454] homologous structures _____

_____

[p.454] analogous structures _____

_____

[p.454] morphological convergence _____

_____

Matching

Choose the most appropriate answer to match with examples of lines of evidence supporting macroevolution.

1. _____ analogous structures [p.454]
2. _____ biogeography [p.453]
3. _____ comparative biochemistry [p.456]
4. _____ comparative embryology [pp.454–455]
5. _____ comparative morphology [p.454]
6. _____ fossilization [p.452]
7. _____ homologous structures [p.454]
8. _____ morphological convergence [p.454]

A. Embryos of all vertebrate lineages go through strikingly similar stages
B. Body parts of close evolutionary relatives that are modified in very different ways as adaptations to different environments
C. Uses decay rate of radioactive isotopes to determine the age of volcanic rocks
D. Why are species of egg-laying mammals found only in Australia, Tasmania, and New Guinea?
E. Involves burial in sediments or volcanic ash, water infiltration, chemical changes, and pressure

9. _____ radiometric dating [p.453]

10. _____ plate tectonics [p.453]

11. _____ preservation [p.452]

12. _____ vestigial structures [p.455]

F. Uses information contained in patterns of body form to reconstruct evolutionary history

G. Gene sequencing and amino acid sequence comparison, especially of neutral mutations

H. Evolutionary pattern in which body parts of different lineages come to resemble one another due to evolution in similar environments

I. More likely with rapid burial in the absence of oxygen

J. Theory stating that all present continents were once part of the supercontinent Pangaea

K. Ear-wagging muscles in humans, pelvic girdle in pythons, and the human coccyx

L. Body parts of organisms without a common recent ancestor that resemble each other due to adaptation to a common environment

True/False

If the statement is true, write a "T" in the blank. If the statement is false, make it correct by writing the word(s) in the blank that should take the place of the underlined word(s).

_____ 13. We will not be able to recover the fossils for most extinct species. [p.452]

_____ 14. Movements of Earth's crust have wiped out parts of the fossil record. [p.452]

_____ 15. Jellyfish and other soft-bodied organisms are well represented in the fossil record. [p.452]

_____ 16. Finding fossils of small populations producing few offspring is less likely than finding fossils of large populations with a high reproductive rate. [p.452]

_____ 17. The environment of a species does not seem to be a factor in how well it is represented in the fossil record. [p.452]

24.7. HOW SPECIES COME AND GO [pp.456–457]

24.8. ENDANGERED SPECIES [p.457]

24.9. EVOLUTION FROM A HUMAN PERSPECTIVE [pp.458–459]

24.10. EMERGENCE OF EARLY HUMANS [pp.460–461]

Selected Words: "background extinction" [p.457], *endemic* [p.457], *prehensile* [p.458], *hominoids* [p.460], *hominids* [p.460], *Sahelanthropus tchadensis* [p.460], *Australopithecus afarensis* or "Lucy" [p.460], *Homo habilis* [p.460], *Homo erectus* [p.460], *Homo sapiens* [p.461], *multiregional model* [p.461], *African emergence* model [p.461], "stone age" [p.461]

Boldfaced, Page-Referenced Terms

[p.456] mass extinction _____

_____

[p.456] adaptive radiation _____

_____

[p.457] endangered species _____

_____

[p.458] genus _____

_____

[p.459] culture _____

_____

[p.459] bipedalism_____

_____

[p.460] humans _____

_____

Matching

Choose the most appropriate description for each term.

1. _____ adaptive radiation [p.456]
2. _____ habitat loss [p.457]
3. _____ endangered species [p.457]
4. _____ mass extinction [p.456]
5. _____ background extinction [p.456]

A. Major threat to over 90 percent of endangered species
B. An abrupt, widespread rise in extinction rates above the background level; a global event in which major groups of organisms are simultaneously wiped out
C. Species living in only one region that is extremely vulnerable to extinction
D. New species of a lineage fill a wide range of habitats during bursts of evolutionary activity
E. A relatively stable rate of species' disappearance over time

Dichotomous Choice

Circle one of the two possible answers given between parentheses in each statement.

6. (Gorillas/Bonobos) are our closest living evolutionary relative. [p.458]

7. (Primates/Hominids) evolved around 60 million years ago. [p.458]

8. Through alterations in hand bones, fingers could be wrapped around objects in (opposable/ prehensile) movements. [p.458]

9. Further, the thumb and tip of each finger could touch in (opposable/prehensile) movements. [p.458]

10. Later primates had (forward-directed eyes/an eye on each side of the head) that allows detecting shapes and movements in three dimensions. [p.459]

11. Through additional modifications, the eyes were able to respond to color variations and light intensity. These visual stimuli are typical of life in the (grasslands/trees). [p.459]

12. Humans have (bow-shaped/rectangular) jaws and (long canine teeth/smaller teeth of about the same length). [p.459]

13. On the road from early primates to humans, there was a shift from eating (insects/a mixed diet) to eating (insects/a mixed diet). [p.459]

14. In many primate lineages, parents started to invest more effort in (fewer/more) offspring. [p.459]

15. As parents formed stronger bonds with their young and maternal care became intense, the learning period grew (shorter/longer). [p.459]

16. The interlocking of brain modifications and behavioral complexity is most evident in the parallel evolution of the human (hand/brain) and culture. [p.459]

17. (Culture/Education) is the sum total of behavior patterns of a social group, passed between generations by learning and symbolic behavior—especially language. [p.459]

18. The capacity for (emotions/language) arose among ancestral humans through changes in the skull bones and expansion of parts of the brain. [p.459]

19. The habitual two-legged gait peculiar to the evolved, reorganized human skeleton is called (opposable movement/bipedalism). [p.459]

20. Compared with monkeys and apes, humans have a (longer/shorter), S-shaped, and somewhat flexible backbone. [p.459]

21. The position and shape of the human backbone, knee and ankle joints, and pelvic girdle are the basis of (bipedalism/prehensile movement). [p.459]

22. By 36 million years ago, tree-dwelling (hominoids/anthropoids) had evolved in tropical forests; they included ancestors of monkeys, apes, and humans. [p.460]

23. From 25 million to 5 million years ago, continents began to assume their current positions, climates became cooler and drier, and an adaptive radiation of apelike forms, the first (hominoids/hominids), occurred. [p.460]

24. A survivor of hominoid extinction led to the great apes and the (chimpanzees/hominids). [p.460]

25. *Australopithecus afarensis*, an early hominid, is nicknamed (Lucy/Linus). [p.460]

26. *Homo habilis* (did/did not) use tools. [p.460]

27. The first *Homo* species to leave Africa was *Homo* (erectus/sapiens). [p.460]

28. Neandertal man was present in Europe and the Near East until about (30,000/100,000) years ago. [p.460]

29. The (multiregional/African emergence) model says that humans evolved in Africa and replaced *Homo erectus* in other areas. [p.461]

30. The (multiregional/African emergence) model says that humans evolved from different dispersed populations of *Homo erectus*. [p.461]

31. Since the emergence of modern humans, evolution has been almost entirely (biological/cultural). [p.461]

24.11. EARTH'S HISTORY AND THE ORIGIN OF LIFE [pp.462–463]

Selected Words: "molecule of life" [p.463]

Boldfaced, Page-Referenced Term

[p.463] chemical evolution_____

_____

True/False

If the statement is true, write a "T" in the blank. If the statement is false, make it correct by writing the word(s) in the blank that should take the place of the underlined word(s).

_____ 1. The first atmosphere <u>contained</u> gaseous oxygen. [p.462]

_____ 2. Liquid <u>water</u> was essential for the formation of cell membranes. [p.462]

_____ 3. Without an <u>oxygen-rich</u> atmosphere, the organic compounds that led to life would never have formed. [p.462]

_____ 4. The first living cells emerged around <u>2.5 billion</u> years ago. [p.463]

_____ 5. The energy to drive chemical reactions that yielded organic molecules might have come from <u>aerobic respiration</u>. [p.463]

_____ 6. Complex compounds may have formed on <u>clay layers</u>. [p.463]

_____ 7. Complex compounds might have formed at <u>deep sea vents</u>. [p.463]

_____ 8. The first "molecule of life" was possibly not DNA but <u>ATP</u>. [p.463]

_____ 9. Membrane-bound sacs resembling cell membranes <u>have</u> been formed in laboratories. [p.463]

Self-Quiz

Are you ready for the exam? Test yourself on key concepts by taking the additional tests linked with your BiologyNow CD-ROM.

___ 1. The term _____ refers to cumulative genetic changes that give rise to new species. [p.449]
   a. macroevolution
   b. mutation
   c. microevolution
   d. variation

___ 2. _____ applies to the large-scale patterns, trends, and rates of change among groups of species. [p.449]
   a. Macroevolution
   b. Natural selection
   c. Microevolution
   d. Variation

___ 3. Evolution occurs when there is change in the genetic makeup of _____. [p.449]
   a. individuals
   b. gametes
   c. populations
   d. ecosystems

___ 4. The idea of natural selection was the principal contribution of _____. [p.448]
   a. Fox
   b. Oparin
   c. Miller
   d. Darwin

___ 5. When the relative numbers of different alleles in a gene pool change randomly through the generations because of chance events alone, that change is called _____. [p.450]
   a. natural selection
   b. genetic drift
   c. gene flow
   d. mutation

___ 6. The source of new alleles for natural selection to work with is _____. [p.450]
   a. natural selection
   b. genetic drift
   c. gene flow
   d. mutation

___ 7. In the model known as _____, new species emerge through many small changes in form over long expanses of time. [p.451]
   a. reproductive isolation
   b. punctuated equilibrium
   c. gradualism
   d. comparative evolution

____ 8. Divergence may be the first stage on the road to _____, the process by which species originate. [p.451]
   a. natural selection
   b. speciation
   c. the founder effect
   d. genetic variation

____ 9. Related species remain alike in many ways; these shared but diverged characteristics of a common genetic plan are _____. [p.454]
   a. analogous
   b. vestigial
   c. homologous
   d. adaptive

____10. The physical record for the long history of life comes from _____. [p.452]
   a. comparative morphology
   b. fossils
   c. comparative embryology
   d. comparative biochemistry

____11. Modern studies of _____ indicate that all present-day continents were once part of the supercontinent called Pangaea. [p.453]
   a. adaptive radiation
   b. plate tectonics
   c. meteor impact
   d. mass extinction

____12. Many new mammal species arose and moved into habitats vacated by dinosaurs during a(n) _____. [pp.456–457]
   a. adaptive radiation
   b. bottleneck
   c. meteor impact
   d. reverse extinction

____13. When gene flow between two populations stops, it leads to _____. [p.451]
   a. adaptive radiation
   b. reproductive isolation
   c. punctuated equilibrium
   d. natural selection

____14. Which of the following is the most ancient of our genus? [pp.460–461]
   a. *Australopithecus afarensis*
   b. *Homo erectus*
   c. *Homo sapiens*
   d. *Homo habilis*

____15. What two molecules that were not present on early Earth are essential for the beginning of life? [p.462]
   a. H_2 and N_2
   b. O_2 and N_2
   c. H_2O and CO_2
   d. O_2 and H_2O

Chapter Objectives/Review Questions

This section lists general and detailed chapter objectives that can be used as review questions. You can make maximum use of these items by writing answers on a separate sheet of paper. Fill in answers where blanks are provided. To check for accuracy, compare your answers with information given in the chapter or glossary.

1. _____ refers to genetic changes in lines of descent over time. [p.448]
2. Briefly describe Charles Darwin's early life and his contribution to evolutionary theory. [p.448]
3. Contrast the definitions of *microevolution* and *macroevolution*. [p.449]
4. List the three types of traits possessed by members of a population. [p.449]
5. Relate the term "gene pool" to the definition of a population. [p.449]
6. _____ are the source of new alleles, hence of the heritable variation in traits. [p.450]
7. Define *natural selection*. [p.450]
8. Define each of the following evolutionary forces: genetic drift, gene flow, reproductive isolating mechanisms. [pp.450–451]
9. The occurrence of chance events in bringing about changes in allele frequencies in small populations is known as _____ _____. [p.450]
10. Name the unit whose individuals can interbreed under natural conditions and produce fertile offspring. [p.451]

11. Why does the explanation of life's history include both the punctuated equilibrium model and the gradualism model? [p.451]
12. Briefly define and cite an example of each of the following lines of evidence supporting macroevolution: the fossil record, biogeography, comparative morphology, homologous and analogous structures, comparative embryology, vestigial structures, and comparative biochemistry. [pp.452–456]
13. A(n) _____ _____ is an abrupt, widespread rise in extinction rates above the background extinction level. [p.456]
14. In a(n) _____ _____, new species of a lineage fill a wide range of habitats during bursts of microevolutionary activity. [p.456]
15. Name and briefly discuss the five trends in human evolution. [pp.458–459]
16. Define *culture*. [p.459]
17. Distinguish between hominoids and hominids. [p.460]
18. Briefly discuss the species of genus Homo that exist in the fossil record, as well as the two ideas for the evolution of modern-day human populations. [pp.460–461]
19. Describe the conditions of early Earth, and explain how they contributed to the origin of life. [p.462]
20. Which possible aspects of the evolution of life have been demonstrated in the lab? [p.463]

Media Menu Review Questions

Questions 1–2 are drawn from the following InfoTrac College Edition article: "Species: Life's Mystery Packages." Jessica Ruvinsky. *U.S. News & World Report*, July 29, 2002.

1. _____ isolation is a good bet for the evolution of new species.
2. _____ that give fruit flies different taste preferences may serve as an isolation mechanism.

Questions 3–5 are drawn from the following InfoTrac College Edition article: "Unlocking the Mystery of Extinction." William Mullen. *Knight Ridder/Tribune News Service*, December 13, 2002.

3. The cause of a mass extinction 206 million years ago may have been a buildup of _____.
4. Jennifer McElwain counted tiny pores (stomata) on the surfaces of ancient _____ and found evidence of a surge in carbon dioxide through the die-off period.
5. The CO_2 level in Earth's atmosphere in 1900 was 300 parts per million (ppm). It is now 360 ppm and will be _____ ppm by the end of the century.

Questions 6–10 are drawn from the following InfoTrac College Edition article: "Fossils Plug Gap in Human Origins." B. Bower. *Science News*, June 14, 2003.

6. Three partial skulls excavated in eastern _____ represent the oldest known fossils of modern people.
7. The Herto fossils show that *Homo sapiens* evolved in Africa independently of _____.
8. Tim White's finds bolster the _____ theory of human evolution.
9. The age estimate of the skulls was generated from measurements of argon gas trapped in _____ above and below the finds.
10. Stone-tool incisions on the _____ probably resulted from removal of flesh after death and other mortuary practices.

Integrating and Applying Key Concepts

1. Imagine that in the next decade three more Chernobyl-type disasters happen, the oceans acquire critical levels of carcinogenic pesticides that work their way up the food chains, and the ozone layer shrinks dramatically in the upper atmosphere. Describe the macroevolutionary events that you believe might follow.
2. As Earth becomes increasingly loaded with carbon dioxide and various industrial waste products, how do you think life forms on Earth will evolve to cope with these changes?

25

ECOLOGY AND HUMAN CONCERNS

Interactive Exercises

Impacts, Issues: The Human Touch [p.467]

25.1. SOME BASIC PRINCIPLES OF ECOLOGY [pp.468–469]

25.2. FEEDING LEVELS AND FOOD WEBS [p.470]

25.3. HOW ENERGY FLOWS THROUGH ECOSYSTEMS [p.472]

For additional practice, use the interactive vocabulary exercises linked with your BiologyNow CD-ROM.

Selected Words: biosphere [p.468], "disturbed" habitats [p.468], "ecological niche" [p.468], *specialist* species [p.468], *generalist* species [p.468], *biotic* [p.468], "climax community" [p.468], *primary succession* [p.468], *secondary succession* [p.468], *primary, secondary,* and *tertiary* consumers [p.470], *trophic levels* [p.470]

Boldfaced, Page-Referenced Terms

[p.468] ecology _____

_____

[p.468] habitat _____

_____

[p.468] community _____

_____

[p.468] niche _____

_____

[p.468] ecosystem _____

_____

[p.468] succession _____

_____

[p.469] producers _____

_____

[p.469] consumers _____

_____

[p.469] herbivores _____

_____

[p.469] carnivores _____

_____

[p.469] omnivores _____

_____

[p.469] decomposers _____

_____

[p.469] food chain _____

_____

[p.469] food webs _____

_____

[p.472] primary productivity _____

_____

[p.472] ecological pyramid _____

_____

Short Answer

1. What is the lesson that ecologists suggest that we learn from Easter Island? [p.468]

_____

_____

Matching

Choose the most appropriate description for each term.

2. _____ biosphere [p.468]

3. _____ ecology [p.468]

4. _____ habitat [p.468]

5. _____ community [p.468]

6. _____ niche [p.468]

7. _____ specialist species [p.468]

8. _____ generalist species [p.468]

9. _____ ecosystem [p.468]

10. _____ biotic [p.468]

11. _____ succession [p.468]

A. In any given habitat, the populations of all species that associate directly or indirectly
B. A community of organisms interacting with one another and with the physical environment
C. Consists of all the physical, chemical, and biological conditions the species requires to live and reproduce in an ecosystem
D. An orderly progression of species replacement in a community until a stable climax community is reached
E. The general type of place where a species normally lives
F. Have broad niches; can live in a range of habitats and eat many types of food
G. Have narrow niches; may be able to use only one or a few types of food or live only in one type of habitat
H. The regions of the Earth's crust, waters, and atmosphere in which organisms live
I. Refers to the living portions of an ecosystem
J. The study of the interactions of organisms with one another and with the physical environment

Choice

For questions 12–15, choose from the following:

a. primary succession   b. secondary succession

12. _____ Successional changes begin when a pioneer species colonizes a newly available habitat. [p.468]

13. _____ A community develops toward the climax state after parts of the habitat have been disturbed. [p.468]

14. _____ This pattern occurs in abandoned fields in which wild grasses and other plants quickly take hold when cultivation stops. [p.468]

15. _____ Might occur in a recently deglaciated region. [p.468]

Fill-in-the-Blanks

Nearly every ecosystem runs on energy from the (16) _____ [p.470]. Plants and other photosyn-

thetic organisms are (17) _____ [p.470] or "self-feeders." All other organisms in the system are

(18) _____ [p.470] that depend directly or indirectly on energy stored in the tissues of producers.

Consumers are "other-feeders" or (19) _____ [p.470]. Within this group are (20) _____

[p.470], such as grazing animals and insects that eat plants; (21) _____ [p.464] such as lions and snakes, which eat animals; (22) _____ [p.470] such as humans and bears that feed on a variety of either plant or animal foods; and (23) _____ [p.470] such as fungi, bacteria, and worms that get energy from the remains or products of organisms. Producers obtain an ecosystem's nutrients and its initial pool of (24) _____ [p.470]. The water, carbon dioxide, and other minerals are used in the ecosystem and ultimately broken down by (25) _____ [p.470], then reused by the producers.

Choice

For questions 26–32, choose from the following:

a. primary producer   b. herbivore   c. primary carnivore   d. secondary carnivore   e. decomposer

26. _____ gain energy directly from sunlight [p.470]

27. _____ a hawk that eats a snake [p.470]

28. _____ fungi and bacteria [p.470]

29. _____ organisms that prey on herbivores [p.470]

30. _____ green plants [p.470]

31. _____ autotrophs [p.470]

32. _____ snails, grasshoppers, and other plant-eaters [p.470]

Dichotomous Choice

Circle one of the two possible answers given between parentheses in each statement.

33. Minerals carried by erosion into a lake represent nutrient (input/output). [p.470]

34. Ecosystems require a continual energy input from the (sun/heterotrophs). [p.470]

35. Nutrients typically (are/are not) recycled. [p.470]

36. Each species in an ecosystem fits somewhere in a hierarchy of feeding relationships called (niche/trophic) levels. [p.470]

37. A linear sequence of who eats whom in an ecosystem is sometimes called a food (chain/web). [p.470]

38. The amount of energy actually stored in an ecosystem depends on how many (plants/animals) are present, and on the balance between energy trapped by photosynthesis and energy used by the plants. [p.472]

39. In a harsh ecosystem environment, productivity would be expected to be (lower/higher) than in a mild ecosystem environment. [p.472]

40. In an ecological pyramid, the primary (producers/consumers) form a base for successive tiers of (producers/consumers) above them. [p.472]

41. A(n) (biomass/energy) pyramid depicts the weight of all an ecosystem's organisms. [p.472]

42. The amount of available energy (increases/decreases) through successive feeding levels of an ecosystem. [p.472]

25.4. CHEMICAL CYCLES—AN OVERVIEW [p.473]

25.5. THE WATER CYCLE [p.474]

25.6. CYCLING CHEMICALS FROM THE EARTH'S CRUST [p.475]

Selected Words: watershed [p.474], *eutrophication* [p.475], "blooms" [p.475]

Boldfaced, Page-Referenced Terms

[p.473] biogeochemical cycle _____

_____

[p.474] water cycle _____

_____

[p.475] phosphorous cycle _____

_____

Analyzing Diagrams

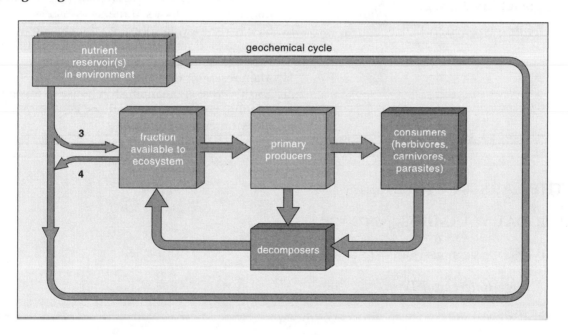

1. _____ The diagram represents the movement of nutrients from the physical environment, through organisms, and back to the environment. What is this known as? [p.473]

2. _____ The amount of nutrient that enters or leaves an ecosystem is far less than what cycles through the ecosystem. What component of the diagram is responsible for this internal cycling? [p.473]

3. _____ In addition to rainfall and snowfall, what else is included in the arrow labeled "3"? [p.473]

4. _____ Give an example of what accounts for the arrow labeled "4." [p.473]

5. _____ Which component of the diagram determines the type of biogeochemical cycle? [p.473]

6. _____ In the water cycle, water is the reservoir of what two substances? [p.473]

   _____

7. _____ In the atmospheric cycle, the reservoir of carbon and nitrogen is in what form? [p.473]

8. _____ What is the major storehouse of nongaseous nutrients such as phosphorus? [p.473]

Matching

Choose the most appropriate phrase or description to match the following.

9. _____ solar energy [p.474]

10. _____ ocean [p.474]

11. _____ watershed [p.474]

12. _____ forms of precipitation [p.474]

13. _____ important role of plants [p.474]

14. _____ forms of atmospheric water [p.474]

15. _____ phosphorus [p.475]

16. _____ uplift and drainage [p.475]

17. _____ eutrophication [p.475]

A. Mostly rain and snow
B. Nutrient enrichment of a body of water resulting in algal blooms and other negative consequences, accelerated by runoff carrying phosphorus into water bodies
C. Take up minerals dissolved by water and prevent their loss in runoff
D. Water vapor, clouds, and ice crystals
E. Where precipitation of a specified region becomes funneled into a single stream or river
F. Process occurring as crustal plates move that makes minerals in seafloor sediments available
G. Slowly drives water through the atmosphere, on or through land mass surface layers, to oceans, and back again
H. Main reservoir of water
I. Earth's crust is the main storehouse of this mineral, calcium, potassium, and others; key component of many organic compounds

25.7. THE CARBON CYCLE [pp.476–477]

25.8. GLOBAL WARMING [pp.478–479]

Selected Words: "greenhouse gases" [p.478]

Boldfaced, Page-Referenced Terms

[p.476] carbon cycle _____

_____

[p.478] greenhouse effect _____

_____

[p.478] global warming _____

_____

Matching

Choose the most appropriate description for each term/phrase.

1. _____ greenhouse gases [p.478]
2. _____ carbon cycle [p.476]
3. _____ how carbon enters the atmosphere [p.476]
4. _____ greenhouse effect [p.478]
5. _____ carbon dioxide (CO_2) [p.476]
6. _____ sediments, gas, petroleum, and coal [p.477]

A. Form of most of the atmospheric carbon
B. Aerobic respiration, fossil fuel burning, and volcanic eruptions
C. CO_2, H_2O, ozone, CFCs, methane, nitrous oxide
D. Buried reservoirs of carbon
E. Carbon moves from reservoirs in the atmosphere and oceans, through organisms, then back
F. Warming of Earth's lower atmosphere due to accumulation of greenhouse gases

Fill-in-the-Blanks

The (7) _____ _____ [p.478] refers to the warming of Earth's surface by (8) _____ [p.478] energy trapped in the lower atmosphere by "greenhouse gases." Human activity has resulted in the high levels of these gases, which is most likely contributing to (9) _____ _____ [p.478]. A warmer planet would mean that (10) _____ [p.478] level would rise so that waterfronts of major coastal cities would be submerged. Regional patterns of (11) _____ [p.478] would be disturbed, reducing crop yields in currently productive regions.

Scientists have found evidence that the atmospheric concentration of (12) _____ _____ [p.479] is increasing. In 2001, the United Nations announced that global warming and long-term (13) _____ [p.479] change are real. Data suggests that we are past the point of being able to (14) _____ [p.479] some impacts of climate change. Unfortunately, the subject of global warming has become a controversial (15) _____ [p.479] topic. The burning of (16) _____ _____ [p.479] is probably the biggest contributor to global warming.

25.9. THE NITROGEN CYCLE [p.480]

25.10. THE DANGER OF BIOLOGICAL MAGNIFICATION [p.481]

Selected Words: "denitrification" [p.480], *Silent Spring* [p.481]

Boldfaced, Page-Referenced Terms

[p.480] nitrogen cycle _____

_____

[p.480] nitrogen fixation _____

_____

[p.480] nitrification _____

_____

[p.481] biological magnification _____

_____

Choice

For questions 1–8, choose from the following:

a. nitrogen cycle   b. nitrogen fixation   c. nitrogen   d. nitrification   e. denitrification

1. _____ Ammonia and ammonium in soil are converted to nitrite (NO_2); other bacteria metabolize nitrite and convert nitrite to nitrate (NO_3). [p.480]

2. _____ Bacteria are its key organisms—they convert nitrogen to forms plants can use, and also release nitrogen to complete the cycle. [p.480]

3. _____ Component of proteins and nucleic acids [p.480]

4. _____ Atmospheric cycle; only certain bacteria can break the bonds of N_2 and put nitrogen into forms that can enter food webs [p.480]

5. _____ Process in which a few kinds of bacteria convert N_2 to ammonia (NH_3), which dissolves quickly in water to form ammonium (NH_4^+) [p.480]

6. _____ Gaseous form makes up 80 percent of the atmosphere [p.480]

7. _____ Plants called legumes have mutually beneficial associations with bacteria that carry out this process. [p.480]

8. _____ Bacteria convert nitrate or nitrite to N_2 (and a bit of nitrous oxide) that escapes into the atmosphere. [p.480]

Fill-in-the-Blanks

During World War II, DDT was sprayed in the tropical Pacific to control (9) _____ [p.481] responsible for transmitting the organisms that cause dangerous malaria. In Europe, it was used to control (10) _____ _____ [p.481] that transmitted typhus. After the war, the use of DDT continued in many applications. DDT spreads through wind and water, and accumulates in the fatty (11) _____ [p.481] of organisms. DDT can show (12) _____ _____ [p.481] in which there is an increase in the concentration of a nondegradable substance in organisms as it is passed through (13) _____ _____ [p.481]. As DDT spread through the environment after the war, it had disastrous effects. Where it was used to control Dutch elm disease, (14) _____ [p.481] began dying, as well as (15) _____ [p.481] in the streams flowing through the forests. In croplands where DDT was used to control one kind of pest, new kinds of (16) _____ [p.481] moved in. In addition, DDT was killing off the natural (17) _____ [p.481] that kept pest insect populations in check. Then, through biological magnification, species at the ends of (18) _____ _____ [p.481] were devastated, including bald eagles, peregrine falcons, ospreys, and brown pelicans. One consequence was that birds produced eggs with (19) _____ _____ [p.481] so that chick embryos didn't hatch, placing some species on the brink of (20) _____ [p.481]. The work of Rachel Carson about the harmful effects of widespread, unregulated (21) _____ [p.481] use became the impetus for the environmental movement in the United States. DDT is now (22) _____ [p.481] in the United States, but DDT from 20 years ago is still contaminating parts of the (23) _____ [p.481].

25.11. HUMAN POPULATION GROWTH [pp.482–483]

25.12. NATURE'S CONTROLS ON POPULATION GROWTH [p.484]

Selected Words: bubonic plague [p.484], *Yersinia pestis* [p.484]

Boldfaced, Page-Referenced Terms

[p.482] total fertility rate (TFR) _____

_____

[p.483] demographics _____

_____

[p.483] population density _____

_____

[p.483] age structure _____

_____

[p.483] reproductive base _____

_____

[p.484] carrying capacity _____

_____

[p.484] logistic growth _____

_____

[p.484] density-dependent controls _____

_____

Dichotomous Choice

Circle one of the two possible answers given between parentheses in each statement.

1. The world population is expected to reach (six/nine) billion by the year 2050. [p.482]
2. When death rate exceeds birth rate, population size (stabilizes/decreases). [p.482]
3. Overall on Earth, (birth/death) rates have been coming down; (birth/death) rates have been falling, too. [p.482]
4. The country expected to show the *most* population growth is (China/India). [p.482]
5. In 2003, the total fertility rate was (below/above) replacement level fertility. [p.482]
6. TFRs are at or below replacement levels for many (third world/industrialized) countries. [p.482]
7. Even if every couple decides to bear no more than two children, the world population will keep growing for (twenty/sixty) years. [p.482]
8. A population's vital statistics—size, age structure, density—are its (national base/demographics). [p.483]
9. The total number of individuals in a given area is population (density/distribution). [p.483]
10. Dividing a population into prereproductive, reproductive, and postreproductive categories characterizes its (age/distribution) structure. [p.483]
11. The reproductive (rate/base) of a population refers to the number of individuals in the prereproductive and reproductive age structure categories. [p.483]

12. The current United States population is growing (faster/slower) than in preceding years. [p.483]

13. The human population has been experiencing (exponential/logistic) growth since the mid-1700s. [p.484]

14. When the course of logistic growth is plotted on a graph, a(n) (J-shaped/S-shaped) curve is obtained. [p.484]

15. Infectious diseases and parasites are a bigger problem in (low-density/high-density) populations. [p.484]

Matching

Select the best description for each term.

16. _____ density-independent controls [p.484]

17. _____ carrying capacity [p.484]

18. _____ limiting factor [p.484]

19. _____ logistic growth [p.484]

20. _____ density-dependent controls [p.484]

21. _____ population size [p.483]

22. _____ exponential growth [p.484]

A. A growth pattern in which a population grows slowly at first, then increases rapidly, but levels off when carrying capacity is reached
B. An essential resource such as food or water that, when in short supply, affects population growth
C. Number of individuals in a population's gene pool
D. An example is overcrowding that results in competition for resources
E. Events such as natural disasters that cause deaths or births regardless of whether crowding exists
F. The number of individuals of a given species that can be sustained indefinitely by the resources in a given area
G. Population growth in doubling increments

Graph Construction and Interpretation

Year	Estimated World Population
1650	500,000,000
1850	1,000,000,000
1930	2,000,000,000
1987	4,000,000,000
1975	5,000,000,000
1993	5,500,000,000
1995	5,700,000,000
1997	5,800,000,000
2000	6,000,000,000

23. Construct a graph of the data in the space provided on the next page.

   a. Estimate the year that the world contained 3 billion humans. _____ [p.482]

   b. Estimate the year that Earth will house 9 billion humans. _____ [p.82]

   c. Do you expect Earth to house 9 billion humans within your lifetime? [Check one] ☐ Yes ☐ No [p.482]

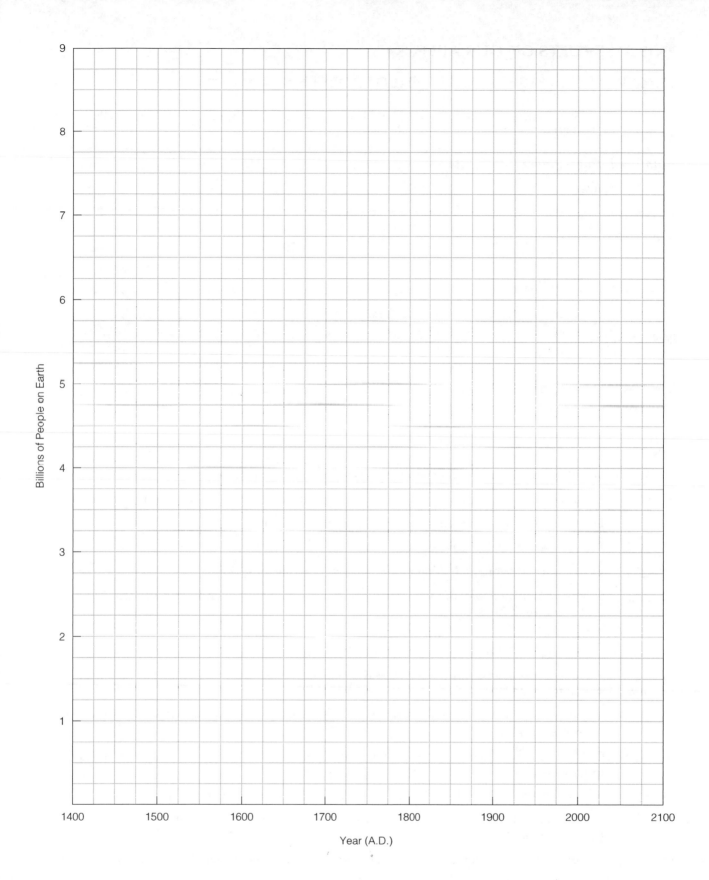

25.13. ASSAULTS ON OUR AIR [p.485]

25.14. WATER, WASTES, AND OTHER PROBLEMS [pp.486–487]

Selected Words: ozone [p.485], ozone "hole" [p.485], *salinization* [p.486], "green revolution" [p.486], *desertification* [p.487]

Boldfaced, Page-Referenced Terms

[p.485] pollutants _____

_____

[p.485] industrial smog _____

_____

[p.485] photochemical smog _____

_____

[p.485] acid rain _____

_____

[p.487] deforestation _____

_____

Choice

For questions 1–12, choose from the following aspects of atmospheric pollution.

a. chlorofluorocarbons   b. industrial smog   c. photochemical smog   d. acid rain   e. ozone layer

1. _____ Develops as a brown, smelly haze over large cities [p.485]

2. _____ Precipitation containing weak sulfuric and nitric acids [p.485]

3. _____ Contributes to ozone reduction more than any other factor [p.485]

4. _____ Where winters are cold and wet, this develops as a gray haze over many cities that burn coal and other fossil fuels. [p.485]

5. _____ Substances from power plants and factories as well as from motor vehicles combine with water, producing a solution with a pH as low as 2.3. [p.485]

6. _____ Each year, from September through mid-October, it thins down by as much as half above Antarctica. [p.485]

7. _____ Today most of this forms in cities of China, India, eastern Europe, and other developing countries. [p.485]

8. _____ Contains airborne pollutants, including dust, smoke, soot, ashes, asbestos, oil, bits of lead and other heavy metals, and sulfur dioxide [p.485]

9. _____ Reduction allows more harmful ultraviolet radiation to reach Earth's surface [p.485]

10. _____ Nitric oxide released from vehicles is exposed to sunlight and reacts with hydrocarbons to form harmful oxidants. [p.485]

11. _____ Widely used as refrigerator coolants, air conditioners, industrial solvents, and plastic foams; enter the atmosphere slowly and resist breakdown [p.485]

12. _____ It will take 100 to 200 years to recover after its main threats are banned. [p.485]

Fill-in-the-Blanks

There is a tremendous amount of water in the world, but most is too (13) _____ [p.486] for human consumption. Today, we produce about one-third of our food supply on land that is (14) _____ [p.486]. Irrigation water is often loaded with (15) _____ [p.486] salts. In regions with poorly draining soils, evaporation may cause buildup, or (16) _____ [p.486]; such soils can stunt growth, decrease yields, and in time kill crop plants. Large-scale (17) _____ [p.486] is also a major factor in groundwater depletion. Farmers take so much water from the Ogallala (18) _____ [p.486] that the overdraft nearly equals the annual flow of the Colorado River. In coastal areas, overuse of groundwater can cause (19) _____ [p.486] intrusion into human water supplies. (20) _____ [p.486] runoff pollutes public water sources with sediments, pesticides, and fertilizers. Power plants and (21) _____ [p.486] pollute water with chemicals, radioactive materials, and excess heat. Many people view the (22) _____ [p.486] as convenient refuse dumps and dump untreated wastes there.

Matching

Choose the most appropriate description for each term.

23. _____ landfills [p.486]

24. _____ green revolution [p.486]

25. _____ deforestation [p.487]

26. _____ recycling [p.486]

27. _____ forested watersheds [p.487]

28. _____ desertification [p.487]

A. More people are participating in this beneficial alternative to landfills

B. Act like giant sponges that absorb, hold, and gradually release water

C. Conversion of large tracts of grasslands, rain-fed cropland, or irrigated cropland to a less productive, more desertlike state

D. Burying wastes in areas that may leak into groundwater supplies

E. Research directed toward improving crop plants for higher yields and exporting modern agricultural practices and equipment to developing countries

F. Removal of all trees from large land tracts; leads to loss of fragile soils and disrupts watersheds

25.15. CONCERNS ABOUT ENERGY [p.488]

25.16. A PLANETARY EMERGENCY: LOSS OF BIODIVERSITY [p.489]

Selected Words: "nuclear power revolution" [p.488], "hybrid" vehicles [p.488]

Boldfaced, Page-Referenced Term

[p.488] fossil fuels _____

_____

Fill-in-the Blanks

(1) _____ [p.488] energy is the energy left over after subtracting the energy used to locate, extract, transport, store, and deliver energy to consumers. Some sources of energy, such as direct (2) _____ [p.488] energy, are renewable. Most of the energy used in (3) _____ [p.488] countries comes from nonrenewable resources. Energy supplies in the form of (4) _____ [p.488] fuels are nonrenewable and dwindling, and their extraction and use come at high environmental cost.

Dichotomous Choice

Circle one of the two possible answers given between parentheses in each statement.

5. Paralleling human population growth is a steep rise in total and per capita energy (conservation/ consumption). [p.488]

6. World coal reserves can meet human energy needs for several centuries, but burning it has been the main source of (ozone depletion/air pollution). [p.488]

7. Compared to coal-burning plants, a nuclear plant emits (less/more) radioactivity and pollutants. [p.488]

8. The greatest hazard in the use of nuclear energy as an energy supply during normal operation is with potential (radioactivity escape/meltdown). [p.488]

9. Use of low-cost (wind harnesses/solar cells) can use sunlight energy to generate electricity and produce hydrogen gas. [p.488]

10. Another energy alternative is (fusion power/wind farms) in which the idea is to bring hydrogen atoms together to form helium atoms, a reaction that releases considerable energy. [p.488]

11. Human actions are leading to the premature extinction of at least six species per (hour/year). [p.489]

12. Globally, (pollution/tropical deforestation) is the greatest killer of species, followed by destruction of coral reefs. [p.489]

13. In the United States, we have cut down 92 percent of old-growth forests and drained (a fourth/half) of the wetlands. [p.489]

14. In (less/more) affluent countries, people use greater than the average amount of resources. [p.489]

15. In (less/more) affluent countries, there is overuse of land and animal resources. [p.489]

Self-Quiz

Are you ready for the exam? Test yourself on key concepts by taking the additional tests linked with your BiologyNow CD-ROM.

____ 1. All the populations of different species that occupy and are adapted to a given habitat are referred to as a(n) _____. [p.468]
  a. biosphere
  b. community
  c. ecosystem
  d. niche

____ 2. The _____ of a species consists of all the physical, chemical, and biological conditions that the species needs to live and reproduce in an ecosystem. [p.468]
  a. habitat
  b. niche
  c. carrying capacity
  d. ecosystem

_____ 3. A network of interactions that involve the cycling of materials and the flow of energy between one or more communities and the physical environment is a(n) _____. [p.468]
   a. population
   b. community
   c. ecosystem
   d. biosphere

_____ 4. _____ get energy from the remains or products of organisms. [p.470]
   a. Herbivores
   b. Parasites
   c. Decomposers
   d. Carnivores

_____ 5. A linear sequence of who eats whom in an ecosystem is sometimes called a(n) _____. [p.470]
   a. trophic level
   b. food chain
   c. ecological pyramid
   d. food web

_____ 6. In a typical food web, the second trophic level includes primary consumers that are _____. [p.470]
   a. herbivores
   b. carnivores
   c. scavengers
   d. decomposers

_____ 7. Succession involves the replacement of species by others as each changes the habitat, finally resulting in a more or less stable _____. [p.468]
   a. climax community
   b. pioneer community
   c. secondary community
   d. pyramid community

_____ 8. A(n) _____ is a way to represent the relative amounts of energy available at each level of an ecosystem. [p.472]
   a. food chain
   b. food web
   c. productivity map
   d. ecological pyramid

_____ 9. In the carbon cycle, carbon enters the atmosphere through _____. [p.476]
   a. carbon dioxide fixation
   b. aerobic respiration, fossil fuel burning, and volcanic eruptions
   c. oceans and accumulation of plant biomass
   d. evaporation

_____10. _____ is the name for the process in which certain bacteria convert N_2 to ammonia, which dissolves to form ammonium, a form that plants can pull into an ecosystem. [p.480]
   a. Nitrification
   b. Ammonification
   c. Denitrification
   d. Nitrogen fixation

_____11. _____ refers to an increase in concentration of a nondegradable (or slowly degradable) substance in organisms as it is passed along food chains. [p.481]
   a. Ecosystem modeling
   b. Nutrient input
   c. Biogeochemical cycle
   d. Biological magnification

_____12. The average number of individuals of the same species per unit area at a given time is the _____. [p.483]
   a. population density
   b. population growth
   c. population birth rate
   d. population size

_____13. The maximum number of individuals of a population (or species) that can be sustained by a given environment defines _____. [p.484]
   a. the carrying capacity
   b. exponential growth
   c. logistic growth
   d. density-independent factors

____14. Which of the following is NOT
characteristic of logistic growth? [p.484]
   a. S-shaped curve
   b. growth levels off as carrying capacity is
      reached
   c. unrestricted growth
   d. slow growth of a low-density
      population followed by rapid growth

____15. Factors that are at work in crowded
populations, and tend to reduce
population size, are called _____.
[p.484]
   a. density-independent controls
   b. density-dependent controls
   c. limiting factors
   d. logistic factors

____16. _____ result(s) when nitrogen
dioxide and hydrocarbons react in the
presence of sunlight. [p.485]
   a. Photochemical smog
   b. Industrial smog
   c. A thermal inversion
   d. Eutrophication

____17. Sulfur and nitrogen dioxides dissolve in
atmospheric water to form a weak solution
of sulfuric acid and nitric acid; this
describes the formation of _____.
[p.485]
   a. photochemical smog
   b. industrial smog
   c. ozone and PANs
   d. acid rain

____18. For every million liters of water in the
world, only about _____ liters
are in a form that can be used for human
consumption or agriculture. [p.486]
   a. 6
   b. 60
   c. 600
   d. 6,000

____19. In theory, world reserves of _____
can meet the energy needs of the human
population for at least several centuries.
[p.488]
   a. carbon monoxide
   b. oil
   c. natural gas
   d. coal

____20. Many biologists believe that extinction is a
more serious problem than ozone
depletion or global warming because it is
happening faster and is _____.
[p.489]
   a. destroying food webs
   b. reducing food supplies for humans
   c. irreversible
   d. causing succession

Chapter Objectives/Review Questions

This section lists general and detailed chapter objectives that can be used as review questions. You can
make maximum use of these items by writing answers on a separate sheet of paper. Fill in answers where
blanks are provided. To check for accuracy, compare your answers with information given in the chapter or
glossary.

1. Define and distinguish between *habitat* and *niche*. [p.468]
2. Distinguish between *primary succession* and *secondary succession*. [p.468]
3. Name and distinguish among the four types of consumers in food webs. [p.470]
4. Biomass pyramids depict the _____ of all the ecosystem's members at each trophic level;
   _____ pyramids reflect the energy losses as energy flows through the trophic levels of an
   ecosystem. [p.472]
5. Give examples of biogeochemical cycles, and what the reservoir for each is. [pp.473–477]

6. In the _____ cycle, minerals move from land to sediments in the seas, and then back to the land. [p.475]
7. Explain the greenhouse effect and the implications this has for life on Earth. [pp.478–479]
8. Define the chemical events that occur during nitrogen fixation, nitrification, ammonification, and denitrification. [p.480]
9. Explain the detrimental effects of biological magnification. [p.481]
10. Define *zero population growth,* and explain why it is attained more often in industrialized countries. [p.482]
11. Define the three stages of a population's age structure. [p.483]
12. Distinguish between exponential growth and logistic growth. [p.484]
13. Define limiting factors and tell how they influence carrying capacity. [p.484]
14. Define *density-dependent controls*, give two examples, and indicate how density-dependent factors act on populations. [p.484]
15. Define *density-independent controls* and give two examples. [p.484]
16. Identify the principal air pollutants, their sources, and their effects. [p.485]
17. Distinguish between industrial smog and photochemical smog. [p.485]
18. Describe the environmental conditions leading to acid rain, and list its negative effects. [p.485]
19. Summarize the major problems contributing to contamination of the world's supply of pure water. [p.486]
20. Describe the process of desertification and its effects. [p.487]
21. Explain the repercussions of deforestation. [p.487]
22. List the disadvantages of the use of fossil fuels as energy sources. [p.488]
23. Explain what a meltdown is. [p.488]
24. Describe how our use of fossil fuels, solar energy, and nuclear energy affects ecosystems. [p.488]
25. Cite major reasons for the continuing loss of biodiversity. [p.489]

Media Menu Review Questions

Questions 1–5 are drawn from the following InfoTrac College Edition article: "Mysterious Island: The More We Learn about Easter Island, the More It Intrigues, as a New Exhibition of Its Art Reminds Us." Paul Trachtman. *Smithsonian*, March 2002.

1. To early travelers, Easter Island's population was too _____, primitive, and isolated to be credited with carving the immense statues there.
2. Rongorongo is a form of _____ writing that has eluded every attempt to decipher it.
3. Veneration of "birdmen" replaced that of stone _____.
4. Legend says that after the stone "moai" were carved, they _____ to their final positions with the help of a chief or priest.
5. By 1877, there were only _____ natives left on Easter Island.

Questions 6–8 are drawn from the following InfoTrac College Edition article: "Spring Forward. Warmer Climates Accelerate Life Cycles of Plants, Animals." Sid Perkins. *Science News*, March 8, 2003.

6. Phenologists report that many creatures are beginning their annual cycles _____, as global temperatures rise; others are expanding their _____.
7. When intimately connected species in an _____ experience changes at different rates, food chains may be destroyed.
8. Analyses by Terry Root and her colleagues show than 80 percent of springtime life cycle changes are in the direction expected if _____ were the culprit.

Questions 9–10 are drawn from the following InfoTrac College Edition article: "Paradise Traded for Canadian Nickel." *Earth Island Journal*, Autumn 2002.

9. A Canadian company's plan to mine Goro Island for nickel risks acid rain and heavy metal contamination of _____.
10. Nickel is so toxic that the World Health Organization will not set a minimum human _____ level for it.

Integrating and Applying Key Concepts

1. Explain why some biologists believe that the endangered species list now includes all species.
2. Is there a *fundamental niche* that is occupied by humans? If you think so, describe the minimal abiotic and biotic conditions required by populations of humans in order to live and reproduce. (Note that "to thrive" and "to be happy" are not requirements.)

ANSWERS TO STUDENT INTERACTIVE WORKBOOK

Chapter 1 Learning About Human Biology

Impacts, Issues: What Am I Doing Here? [p.1]

1.1. The Characteristics of Life [p.2]
1. b; 2. a; 3. a; 4. c; 5. e; 6. e; 7. d; 8. b; 9. d; 10. f; 11. f; 12. a; 13. c; 14. f.

1.2. Where Do We Fit in the Natural World? [p.3]
1. animals; 2. vertebrates; 3. mammals; 4. primates; 5. humans; 6. Protistans, Plants, Fungi, Animals; 7. Archaebacteria, Eubacteria.

1.3. Life's Organization [pp.4–5]
1. D; 2. C; 3. E; 4. F; 5. G; 6. I; 7. J; 8. B; 9. H; 10. A; 11. E; 12. K; 13. I; 14. J; 15. C; 16. A; 17. F; 18. G; 19. H; 20. D; 21. B; 22. sun; 23. producers; 24. consumers; 25. decomposing; 26. web.

1.4. Science as a Way of Learning about the Natural World [pp.6–7]

1.5. Science in Action: Cancer, Broccoli, and Mighty Mice [p.8]
1. a. hypothesis; b. prediction; c. experiment; d. variable; e. control group; 2. F; 3. A; 4. D; 5. B; 6. C; 7. E; 8.E—question; 9.D—hypothesis; 10.B—prediction; 11.F—test; 12.C—repeat testing; 13.A—conclusion; 14. inductive; 15. deductive.

1.6. Science in Perspective [p.9]

1.7. Critical Thinking in Science and Life [p.10]

1.8. What Is the Truth about Herbal Food Substances? [p.11]
1. F; 2. F; 3. T; 4. F; 5. T; 6. T; 7. F; 8. correlated; 9. cause; 10. critical; 11. opinion; 12. fact; 13. b, d; 14. a, c, e.

Self-Quiz
1. b; 2. d; 3. d; 4. c; 5. a; 6. b; 7. d; 8. b; 9. c; 10. d.

Chapter Objectives/Review Questions
1. DNA; 3. homeostasis; 4. cell; 11. inductive, deductive.

Media Menu Review Questions
1. confirming, rationalize; 2. d; 3. a lot, succeed; 4. external; 5. confirmation bias; 6. check; 7. genetic code; 8. bioweapons; 9. biotechnology; 10. controls.

Chapter 2 Molecules of Life

Impacts, Issues: It's Elemental [p.15]

2.1. The Atom [p.16]

2.2. Saving Lives with Radioisotopes [p.17]
1. A; 2. B; 3. K; 4. M; 5. N; 6. I; 7. C; 8. O; 9. D; 10. G; 11. J; 12. E; 13. F; 14. H; 15. L; 16. a. hydrogen, H; b. carbon, C; c. nitrogen, N; d. oxygen, O; e. sodium, Na; f. chlorine, Cl; 17. atomic number; 18. mass numbers; 19. protons;

20. electrons; 21. neutrons.

2.3. What Is a Chemical Bond? [pp.18–19]

2.4. Important Bonds in Biological Molecules [pp.20–21]

2.5. Antioxidants—Fighting Free Radicals [p.22]
1. Both oxygen gas and water are molecules, since they consist of two or more atoms joined together. Only water is a compound, because the atoms are different elements.
2.

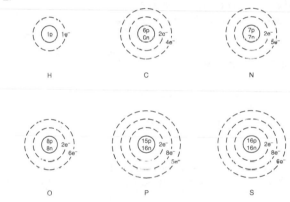

3. Helium will not form chemical bonds because it is *inert*, with no electron vacancies in its outer shell.
4. oxygen, carbon, hydrogen, and nitrogen; each has one or more electron vacancies in its outer shell
5.

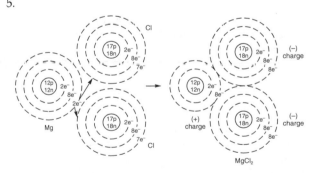

6. In a nonpolar covalent bond, atoms attract shared electrons equally. An example of a nonpolar covalent bond is that found in molecular hydrogen—H_2. In a polar covalent bond, atoms do not share electrons equally, so the bond is slightly positive at one end and slightly negative at the other. An example is water—H_2O.

7.

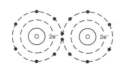

8. orbital; 9. mixture; 10. hydrogen; 11. chemical equations; 12. reactants; 13. products, molecules, and compounds; 14. balanced; 15. donated, accepted; 16. opposite; 17. covalent, polar; 18. covalent; 19. hydrogen; 20. Free radicals form by oxidation—the loss of one or more electrons to another atom. These highly unstable molecules have the potential to "steal" electrons from stable molecules, disrupting their structure and function. The effects of such damage to essential molecules in the body can be fatal; 21. An antioxidant is a chemical that can give up an electron to a free radical before it causes damage. Antioxidants in our diet include ascorbic acid (vitamin C) and vitamin E, as well as carotenoids found in orange and leafy green vegetables.

2.6. Life Depends on Water [p.23]

2.7. Acids, Bases, and Buffers [pp.24–25]

1. two-thirds; 2. hydrogen; 3. hydrophilic; 4. hydrophobic; 5. heat; 6. do not re-form; 7. losing; 8. solvent; 9. solutes; 10. dissolve; 11. acidic; 12. alkaline; 13. 7; 14. 100 times; 15. 100,000 times; 16. basic; 17. buffer; 18. H_2CO_3, HCO_3^-; 19. right, remove; 20. left, add; 21. acidosis, alkalosis; 22. acid; 23. base; 24. salt; 25. salt; 26. It raises the pH as magnesium ions and hydroxide ions are released. The hydroxide ions (OH^-) combine with the excess hydrogen ions (H^+); 27. a. 7.3–7.5, slightly basic; b. 5.0–7.0, slightly acid to neutral; c. 1.0–3.0, acid.

2.8. Organic Compounds: Building on Carbon Atoms [pp.26–27]

1. T; 2. water—nucleic acids; 3. two—four; 4. T; 5. carbohydrate—hydrocarbon; 6. enzymes—functional groups; 7. hydroxyl (sugars); 8. ketone (sugars); 9. amino (amino acids, proteins); 10. phosphate (phosphate compounds); 11. carboxyl (fats, sugars, amino acids); 12. aldehyde (sugars); 13. Enzymes are proteins that speed up metabolic reactions. Each enzyme facilitates a specific reaction required for either building or breaking apart large organic compounds.

14.

amino acid   amino acid   dipeptide

15.a. hydrolysis; b. yes; c. water.

2.9. Carbohydrates: Plentiful and Varied [pp.28–29]

1.

glucose (a monosaccharide)   glucose (a monosaccharide)   enzyme (condensation) / (hydrolysis) enzyme   maltose (a disaccharide)   water + H_2O

2. a. sucrose; b. glucose; c. cellulose; d. deoxyribose; e. lactose; f. glycogen; g. starch.

2.10. Lipids: Fats and Their Chemical Kin [pp.30–31]

1.

glycerol   three fatty acids   yields   triglyceride (a complete fat molecule)

2. a. unsaturated; b. saturated; 3. A phospholipid has two hydrophobic fatty acid tails attached to a glycerol backbone. The hydrophilic head consists of a phosphate group and another polar group. Phospholipids are the main structural component of cell membranes; 4. Unlike fats, sterols have no fatty acid tails. Although they differ from one another in number, position, and type of functional groups, all sterols have four carbon rings bonded together; 5. B; 6. E; 7. D; 8. A; 9. C.

2.11. Proteins: Biological Molecules with Many Roles [pp.32–33]

2.12. A Protein's Function Depends on its Shape [pp.34–35]

1. amino group; 2. R group; 3. carboxyl group; 4. R group.

5.

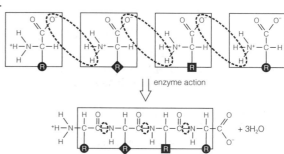

enzyme action

+ 3H_2O

6. C; 7. A; 8. D; 9. B; 10. They differ by what is attached to the polypeptide chain—in lipoproteins, cholesterol, a triglyceride, or a phospholipid is attached; oligosaccharides are attached to glycoproteins; 11. Protein denaturation is a change in the three-dimensional structure of a protein occurring as a result of a change in pH or an increase in temperature that breaks weak hydrogen bonds within the molecule.

2.13. Nucleotides and Nucleic Acids [p.36]

2.14. Food Production and a Chemical Arms Race [p.37]

1. phosphate group; 2. 5-carbon sugar; 3. base; 4.a. deoxyribose, b. ribose.

5.

6. B; 7. A; 8. D; 9. C; 10. Benefits are protecting crop yields, stored grains, public water supplies, pets, and ornamental plants; Drawbacks are that pesticides kill birds and other natural predators that help control pests, and may be dangerous to people.

Self-Quiz

1. a. lipids; b. proteins; c. proteins; d. nucleic acids; e. carbohydrates; f. lipids; g. proteins; h. lipids; i. nucleic acids; j. lipids; k. proteins; l. carbohydrates; 2. a; 3. d; 4. c; 5. e; 6. e; 7. b; 8. c; 9. c; 10. b; 11. b.

Chapter Objectives/Review Questions

1. element; 4. isotopes; 5. radioisotopes; 6. positron-emission tomography; 8. bonds; 11. ionic; 12. covalent; 15. hydrophilic; 16. hydrophobic; 18. acid, base; 21. salts; 22. salts; 24. hydrocarbon; 25. functional; 31. triglycerides; 33. phospholipids; 34. amino; 35. enzymes; 38. denaturation.

Media Menu Review Questions

1. hydrogenated; 2. semi-solids; 3. trans; 4. shorter; 5. saturated; 6. L; 7. bound; 8. margarine; 9. non-caloric; 10. less.

Chapter 3 Cells and How They Work

Impacts, Issues: When Mitochondria Spin Their Wheels [p.41]

3.1. Cells: Organized for Life [pp.42–43]

3.2. The Parts of a Eukaryotic Cell [pp.44–45]

1. Every organism is composed of one or more cells. The cell is the smallest unit having the properties of life. All cells come from pre-existing cells; 2.A. DNA B. cytoplasm C. plasma membrane; 3. Prokaryotic cells have no distinctly separated internal compartments; the DNA is loose in the cytoplasm. In the eukaryotic cell, there is distinct compartmentalization; the DNA is kept separate from the other cellular components by the nuclear envelope. The cell contains several membrane-bound organelles besides the nucleus; 4. Most cells are small because of the surface-to-volume ratio. If the cell were large, there would not be enough surface area for the necessary high rate of import and export of nutrients and wastes. The cell would die. To compensate, large cells are either long and thin or have folds to increase surface area; 5. The lipid bilayer has the hydrophilic heads of the phospholipids to the outside because the

intracellular fluid and extracellular fluid are aqueous and therefore polar, like water. The tails are nonpolar and therefore hydrophobic and so tend to be attracted to each other. They form a "sandwich" of tails inside with the heads outside; 6. Compartments separate cellular reactions and concentrate the enzymes and other molecules needed for a specific type of reaction; 7.a. microfilaments; b. microtubules; c. plasma membrane; d. mitochondrion; e. nuclear envelope; f. nucleolus; g. DNA and nucleoplasm; h. vesicle; i. lysosome; j. rough endoplasmic reticulum; k. ribosomes—attached and free; l. smooth endoplasmic reticulum; m. Golgi vesicle; n. Golgi body; o. centrioles; 8. E; 9. B; 10. F; 11. A; 12. C; 13. D.

3.3. The Plasma Membrane: A Lipid Bilayer [p.46]

3.4. Sugar Wars [p.47]

3.5. The Nucleus [pp.48–49]

3.6. The Endomembrane System [pp.50–51]

1. The cell membrane must be "fluid" rather than a solid barrier so that substances can enter and leave the cell. The "mosaic" quality refers to the mixture of proteins and the lipid bilayer; 2. phospholipids; 3. proteins; 4.a. recognition proteins; b. lipid bilayer; c. transport proteins; d. adhesion proteins; e. receptor proteins; 5. lipids is crossed out; 6. providing energy for transport is crossed out; 7. smallpox is crossed out; 8. as an energy supply is crossed out; 9. F; 10. D; 11. E; 12. B; 13. A; 14. C; 15. E; 16. F; 17. B; 18. D; 19. C; 20. A; 21.a. ribosome; b. rough ER; c. rough ER vesicle; d. Golgi body; e. Golgi body vesicle; f. exocytosis into small intestine.

3.7. Mitochondria: The Cell's Energy Factories [p.52]

3.8. Microscopes: Windows into the World of Cells [p.53]

3.9. The Cytoskeleton: Cell Support and Movement [pp.54–55]

1. ATP; 2. organic; 3. carbon dioxide; 4. water; 5. eukaryotic; 6. oxygen; 7. membrane; 8. cristae; 9. compartments; 10. Mitochondria resemble bacteria in size. They have their own DNA and some ribosomes. They reproduce on their own; 11. B; 12. A; 13. C; 14. Unlike your skeleton, the cytoskeleton is made of elements that are not permanently rigid—they assemble and disassemble as needed; 15. microtubules, microfilaments, intermediate filaments; 16. motor proteins; 17. basal bodies; 18. microfilaments; 19. microtubules, 20. flagella; 21. intermediate filaments; 22. cytoskeleton.

3.10. Moving Substances Across Membranes by Diffusion and Osmosis [pp.56–57]

3.11. Other Ways Substances Cross Cell Membranes [pp.58–59]

3.12. Revenge of El Tor [p.59]

1. E; 2. A; 3. K; 4. B; 5. G; 6. C; 7. D; 8. I; 9. F; 10. L; 11. H; 12. J; 13. N; 14. M; 15. T; 16. T; 17. greater—equal to; 18. T; 19. T; 20. swell and burst—shrink; 21. hypertonic—isotonic; 22. T; 23. hypotonic—hypertonic; 24. fungus—bacterium; 25. phagocytosis—exocytosis; 26. exocytosis—endocytosis; 27. diffusion; 28. lipid bilayer; 29. water; 30. lipid bilayer; 31. passive transport; 32. active transport; 33. transport proteins.

3.13. Metabolism: Doing Cellular Work [pp.60–61]
1. C; 2. D; 3. E; 4. G; 5. I; 6. H; 7. F; 8. B; 9. A;
10. Enzymes speed up reactions by hundreds to millions of times. They are not permanently altered or used up during reactions. Usually they work for both anabolic reactions and catabolic reactions. Each enzyme reacts with a specific substrate; 11. C; 12. A; 13. D; 14. B;
15. anabolic; 16. 7.35–7.4; 17. anabolic; 18. proteins;
19. coenzymes.

3.14. How Cells Make ATP [pp.62–63]
3.15. Summary of Aerobic Respiration [p.64]
3.16. Alternative Energy Sources in the Body [p.65]
1. covalent; 2. electrons; 3. aerobic; 4. glycolysis; 5. cytoplasm; 6. pyruvate; 7. two; 8. energy requiring; 9. glucose; 10. PGAL; 11. four; 12. two; 13. intermediates;
14. mitochondria; 15. inner; 16. coenzyme A; 17. oxaloacetate; 18. CO_2; 19. two; 20. coenzyme; 21. electron transport system; 22. inner; 23. H^+; 24. ATP; 25. water;
26. 36; 27. glycolysis; 28. two; 29. two; 30. mitochondrion; 31. two; 32. 8 NADH + 2 $FADH_2$; 33. electron transport phosphorylation; 34. mitochondrion; 35. 32;
36. 4, 2, 1, 5, 3, 6; 37. 78%, triglycerides; 38. between meals as an alternative to glucose; 39. The energy in fatty acids is stored in the covalent bonds that bond the hydrogen atoms to the carbon chain. Glucose doesn't have all those bonds; 40. 21%, urea in urine.

Self-Quiz
1. c; 2. c; 3. a; 4. b; 5. d; 6. c; 7. d; 8. c; 9. a; 10. c; 11. c;
12. d; 13. d; 14. d.

Chapter Objectives/Review Questions
4. Organelles; 5. hydrophilic, hydrophobic; 8. envelope;
9. nucleolus; 10. chromosome; 12. Lysosomes; 13. Mitochondria; 14. cytoskeleton; 15. more, less; 18. isotonic, hypotonic, hypertonic; 21. metabolism; 25. Substrates;
26. active; 27. ATP.

Media Menu Review Questions
1. thought; 2. axon; 3. organism; 4. mitochondria; 5. electrons; 6. neurons; 7. calcium; 8. Alzheimer's; 9. proteins;
10. ATP.

Chapter 4 Tissues, Organ Systems, and Homeostasis

Impacts, Issues: Open or Close the Stem Cell Factories? [p.69]
4.1. Epithelium: The Body's Covering and Linings [pp.70–71]
4.2. Connective Tissue: Binding, Support, and Other Roles [pp.72–73]
4.3. Muscle Tissue: Movement [p.74]
4.4. Nervous Tissue: Communication [p.75]
4.5. The Human Body Shop: Engineering New Tissues and Organs [pp.76–77]
1. T; 2. F; 3. T; 4. T; 5. F; 6. B; 7. C; 8. A; 9. D; 10. stable operating conditions in the internal environment;
11. Epithelial; 12. layers; 13. microvilli; 14. basement;
15. simple; 16. stratified; 17. pseudostratified; 18. a. stratified columnar; b. simple cuboidal; c. stratified cuboidal;
d. pseudostratified columnar; e. simple squamous;

f. stratified squamous; g. simple columnar; 19. secretes;
20. Exocrine; 21. mucus; 22. saliva; 23. earwax; 24. oil;
25. milk; 26. digestive enzymes; 27. Endocrine; 28. hormones; 29. thyroid; 30. adrenal; 31. pituitary; 32. connective; 33. soft; 34. specialized; 35. ground substance;
36. collagen; 37. elastin; 38. matrix; 39. b; 40. c; 41. a;
42. b; 43. a; 44. c; 45. Cartilage is solid but pliable and resistant to compression. Chondroblasts make the cartilage matrix; 46. There are no blood vessels in cartilage;
47. a; 48. c; 49. a; 50. a; 51. b; 52. c; 53. b; 54. collagen;
55. calcium; 56. protect; 57. muscles; 58. blood; 59. fat;
60. skin; 61. connective; 62. transport; 63. plasma;
64. platelets; 65. Muscle cells contract when stimulated, and relax; by working in layers, they work in a coordinated manner; 66. b; 67. c; 68. a; 69. a; 70. b; 71. c; 72. a;
73. c; 74. c; 75. b; 76. a; 77. Cartilage or bone cells are seeded into tiny spaces in a biodegradable plastic model and provided with nutrients—after awhile the plastic model biodegrades, leaving usable tissue; 78. Stem cells from adults are already somewhat specialized so they don't have the same developmental potential that embryonic ones do.

4.6. Cell Junctions: Holding Tissues Together [p.78]
4.7. Membranes: Thin, Sheetlike Covers [p.79]
1. tight junction (B); 2. adhering junction (A); 3. gap junction (C); 4. a. serous; b. synovial; c. cutaneous;
d. mucous; 5. C; 6. D; 7. C; 8. B; 9. D; 10. A.

4.8. Organ Systems: Built from Tissues [pp.80–81]
4.9. The Skin—Example of an Organ System [p.82]
4.10. Sun, Skin, and the Ozone Layer [p.83]
1. cranial; 2. spinal; 3. thoracic; 4. abdominal; 5. pelvic;
6. midsagittal; 7. frontal; 8. anterior; 9. transverse;
10. inferior; 11.a. ectoderm—outer skin and nervous system, b. mesoderm—muscles, bones, most organ systems; endoderm—lining of digestive tract and organs derived from it. 12. E; 13. C; 14. G; 15. I; 16. F; 17. B; 18. H; 19. J;
20. D; 21. K; 22. A; 23. integument; 24. sun; 25. bacteria;
26. moisture; 27. temperature; 28. sensory; 29. D; 30. epidermis; 31. dermis; 32. connective; 33. hypodermis;
34. anchor; 35. Fat; 36. stratified squamous epithelium;
37. keratin; 38. stratum corneum; 39. melanocytes;
40. ultraviolet; 41. skin color; 42. Langerhans;
43. Granstein; 44. dermis; 45. blood; 46. lymph; 47. sebaceous; 48. melanin; 49. elastin; 50. immune; 51. protooncogenes; 52. cancer; 53. increasing; 54. 100; 55. sebaceous gland; 56. smooth muscle; 57. melanocyte; 58. hair shaft; 59. adipose (fat) cells; 60. sweat gland; 61. pressure receptor; 62. hair follicle; 63. nerve fiber; 64. hypodermis;
65. dermis; 66. epidermis.

4.11. Homeostasis: The Body in Balance [pp.84–85]
1. Extracellular; 2. interstitial; 3. cells; 4. ions; 5. Homeostasis; 6. stability; 7. receptors; 8. stimulus; 9. integrator;
10. effectors; 11. "set points"; 12. feedback; 13. negative;
14. positive; 15. stimulus (B); 16. receptor (E); 17. integrator (A); 18. effector (C); 19. response (D) 20. b; 21. a;
22. b; 23. b; 24. a; 25. a.

Self-Quiz
1. connective (D) (g); 2. epithelial (G) (b); 3. muscle (I) (e); 4. muscle (J) (l); 5. connective (E) (d); 6. connective

374 Answers to Student Interactive Workbook

(L) (i); 7. connective (B) (k); 8. epithelial (H) (j); 9. connective (K) (f); 10. nervous (M) (c); 11. muscle (C) (a); 12. epithelial (F) (m); 13. connective (A) (h); 14. d; 15. c; 16. c; 17. a; 18. d; 19. c; 20. d; 21. b; 22. c.

Chapter Objectives/Review Questions
1. tissues; 3. Epithelium; 4. Endocrine; 7. Dense; 8. hyaline; 9. elastic; 10. fibrocartilage; 11. energy; 13. contract; 14. communication; 15. dendrites, axons; 16. nerve; 18. epithelium.

Media Menu Review Questions
1. tissue engineering; 2. 20; 3. Capillaries; 4. bladder; 5. layer; 6. cartilage; 7. mice; 8. angioplasty; 9. medication; 10. 10%.

Chapter 5 The Skeletal System

Impacts, Issues: Hold the Hype [p.89]
5.1. Bone—Mineralized Connective Tissue [pp.90–91]
5.2. The Skeleton: The Body's Bony Framework
    [pp.92–93]
1. Compact bone is dense, smooth tissue forming the walls of long bones. Spongy bone is found inside a long bone and appears lacy and delicate but is quite strong; 2. 3, 1, 4, 2; 3. Q; 4. D; 5. F; 6. N; 7. C; 8. G; 9. P; 10. J; 11. E; 12. O; 13. M; 14. A; 15. L; 16. H; 17. B; 18. K; 19. I; 20. nutrient canal; 21. marrow cavity; 22. compact bone; 23. spongy bone; 24. dense connective tissue; 25. osteon; 26. blood vessel; 27. lacuna; 28. yellow marrow; 29. irregular; 30. ligaments; 31. tendons; 32. movement, support, protection, mineral storage, blood cell formation; 33. cranial bones; 34. clavicle; 35. sternum; 36. scapula; 37. radius; 38. carpals; 39. femur; 40. tibia; 41. tarsals; 42. metatarsals.

5.3. The Axial Skeleton [pp.94–95]
5.4. The Appendicular Skeleton [pp.96–97]
1. parietal; 2. temporal; 3. occipital; 4. frontal; 5. maxilla; 6. mandible; 7. palatine; 8. zygomatic; 9. foramen magnum; 10. cervical; 11. thoracic; 12. lumbar; 13. sacrum; 14. coccyx; 15. axial; 16. sinuses; 17. Temporal; 18. sphenoid; 19. nose; 20. occipital; 21. mandible; 22. zygomatic; 23. lacrimal; 24. maxillary; 25. vomer; 26. cervical; 27. chest; 28. lumbar; 29. coccyx; 30. intervertebral disks; 31. herniated; 32. sternum; 33. pectoral; 34. carpals; 35. pubic; 36. fibula; 37. tarsals; 38. metatarsals.
5.5. Joints—Connections Between Bones [pp.98–99]
5.6. Skeletal Diseases, Disorders, and Injuries
    [pp.100–101]
5.7. Replacing Joints [p.101]
1. Synovial; 2. synovial; 3. hingelike; 4. ball-and-socket; 5. Cartilaginous; 6. fibrous; 7. abduction; 8. adduction; 9. circumduction; 10. rotation; 11. flexion; 12. extension; 13. supination; 14. pronation; 15. cartilage; 16. osteoarthritis; 17. rheumatoid arthritis; 18. Tendinitis; 19. carpel tunnel syndrome; 20. strain; 21. sprain; 22. dislocated; 23. simple; 24. complete; 25. compound; 26. prosthesis.
Self-Quiz
1. d; 2. e and g; 3. a; 4. j; 5. i; 6. c; 7. h and k; 8. f; 9. b; 10. i; 11. c; 12. d; 13. b; 14. f; 15. a; 16. c; 17. a; 18. a; 19. b.

Media Menu Review Questions
1. weight-bearing, resistance; 2. weight-bearing; 3. resistance; 4. 30; 5. Osteoarthritis; 6. chondrocytes; 7. genetic; 8. inflammatory; 9. destructive; 10. exercise.

Chapter 6 The Muscular System

Impacts, Issues: Pumping up Muscles [p.105]
6.1. The Structure and Functioning of Skeletal Muscles
    [pp.106–107]
6.2. How Muscles Contract [pp.108–109]
1. Only skeletal muscle interacts with the skeleton to move the body; 2. origin 3. insertion; 4. pulls; 5. fibers; 6. connective; 7. tendon; 8. bone; 9. pairs; 10. opposition; 11. relax; 12. bend; 13. relax; 14. straighten; 15. deltoid (C); 16. trapezius (H); 17. gluteus maximus (G); 18. biceps femoris (J); 19. gastrocnemius (A); 20. pectoralis major (F); 21. rectus abdominis (D); 22. sartorius (E); 23. quadriceps femoris (I); 24. tibialis anterior (B); 25. parallel; 26. bands; 27. striped; 28. Z; 29. sarcomere; 30. actin; 31. myosin; 32. sliding filament; 33. myosin; 34. actin; 35. sarcomere; 36. ATP; 37. outer sheath (connective tissue); 38. muscle cell bundles; 39. muscle cell; 40. myofibril; 41. sarcomere; 42. Z bands.
6.3. Energy for Muscle Cells [p.109]
6.4. How the Nervous System Controls Muscle
    Contraction [pp.110–111]
6.5. Properties of Whole Muscles [pp.112–113]
6.6. Muscle Matters [pp.114–115]
1. Phosphate is transferred from creatine phosphate to ADP in order to quickly provide ATP needed for a muscle contraction; 2. Glycogen stored in the muscle is used for the first 5–10 minutes. Then, glucose and fatty acids from the blood are used for about half an hour. Beyond that, fatty acids are used; 3. Glycolysis produces a small amount of ATP when there is not enough oxygen for aerobic respiration to continue, such as during hard exercise; 4. Muscle fatigue and heavy breathing result from the oxygen debt that occurs when the muscles produce ATP by glycolysis, not aerobic respiration; 5. motor; 6. T tubules; 7. sarcoplasmic reticulum; 8. calcium; 9. troponin; 10. actin; 11. myosin cross-bridges; 12. calcium; 13. myosin; 14. neuromuscular junction; 15. axons; 16. synapse; 17. neurotransmitter; 18. contract; 19. motor neuron; 20. neuromuscular junction; 21. motor unit; 22. spinal cord; 23. C; 24. H; 25. E; 26. M; 27. B; 28. L; 29. I; 30. F; 31. J; 32. G; 33. A; 34. D; 35. K; 36. wasting away; 37. aerobic; 38. strength-training; 39. tension; 40. increase in the number and size of mitochondria, increase in number of capillaries supplying the muscle, and more myoglobin; 41. endurance; 42. An anabolic steroid mimics the hormone testosterone. In men, side effects include acne, baldness, shrinking testes, infertility, atherosclerosis, and "roid rage." In women, side effects include deepening of the voice, obvious facial hair, irregular menstrual periods, shrinking breasts, and enlarged clitoris.
Self-Quiz
1. d; 2. e; 3. a; 4. b; 5. c; 6. b; 7. d; 8. a; 9. e.

1. amino acid, muscle; 2. ATP; 3. repeated, high-intensity; 4. Lou Gehrig's, dystrophy; 5. physician (doctor), dehydration; 6. excreted, side effects.

Chapter 7 Digestion and Nutrition

Impacts, Issues: Hips and Hunger [p.119]

7.1. The Digestive System: An Overview [pp.120–121]

7.2. Chewing and Swallowing: Food Processing Begins [pp.122–123]

7.3. The Stomach: Food Storage, Digestion, and More [p.124]

1. mouth (oral) cavity; 2. salivary glands; 3. stomach; 4. small intestine; 5. anus; 6. large intestine; 7. pancreas; 8. gallbladder; 9. liver; 10. esophagus; 11. pharynx; 12. a. mouth cavity; b. salivary glands; c. pharynx; d. esophagus; e. stomach; f. small intestine; g. pancreas; h. liver; i. gallbladder; j. large intestine; k. rectum; l. anus; 13. Mechanical processing and motility (of food along the digestive tract), secretion (of digestive enzymes and other substances), digestion (or chemical breakdown of food into small molecules), absorption (of nutrients and fluids), and elimination (of undigested matter); 14. The mucosa is the inner lining of the tube. Surrounding the mucosa is the submucosa, a layer of connective tissue containing blood and lymph vessels and networks of nerve cells. Next is the smooth muscle, one circular and one lengthwise layer. The outermost layer is the serosa, a thin, serous membrane; 15. C; 16. E; 17. B; 18. I; 19. H; 20. J; 21. A; 22. D; 23. G; 24. F; 25. pharynx; 26. involuntary; 27. epiglottis; 28. Heimlich; 29. stomach; 30. small intestine; 31. Gastric; 32. pepsins; 33. intrinsic; 34. chyme; 35. Protein; 36. denatures; 37. pepsins; 38. gastrin; 39. gastric mucosal barrier; 40. bacterium; 41. peptic ulcer; 42. peristalsis; 43. pyloric; 44. close; 45. small intestine; 46. rugae; 47. alcohol.

7.4. The Small Intestine: A Huge Surface Area for Digestion and Absorption [p.125]

7.5. How Are Nutrients Digested and Absorbed? [pp.126–127]

7.6. The Multipurpose Liver [p.128]

7.7. The Large Intestine [p.129]

7.8. Digestion Controls and Disruptions [p.130]

1. wall; 2. surface area; 3. villi; 4. microvilli; 5. duodenum; 6. pancreas; 7. liver; 8. gallbladder; 9. peptide; 10. amino acids; 11. bicarbonate; 12. enzymes; 13. bile; 14. gallbladder; 15. emulsification; 16. Absorption; 17. segmentation; 18. Transport; 19. Bile; 20. micelles; 21. epithelial; 22. triglycerides; 23. exocytosis; 24. blood; 25. lacteals; 26. hepatic portal vein; 27. hepatic vein; 28. glycogen; 29. bile; 30. gallbladder; 31. cholesterol; 32. gallstones; 33. urea; 34. I; 35. L; 36. H; 37. B; 38. N; 39. A; 40. G; 41. J; 42. M; 43. K; 44. F; 45. D; 46. C; 47. E; 48. G; 49. D; 50. E; 51. H; 52. B; 53. I; 54. F; 55. A; 56. C.

7.9. Nutrient Processing After and Between Meals [p.131]

7.10. The Body's Nutritional Requirements [pp.132–133]

7.11. Diet Alternatives [p.134]

7.12. Malnutrition and Undernutrition [p.135]

7.13. Vitamins and Minerals [pp.136–137]

7.14. Calories Count: Food Energy and Body Weight [pp.138–139]

1. break apart; 2. fats; 3. glucose; 4. carbohydrates; 5. fiber; 6. vitamins; 7. membranes; 8. energy; 9. cushion; 10. insulation; 11. vitamins; 12. liver; 13. saturated; 14. 8; 15. Complete; 16. incomplete; 17. If you eat less fat and limit sweets, you will maintain a healthy weight and live longer. 18. lower risk of heart disease; 19. grain products; 20. 2–3 ounces; 21. 3 to 5; 22. teenager; 23. a few times a month; 24. They should take supplemental vitamins B$_{12}$ and B$_2$; 25. low-carb; 26. glycemic; 27. insulin; 28. glucose; 29. fat; 30. hungry; 31. triglycerides; 32. diabetes; 33. insulin; 34. fat; 35. kidney; 36. A; 37. H; 38. D; 39. B; 40. F; 41. I; 42. G; 43. E; 44. C; 45. vitamins; 46. minerals; 47. vitamins; 48. can; 49. C; 50. B; 51. A; 52. D; 53. 2100, 2700, 1100; 54. 25-year-old female; 55. male; 56. 59-year-old female; 57. about 6 hours; 58. age, emotional state, hormones; 59. Ghrelin increases appetite while leptin and PYY3-36 diminish it; 60. An anorexic eats much too little and exercises too much. A bulimic eats large quantities of food but then vomits afterward.

Self-Quiz

1. b; 2. b; 3. a; 4. e; 5. c; 6. b; 7. b; 8. c; 9. d; 10. d; 11. E; 12. D; 13. A; 14. G; 15. B; 16. C; 17. F.

Media Menu Review Questions

1. ghrelin; 2. brain; 3. stomach; 4. starving; 5. 60; 6. inactivity, smoking; 7. weight, fat; 8. body mass index; 9. abdominal; 10. breakfast; 11. one to two; 12. 30, 10; 13. percentage; 14. thirty; 15. 30; 16. protein; 17. dehydration, kidneys, 18. nitrogen, urea, urine.

Chapter 8 Blood

Impacts, Issues: Chemical Queries [p.143]

8.1. Blood: Plasma, Blood Cells, and Platelets [pp.144–145]

8.2. How Blood Transports Oxygen [p.146]

8.3. Life Cycle of Red Blood Cells [p.147]

1. a. 91%–92%; b. Proteins; c. Neutrophils; d. Lymphocytes; e. Phagocytosis; f. Red blood; g. Platelets; 2. 4–5; 3. plasma; 4. water; 5. albumin; 6. water; 7. proteins; 8. volume; 9. pH; 10. Red blood; 11. hemoglobin; 12. stem cells; 13. white blood; 14. foreign; 15. tissues; 16. bone marrow; 17. granulocytes; 18. neutrophils; 19. eosinophils; 20. basophils; 21. agranulocytes; 22. monocytes; 23. lymphocytes; 24. platelets; 25. clotting; 26. red blood (E); 27. neutrophil (A); 28. lymphocytes (D); 29. macrophage (B); 30. platelets (C); 31. monocyte; 32. stem cell; 33. a. hemoglobin; b. oxyhemoglobin; c. more O$_2$, cooler, less acidic; d. lungs; e. less O$_2$, warmer, more acidic; f. tissues; g. hemoglobin; h. red; i. four; j. iron; k. binds oxygen; l. oxyhemoglobin; 34. F; 35. G; 36. A; 37. B; 38. C; 39. D; 40. H; 41. E.

8.4. Blood Types—Genetically Different Red Blood Cells [pp.148–149]

8.5. Hemostasis and Blood Clotting [p.150]

8.6. Blood Disorders [p.151]

1. I; 2. J; 3. B; 4. A; 5. F; 6. G; 7. H; 8. C; 9. D; 10. E; 11. B; 12. E; 13. C; 14. D; 15. A; 16. F; 17. The series of reactions

is triggered outside the blood by substances coming from damaged blood vessels or from surrounding tissues; 18. Both are clots that form in an unbroken blood vessel. A thrombus stays where it forms, while an embolus breaks free and circulates through the bloodstream. An embolus is more dangerous because it may shut down an organ's blood supply; 19. C; 20. D; 21. F; 22. B; 23. E; 24. A.

Self-Quiz

1. d; 2. e; 3. a; 4. e; 5. e; 6. a; 7. b; 8. d; 9. a; 10. b.

Chapter Objectives/Review Questions

5. erythropoietin; 9. iron deficiency anemia

Media Menu Review Questions

1. plasma; 2. 500; 3. blood serum; 4. tripled; 5. stem; 6. red, platelets; 7. chemotherapy; 8. immune; 9. kill; 10. inflammatory.

Chapter 9 Circulation—The Heart and Blood Vessels

Impacts, Issues: The Breath of Life [p.153]

9.1. The Cardiovascular System—Moving Blood through the Body [pp.154–155]

9.2. The Heart: A Double Pump [pp.156–157]

9.3. The Two Circuits of Blood Flow [pp.158 159]

9.4. Heart-Saving Drugs [p.159]

9.5. How Does Cardiac Muscle Contract [p.160]

1. cardiovascular; 2. heart; 3. blood vessels; 4. arteries; 5. arterioles; 6. capillaries; 7. venules; 8. veins; 9. rapidly; 10. slowly; 11. oxygen; 12. nutrients; 13. wastes; 14. homeostasis; 15. interstitial; 16. lymphatic; 17. jugular; 18. superior; 19. pulmonary; 20. hepatic portal; 21. renal; 22. inferior; 23. iliac; 24. femoral; 25. femoral; 26. iliac; 27. abdominal; 28. renal; 29. brachial; 30. coronary; 31. pulmonary; 32. ascending; 33. carotid; 34. D; 35. K; 36. G; 37. E; 38. J; 39. F; 40. H; 41. B; 42. I; 43. C; 44. L; 45. M; 46. A; 47. N; 48. superior vena cava; 49. right semilunar valve; 50. right atrioventricular (or tricuspid) valve; 51. chordae tendineae; 52. inferior vena cava; 53. myocardium; 54. left ventricle; 55. arch of aorta; 56. 1, 5, 2, 7, 8, 3, 6, 4; . I; 57. D; 58. F; 59. C; 60. J; 61. A; 62. H; 63. E; 64. G; 65. B; 66. I; 67. D; 68. pulmonary; 69. systemic; 70. arteries; 71. organs; 72. arteries; 73. arterioles; 74 capillaries; 75. venules; 76. veins; 77. superior vena cava; 78. inferior vena cava; 79. capillary beds; 80. arterioles; 81. venules; 82. sphincters; 83. relaxes; 84. contracts; 85. hepatic portal vein; 86. impurities; 87. hepatic; 88. oxygenated; 89. statins; 90. LDL; 91. HDLs; 92. triglycerides; 93. heart attacks; 94. brain strokes; 95. intercalated discs; 96. cardiac conduction; 97. sinoatrial; 98. atrioventricular; 99. Purkinje fibers; 100. SA.; 101. nervous.

9.6. Blood Pressure [p.161]

9.7. The Structure and Functions of Blood Vessels [pp.162–163]

9.8. Heart-Healthy Exercise [p.164]

9.9. Exchanges at Capillaries [pp.164–165]

9.10. Cardiovascular Disorders [pp.166–167]

9.11. The Multipurpose Lymphatic System [pp.168–169]

1. Blood pressure; 2. aorta; 3. systemic; 4. systolic; 5. aorta; 6. diastolic; 7. aorta; 8. hypertension; 9. hypotension; 10. a; 11. c; 12. b; 13. c; 14. a; 15. b; 16. c; 17. e; 18. c; 19. e; 20. a; 21. e; 22. e; 23. d; 24. artery, outer coat; 25. arteriole, endothelium; 26. capillary, basement membrane; 27. vein, smooth muscle with elastic fibers; 28. E; 29. C; 30. A; 31. B; 32. D; 33. The heart muscle becomes stronger and has increased blood flow, the number of red blood cells increases and cholesterol levels fall as does blood pressure; 34. higher; 35. T; 36. capillary beds; 37. T; 38. T; 39. lymphatic; 40. scratch through angina pectoralis; 41. scratch through angioplasty and hypotention; 42. scratch through angioplasty and angina pectoralis; 43. scratch through organ failure; 44. scratch through hypertension, stroke; 45. scratch through drinking; 46. scratch through sugar, starch; 47. scratch through comes after, has nothing to do with; 48. scratch through angioplasty, heart failure; 49. scratch through coronary bypass surgery, angioplasty; 50. scratch through CDLs, HDLs; 51. scratch through triglycerides, steroids; 52. scratch through the liver and the heart; 53. scratch through arrhythmia; 54. scratch through inflammation, arrhythmia; 55. scratch through bradycardia, tachycardia; 56. scratch through bradycardia, angina pectoralis; 57. tonsils; 58. thymus gland; 59. thoracic duct; 60. spleen; 61. lymph vessels; 62. lymph nodes; 63. lymphocytes; 64. G; 65. A; 66. E; 67. F; 68. B; 69. D; 70. C; 71. Vessels act as drainage channels, collecting fluids that have leaked from capillary beds and returning them to the bloodstream. Vessels deliver fats absorbed from the small intestine to the bloodstream. The system transports foreign cells and cellular debris from body tissues to lymph nodes for disposal.

Self-Quiz

1. e; 2. c; 3. b; 4. b; 5. a; 6. a; 7. a; 8. e; 9. e; 10. b.

Chapter Objectives/Review Questions

3. systole, diastole; 5. pulmonary, systemic

Media Menu Review Questions

1. fatty arterial plaque; 2. HDL; 3. lower; 4. 5 %; 5. 11; 6. potassium; 7. three; 8. 9,608; 9. 27.

Chapter 10 Immunity

Impacts, Issues: Viral Villains [p.173]

10.1. Three Lines of Defense [p.174]

10.2. Complement Proteins: "Defense Team" Partners [p.175]

10.3. Inflammation—Responses to Tissue Damage [pp.176–177]

1. a; 2. c; 3. b; 4. a; 5. b; 6. a; 7. c; 8. c; 9. b; 10. white; 11. neutrophils; 12. eosinophils; 13. basophils; 14. macrophages; 15. foreign; 16. damaged; 17. cells, tissue; 18. capillaries; 19. mast; 20. vasodilate; 21. permeability; 22. neutrophils; 23. debris; 24. chemotaxins; 25. interleukins; 26. bacteria; 27. prostaglandin; 28. fever; 29. complement; 30. clotting; 31. clots; 32. C; 33. B; 34. D; 35. A; 36. E.

10.4. An Immune System Overview [pp.178–179]

10.5. How Lymphocytes Form and Do Battle [pp.180–181]

10.6. A Closer Look at Antibodies in Action [pp.182–183]

10.7. Cell-Mediated Responses—Countering Threats Inside Cells [pp.184–185]

10.8. Organ Transplants: Beating the (Immune) System [p.185]

1. B and T cells are part of the body's third line of defense. It is called upon when the first two lines of defense—physical barriers and inflammation—don't stop an invader; 2. Self/nonself recognition—B and T cells attack only "nonself" substances or cells, Specificity—B and T cells attack only specific invaders, Diversity—B and T cells can respond to over a billion specific threats, Memory—Some B and T cells are reserved for a future battle with a specific invader; 3. Our own cells have "self" markers recognized by B and T cells, but foreign agents have "nonself" markers that alert the T and B cells. 4. D; 5. A; 6. C; 7. B; 8. d; 9. a; 10. d; 11. b; 12. e; 13. c; 14. a; 15. c; 16. d; 17. a; 18. b; 19. T; 20. T; 21. cell-mediated; 22. B cells; 23. T; 24. antigens; 25. antigens; 26. variable regions; 27. shuffled; 28. antibody; 29. project; 30. naive; 31. thymus; 32. TCRs; 33. antigen-MHC; 34. antigen receptor; 35. clonal selection; 36. clone; 37. Immunological memory; 38. Memory cells; 39. terminated; 40. macrophages; 41. lymph; 42. B and T; 43. naïve; 44. lymph; 45. lymph nodes; 46. spleen; 47. helper T; 48. immune response; 49. F; 50. C; 51. G; 52. D; 53. E; 54. B; 55. A; 56. E; 57. D; 58. B; 59. A; 60. C; 61. cells; 62. cell-mediated; 63. antigens; 64. general; 65. interferons; 66. MHC; 67. "touch-kill"; 68. perforins; 69. apoptosis; 70. rejection; 71. interleukins; 72. first; 73. foreign; 74. 75; 75. relatives; 76. suppress; 77. antibiotics; 78. pigs; 79. xenotransplantation; 80. eye; 81. testicles.

10.9. Practical Applications of Immunology [pp.186–187]

10.10. A Can't Win Proposition? [p.187]

10.11. Disorders of the Immune System [pp.188–189]

1. G; 2. H; 3. A; 4. E; 5. K; 6. I; 7. F; 8. B; 9. C; 10. J; 11. D; 12. L; 13. Strains of influenza have different surface antigens. In addition, "gene swapping" increases variation within a strain; 14. allergy; 15. allergen; 16. predisposed; 17. antibodies; 18. mast; 19. histamines; 20. constrict; 21. invader; 22. anaphylactic shock; 23. cardiovascular; 24. epinephrine; 25. Antihistamines; 26. IgG; 27. autoimmune; 28. rheumatoid arthritis; 29. diabetes; 30. systemic lupus erythematosus; 31. lymphocytes; 32. severe combined immune deficiency; 33. human immunodeficiency virus; 34. acquired immunodeficiency syndrome; 35. body fluids; 36. helper T; 37. macrophages; 38. cancer.

Self-Quiz

1. d; 2. b; 3. b; 4. a; 5. a; 6. e; 7. e; 8. e; 9. a; 10. a; 11. H; 12. D; 13. E; 14. J; 15. B; 16. G; 17. C; 18. I; 19. F; 20. A.

Media Menu Review Questions

1. sanitation; 2. childhood; 3. T lymphocytes (T cells); 4. major histocompatibility complex, healthy; 5. mycobacterial; 6. memory, spleen; 7. autoimmune; 8. evolution; 9. sickle cell anemia, Tay-Sachs; 10. beneficial.

Chapter 11 The Respiratory System

Impacts, Issues: Up in Smoke [p.193]

11.1. The Respiratory System—Built for Gas Exchange [pp.194–195]

1. cilia; 2. white blood cells; 3. bronchitis; 4. carcinogens; 5. blood pressure; 6. bad cholesterol; 7. good cholesterol; 8. oral cavity (H); 9. pleural membrane (J); 10. intercostal muscles (F); 11. diaphragm (B); 12. nasal cavity (E); 13. pharynx (K); 14. epiglottis (D); 15. larynx (A); 16. trachea (G); 17. lung (L); 18. bronchial tree (C); 19. bronchiole; 20. alveolar duct; 21. alveoli (I); 22. alveolar sac; 23. pulmonary capillaries; 24. B; 25. H; 26. E; 27. K; 28. J; 29. I; 30. N; 31. F; 32. O; 33. M; 34. C; 35. A; 36. D; 37. G; 38. L.; 39. D; 40. E; 41. B; 42. G; 43. C; 44. F; 45. A.

11.2. The "Rules" of Gas Exchange [p.196]

11.3. Breathing as a Health Hazard [p.197]

11.4. Breathing—Air In, Air Out [pp.198–199]

1. Gas exchange; 2. pressure; 3. 21; 4. carbon dioxide; 5. respiratory; 6 dissolved; 7. pressure gradient; 8. alveoli; 9. hemoglobin; 10. four; 11. releases; 12. gradient; 13. decreases; 14. hyperventilation; 15. hypoxia; 16. c; 17. b; 18. a; 19. b; 20. c; 21. a; 22. b; 23. c; 24. b; 25. left; 26. right; 27. intercostals; 28. diaphragm; 29. down; 30. upward and outward; 31. expands; 32. up; 33. inward and downward; 34. They recoil or relax; 35. thoracic cavity; 36. enters; 37. expiration; 38. out of; 39. thoracic; 40. pleural; 41. pleural sac; 42. tidal volume; 43. inspiratory; 44. expiratory; 45. vital capacity; 46. one-half; 47. residual; 48. 350.

11.5. How Gases Are Exchanged and Transported [pp.200–201]

11.6. Homeostasis Depends on Controls over Gas Exchange [pp.202–203]

11.7. Tobacco and Other Threats to the Respiratory System [pp.204–205]

1. H; 2. J; 3. B; 4. D; 5. K; 6. G; 7. C; 8. A; 9. E; 10. F; 11. L; 12. I; 13. more; 14. decreases; 15. more; 16. bicarbonate; 17. water; 18. falls; 19. lower; 20. buffer; 21. nervous; 22. rhythm; 23. magnitude; 24. diaphragm; 25. inspiration; 26. expiration; 27. pons; 28. oxygen; 29. carbon dioxide; 30. arteries; 31. diaphragm; 32. medulla; 33. cerebrospinal; 34. respiratory; 35. carbon dioxide (in response to increase in rate and depth of breathing); 36. carotid; 37. aortic; 38. ventilation; 39. increase; 40. dilate; 41. constrict; 42. involuntary; 43. apnea; 44. elderly; 45. secondhand; 46. 80; 47. Each individual will have his or her own responses.

Self-Quiz

1. c; 2. e; 3. a; 4. b; 5. a; 6. a; 7. a; 8. a; 9. c; 10. d.

Chapter Objectives/Review Questions

5. inspiration, expiration; 10. oxyhemoglobin, carbaminohemoglobin.

Media Menu Review Questions

1. hypoxia; 2. fewer; 3. months; 4. hemoglobin, ventilation; 5. nitric oxide; 6. acetylcholine; 7. apoptosis; 8. blood vessels; 9. speeds; 10. protein.

Chapter 12 The Urinary System

Impacts, Issues: Double-Edged Sword [p.209]

12.1. The Challenge: Shifts in Extracellular Fluid [pp.210–211]

12.2. The Urinary System—Built for Filtering and Waste Disposal [pp.212–213]

1. The urinary system maintains stable conditions in the extracellular fluid by removing substances that enter the ECF from the intracellular fluid and adding needed substances that move from the ECF into the intracellular fluid; 2. a. absorption from liquids and solid food; b. metabolic reactions; c. excretion in urine; d. evaporation from lungs and skin; e. sweating; f. elimination in feces; g. absorption from liquids and solid food; h. secretion from cells; i. respiration; j. metabolism; k. urinary excretion; l. respiration; m. sweating; 3. brain; 4. urinary excretion; 5. T; 6. respiratory; 7. T; 8. urine; 9. T; 10. protein; 11. urea; 12. T; 13. kidney; 14. ureter; 15. urinary bladder; 16. urethra; 17. kidney cortex; 18. kidney medulla; 19. ureter; 20. kidneys; 21. nephrons; 22. solutes; 23. blood; 24. million; 25. afferent; 26. renal corpuscle; 27. glomerulus; 28. Bowman's; 29. proximal; 30. Henle; 31. distal; 32. efferent; 33. peritubular; 34. reabsorbed; 35. kidney cortex; 36. kidney medulla; 37. glomerular capillaries; 38. proximal tubule; 39. distal tubule; 40. peritubular capillaries; 41. collecting duct; 42. Bowman's capsule.

12.3. How Urine Forms [pp.214–215]

12.4. Keeping Water and Sodium in Balance [pp.216–217]

12.5. Here's to the Health of Your Urinary Tract! [p.217]

12.6. Adjusting Reabsorption: Hormones Are the Key [pp.218–219]

1. a. glomerulus; b. water, solutes; c. Bowman's, proximal; d. nephron; e. out; f. capillaries; g. nephron; h. capillaries; i. urine; 2. urinalysis; 3. urinary bladder; 4. relaxes; 5. external; 6. Kidney; 7. lungs; 8. higher; 9. very permeable; 10. vasoconstriction; 11. proximal tubule; 12. sodium "pumps"; 13. tissue fluid; 14. amino acids; 15. out; 16. peritubular capillaries; 17. medulla; 18. descending; 19. out; 20. sodium; 21. dilute; 22. D; 23. H; 24. C; 25. E; 26. A; 27. F; 28. G; 29. B; 30. hypothalamus; 31. pituitary gland; 32. distal; 33. more; 34. loss; 35. ADH; 36. sodium; 37. renin; 38. angiotensin II; 39. faster; 40. rise; 41. saliva.

12.7. Maintaining the Body's Acid–Base Balance [p.220]

12.8. Kidney Disorders [p.221]

12.9. Maintaining the Body's Core Temperature [pp.222–223]

1. 7.37, 7.43; 2. eliminate; 3. bicarbonate; 4. blood; 5. nephron tubule 6. E; 7. A; 8. B; 9. C; 10. D; 11. B; 12. H; 13. E; 14. D; 15. G; 16. I; 17. C; 18. K; 19. F; 20. J; 21. A; 22. High temperatures denature enzymes, which can lead to death; very low temperatures result in reduced enzyme function, which causes the heart rate to fall and even stop, especially when heat-generating mechanisms stop; 23. Sweating during exercise has the purpose of cooling you off, but electrolytes are lost along with the water in sweat; sports drinks have the purpose of replacing these electrolytes; 24. Shivering results from the body's efforts to restore core temperature; the rapid shift in temperature involves peripheral vasodilation and possibly sweating.

Self-Quiz

1. d; 2. c; 3. e; 4. b; 5. d; 6. d; 7. d; 8. b; 9. e; 10. a.

1. dehydrated; 2. blood; 3. nitrogen; 4. scarring.

Chapter 13 The Nervous System

Impacts, Issues: In Pursuit of Ecstasy [p.227]

13.1. Neurons—The Communication Specialists [pp.228–229]

13.2. A Closer Look at Action Potentials [pp.230–231]

13.3. Chemical Synapses: Communication Junctions [pp.232–233]

1. E; 2. J; 3. B; 4. G; 5. C; 6. D; 7. A; 8. F; 9. H; 10. I; 11. The neuron can respond to stimuli by producing an electrical signal; 12. dendrites; 13. cell body; 14. axon hillock; 15. axon; 16. axon ending; 17. concentrations; 18. closed; 19. potassium; 20. in, out; 21. negatively; 22. potential; 23. reverse; 24. open; 25. positive; 26. trigger zone; 27. away from; 28. T; 29. active transport; 30. into; 31. T; 32. potassium; 33. resting level; 34. threshold level; 35. action potential; 36. neurotransmitters; 37. chemical synapse; 38. presynaptic; 39. calcium; 40. vesicles; 41. postsynaptic; 42. action potential; 43. inhibitory; 44. acetylcholine; 45. neurotransmitter; 46. endorphins; 47. pain; 48. synapses; 49. billion; 50. graded; 51. EPSPs; 52. IPSPs; 53. Synaptic integration; 54. summation; 55. Integration; 56. removal; 57. enzymes, 58. pumped; 59. reuptake; 60. serotonin.

13.4. Information Pathways [pp.234–235]

13.5. The Nervous System: An Overview [pp.236–237]

13.6. An Environmental Assault on the Nervous System [p.237]

1. nerve; 2. myelin sheath; 3. action potentials; 4. Schwann; 5. node; 6. jump; 7. propagate; 8. central nervous system; 9. oligodendrocytes; 10. stretch reflex; 11. reflex arc; 12. contracts; 13. interneurons; 14. circuits; 15. diverge; 16. reverberating; 17. axon; 18. myelin sheath; 19. nerve fascicle; 20. blood vessels; 21. F; 22. E; 23. G; 24. D; 25. B; 26. C; 27. A; 28. central nervous system; 29. peripheral nervous system; 30. somatic; 31. autonomic; 32. spinal; 33. cranial; 34. ganglia; 35. cranial; 36. spinal cord; 37. cervical; 38. thoracic; 39. lumbar; 40. sacral; 41. coccygeal; 42. Pregnant women should avoid eating fish species that accumulate excessive mercury as it can cause brain damage in the developing fetus; mercury contamination in fish is rising 3 to 5 percent each year.

13.7. Major Expressways: Peripheral Nerves and the Spinal Cord [pp.238–239]

13.8. The Brain—Command Central [pp.240–241]

13.9. A Closer Look at the Cerebrum [pp.242–243]

13.10. Memory and States of Consciousness [pp.244–245]

1. A; 2. S; 3. S; 4. A; 5. A; 6. S; 7. A; 8. Parasympathetic; 9. Sympathetic; 10. Parasympathetic; 11. Sympathetic; 12. norepinephrine; 13. gray matter; 14. gray matter; 15. meninges; 16. do not; 17. autonomic; 18. spinal cord; 19. ganglion; 20. nerve; 21. vertebra; 22. intervertebral disk; 23. meninges; 24. gray matter; 25. white matter; 26. brain; 27. cranium; 28. meninges; 29. dura; 30. hemispheres; 31. cushion; 32. brain stem; 33. gray matter; 34. F, B; 35. F, A; 36. H, G; 37. H, H; 38. M, F; 39. F, D;

40. H, E; 41. F, C; 42. reticular formation; 43. posture;
44. cerebral cortex; 45. cerebrospinal; 46. ventricles;
47. meninges; 48. blood-brain barrier; 49. lipid; 50. cerebrum; 51. T; 52. right; 53. T; 54. nerve tracts; 55. T;
56. cerebral cortex; 57. T; 58. learned; 59. frontal; 60. parietal; 61. T; 62. T; 63. T; 64. self-gratifying; 65. thalamus;
66. hypothalamus; 67. corpus callosum; 68. midbrain;
69. pons; 70. medulla oblongata; 71. cerebellum;
72. short-term; 73. long-term; 74. skills; 75. amygdala;
76. hippocampus; 77. prefrontal; 78. corpus striatum;
79. cerebellum; 80. Amnesia; 81. skills; 82. Parkinson's;
83. learning; 84. Alzheimer's; 85. PET; 86. reticular formation; 87. serotonin.

13.11. Disorders of the Nervous System [p.246]
13.12. The Brain on Drugs [p.247]
1. C; 2. D; 3. H; 4. E; 5. G; 6. B; 7. A; 8. F; 9. C; 10. E;
11. G; 12. A; 13. F; 14. D; 15. B; 16. The addict must
increase intake to stay ahead of the liver's ability to
break down the drug; 17. Habituation develops when a
user cannot feel good or function normally without a
steady supply of the drug being abused.

Self-Quiz
1. a; 2. d; 3. a; 4. b; 5. a; 6. d; 7. e; 8. b; 9. d; 10. c.

Media Menu Review Questions
1. myelin; 2. more; 3. silent; 4. interferon; 5. genetics;
6. survival; 7. dopamine; 8. conditioned; 9. glutamate;
10. hereditary.

Chapter 14 Sensory Systems

Impacts, Issues: Private Eyes [p.251]
14.1. Sensory Receptors and Pathways—An Overview
   [pp.252–253]
14.2. Somatic "Body" Sensations [pp.254–255]
1. A sensation is conscious awareness of a stimulus,
while perception is understanding what it means;
2. Instead of involving one stimulus, a compound sensation integrates different stimuli; 3. The brain's assessment of the nature of a stimulus depends on (1) which
sensory area receives signals from nerves, (2) the frequency of signals, and (3) the number of axons that
respond; 4. Sensory adaptation occurs with pressure
sensations, such as the pressure from clothing on the
skin. It does not occur with stretch receptors, as when
the arm is holding weight; 5. Somatic sensations involve
receptors present at more than one location in the body,
while special senses depend on receptors restricted to
specific sense organs; 6. D; 7. C; 8. A; 9. B; 10. A; 11. E;
12. E; 13. B; 14. D; 15. B; 16. F; 17. B; 18. cerebrum;
19. greatest; 20. mechanoreceptors; 21. Meissner's
corpuscles; 22. Ruffini endings; 23. visceral pain;
24. activate pain receptors; 25. referred pain.

14.3. Taste and Smell—Chemical Senses [pp.256–257]
14.4. A Tasty Morsel of Sensory Science [p.257]
14.5. Hearing: Detecting Sound Waves [pp.258–259]
14.6. Balance: Sensing the Body's Natural Position
   [pp.260–261]
14.7. Noise Pollution: An Attack on the Ears [p.261]
1. chemoreceptors; 2. thalamus; 3. taste buds; 4. five;
5. salty; 6. umami; 7. olfactory; 8. olfactory; 9. cerebral

cortex; 10. vomeronasal; 11. pheromones; 12. tastants;
13. two; 14. sensitive; 15. amplitude; 16. frequency;
17. mechanoreceptors; 18. hairs; 19. action potential;
20. sound; 21. D; 22. B; 23. J; 24. G; 25. L; 26. K; 27. H;
28. E; 29. A; 30. I; 31. C; 32. F; 33. I; 34. H; 35. D; 36. G;
37. K; 38. F; 39. J; 40. C; 41. B; 42. E; 43. A; 44. tympanic
membrane; 45. middle ear bones; 46. oval window;
47. cochlea; 48. auditory nerve; 49. scala vestibuli;
50. cochlear duct; 51. scala tympani; 52. basilar membrane; 53. tectorial membrane; 54. equilibrium position;
55. receptors; 56. vestibular; 57. semicircular canals;
58. dynamic equilibrium; 59. deceleration; 60. cupula;
61. static; 62. vestibular; 63. otolith organ; 64. otoliths;
65. medulla oblongata; 66. balance; 67. Motion sickness.

14.8. Vision: An Overview [pp.262–263]
14.9. From Visual Signals to "Sight" [pp.264–265]
14.10. Disorders of the Eye [pp.266–267]
1. photoreceptors; 2. brain; 3. vision; 4. eyes; 5. cornea;
6. iris; 7. retina; 8. H; 9. K; 10. A; 11. B; 12. E; 13. G; 14. J;
15. F; 16. D; 17. I; 18. C; 19. choroid; 20. vitreous humor;
21. ciliary muscle; 22. iris; 23. pupil; 24. lens; 25. cornea;
26. aqueous humor; 27. sclera; 28. retina; 29. fovea;
30. optic disk (blind spot); 31. optic nerve; 32. c; 33. e;
34. d; 35. d; 36. cones; 37. T; 38. T; 39. a few; 40. greater;
41. ganglion cells; 42. T; 43. prevents; 44. "visual field";
45. T; 46. I; 47. D; 48. F; 49. H; 50. G; 51. C; 52. E; 53. A;
54. B; 55. K; 56. J.

Self-Quiz
1. a; 2. c; 3. d; 4. a; 5. e; 6. d; 7. a; 8. c; 9. b; 10. b.

Media Menu Review Questions
1. retina; 2. rods, cones; 3. color; 4. fovea; 5. migraines;
6. epilepsy; 7. tension; 8. repeat; 9. migraine; 10. result;
11. trigeminal; 12. fluorescent; 13. soften.

Chapter 15 The Endocrine System

Impacts, Issues: Hormones in the Balance [p.271]
15.1. The Endocrine System: Hormones [pp.272–273]
15.2. Hormone Categories and Signaling [pp.274–275]
1. E; 2. D; 3. G; 4. F; 5. A; 6. B; 7. C; 8. a. (9) PTH; b.
(4) cortisol, aldosterone; c. (7) melatonin; d. (11) insulin,
glucagon, somatostatin; e. (5) estrogens, progesterones;
f. (1) produces and secretes releasing and inhibiting hormones and synthesizes ADH and oxytocin; g. (2) ACTH,
TSH, FSH, LH, PRL, GH; h. (8) thyroxine, triiodothyronine, calcitonin; i. (10) thymosins; j. (6) androgens
(including testosterone); k. (3) stores and secretes ADH
and oxytocin, both from the hypothalamus;
l. (4) epinephrine, norepinephrine; 9. a; 10. b; 11. a; 12. b;
13. b; 14. b; 15. a; 16. b.

15.3. The Hypothalamus and Pituitary Gland—Major
   Controllers [pp.276–277]
15.4. When Pituitary Signals Go Awry [p.278]
1. A (G); 2. P (F); 3. A (A); 4. A (H); 5. A (C); 6. P (D);
7. A (B); 8. A (E); 9. hypothalamus; 10. posterior;
11. anterior; 12. releaser; 13. inhibitor; 14. ACTH;
15. FSH; 16. Pituitary dwarfism; 17. gigantism;
18. acromegaly.

15.5. Sources and Effects of Other Hormones [p.279]

15.6. Hormones and Feedback Controls—The Adrenals and Thyroid [pp.280–281]

15.7. Fast Responses to Local Changes—Parathyroids and the Pancreas [pp.282–283]

15.8. Sweet Treachery—The Diabetes Epidemic [pp.284–285]

15.9. Some Final Examples of Integration and Control [pp.286–287]

1. D, c; 2. I, e; 3. F, h; 4. A, g; 5. H, k; 6. E, d; 7. C, a; 8. K, j; 9. B, i; 10. J, f; 11. G, b; 12. F, l; 13. A, m; 14. B, o; 15. C, n; 16. E; 17. H; 18. F; 19. A; 20. B; 21. G; 22. D; 23. I; 24. C; 25. thyroid; 26. metabolic; 27. calcitonin; 28. iodine; 29. TSH; 30. goiter; 31. hypothyroidism; 32. overweight; 33. iodized salt; 34. Graves; 35. toxic; 36. hyperthyroidism; 37. increased; 38. parathyroid; 39. parathyroid; 40. calcium; 41. remodeling; 42. kidney; 43. D; 44. rickets; 45. bones; 46. exocrine, endocrine; 47. islets; 48. alpha; 49. beta; 50. inhibit; 51. rises; 52. excessive; 53. energy; 54. decreased; 55. insulin; 56. Glucagon; 57. type 1 diabetes; 58. type 1 diabetes; 59. type 2 diabetes; 60. Type 2 diabetes; 61. a viral infection; 62. obesity; 63. is; 64. capillaries; 65. edema; 66. is at risk for; 67. can; 68. c; 69. f; 70. a; 71. g; 72. b; 73. e; 74. b; 75. d; 76. g; 77. a; 78. e; 79. b; 80. f; 81. a.

Self-Quiz

1. a; 2. e; 3. d; 4. d; 5. c; 6. e; 7. b; 8. a; 9. c; 10. a; 11. H; 12. F; 13. O; 14. N; 15. D; 16. E; 17. A; 18. G; 19. B; 20. L; 21. I; 22. M; 23. Q; 24. C; 25. K; 26. J; 27. P.

Chapter Objectives/Review Questions

1. Target; 3. endocrine; 10. releasers, inhibitors; 11. growth hormone; 13. cortex; 15. aldosterone; 16. thymus, immune; 17. adrenal medulla; 26. melatonin; 28. prostaglandins.

Media Menu Review Questions

1. diet, exercise; 2. 2; 3. 20; 4. chemicals; 5. endocrine; 6. reproductive; toxicants; 7. mechanisms; 8. sperm, cancer, neurological; 9. neurobehavioral; 10. Delaney; 11. 2005; 12. 60.

Chapter 16 Reproductive Systems

Impacts, Issues: Sperm with a Nose for Home? [p.291]

16.1. The Male Reproductive System [pp.292–293]

16.2. How Sperm Form [pp.294–295]

1. T; 2. before; 3. T; 4. T; 5. epididymis; 6. sperm + glandular secretions; 7. T; 8. T; 9. one; 10. T; 11. haploid; 12. T; 13. T; 14. e; 15. a; 16. d; 17. c; 18. b; 19. b; 20. c; 21. a; 22. seminiferous tubules; 23. spermatogonia; 24. mitosis; 25. meiosis; 26. primary spermatocytes; 27. meiosis I; 28. secondary spermatocytes; 29. meiosis II; 30. spermatids; 31. sperm; 32. tail; 33. Sertoli; 34. head; 35. acrosome; 36. mitochondria; 37. Leydig cells; 38. Testosterone; 39. testosterone; 40. anterior; 41. hypothalamus; 42. decrease; 43. LH; 44. FSH; 45. increase.

16.3. The Female Reproductive System [pp.296–297]

16.4. The Ovarian Cycle: Oocytes Develop [pp.298–299]

16.5. Visual Summary of the Menstrual and Ovarian Cycles [p.300]

1. ovary; 2. oviduct; 3. uterus; 4. myometrium; 5. endometrium; 6. cervix; 7. vagina; 8. labia majora; 9. labia minora; 10. clitoris; 11. urethra; 12. three; 13. menstruation; 14. endometrium; 15. Proliferative; 16. Endometrium; 17. Ovulation; 18. Progestational; 19. Corpus luteum; 20. Endometrium; 21. Menarche; 22. Menopause; 23. Endometriosis; 24. Ovarian; 25. 300; 26. ovary; 27. granulosa; 28. FSH; 29. zona pellucida; 30. estrogens; 31. secondary; 32. polar; 33. follicle; 34. Secondary oocyte; 35. oviduct; 36. Fertilization; 37. ovum; 38. endometrium; 39. progesterone; 40. cervix; 41. corpus luteum; 42. endometrium; 43. follicles; 44. implant; 45. endometrium; 46. estrogen; 47. grows; 48. LH; 49. decrease; 50. increase; 51. increase; 52. remain stable; 53. fully developed; 54. luteal; 55. menstruation; 56. be maintained; 57. corpus luteum; 58. around the middle of.

16.6. Sexual Intercourse, etc. [p.301]

16.7. Controlling Fertility [p.302]

1. scratch through fertilization; 2. scratch through lubrication; 3. scratch through the male's bladder; 4. scratch through pancreas; 5. scratch through pregnancy; 6. scratch through orgasm; 7. scratch through at any time in the menstrual cycle; 8. F (a); 9. L (d); 10. B (e); 11. D (e); 12. O (c); 13. N (c); 14. P (b); 15. J (c); 16. C (b); 17. M (b); 18. Q (c); 19. K (b); 20. H (b); 21. E (a); 22. I (a); 23. A (a); 24. G (a); 25. Biodegradable implants, contraceptives for males, chemical "sterilizers," reversible sterilization for both males and females.

16.8. Options for Coping with Infertility [pp.304–305]

16.9. Dilemmas of Fertility Control [p.305]

1. c; 2. a; 3. b; 4. a; 5. b; 6. a; 7. c; 8. b; 9. a; 10. c; 11. a; 12. b; 13. a; 14. later trimesters; 15. Abortion is more accepted and even encouraged in some countries with population problems.

Self-Quiz

1. a; 2. b; 3. d; 4. c; 5. a; 6. b; 7. c; 8. e; 9. e; 10. c; 11. d; 12. a; 13. d; 14. a; 15. c; 16. e; 17. b; 18. c; 19. b; 20. a.

Chapter Objectives/Review Questions

1. testes, ovaries; 2. sex hormones, secondary; 3. testes, epididymis; 8. oocyte; 11. FSH, LH, estrogen, progesterone; 12. GnRH; 15. tubal ligation, vasectomy; 16 condom.

Media Menu Review Questions

1. ovulation; 2. progesterone; 3. endometrium; 4. menstrual bleeding; 5. normal, 6. endometriosis; 7. preferable; 8. receptors; 9. spearmint; 10. water.

Chapter 17 Development and Aging

Impacts, Issues: Fertility Factors and Mind-Boggling Births [p.309]

17.1. The Six Stages of Early Development: An Overview [pp.310–311]

17.2. The Beginnings of You—Early Steps in Development [pp.312–313]

17.3. Vital Membranes Outside the Embryo [pp.314–315]

1. F; 2. B; 3. C; 4. A; 5. E; 6. D; 7. a. mesoderm; b. ectoderm; c. endoderm; d. mesoderm; e. ectoderm; f. mesoderm; g. endoderm; h. mesoderm; i. ectoderm;

8. organogenesis; 9. specialization; 10. determination; 11. differentiation; 12. morphogenesis; 13. movements; 14. paddle; 15. died; 16. complex; 17. G; 18. D; 19. J; 20. B; 21. F; 22. A; 23. K; 24. H; 25. I; 26. C; 27. E; 28. I; 29. M; 30. C; 31. J; 32. N; 33. E; 34. B; 35. H; 36. K; 37. A; 38. L; 39. F; 40. G; 41. D; 42. O; 43. 31; 44. 30; 45. 29; 46. 40; 47. 33; 48. f; 49. e; 50. e; 51. d; 52. a; 53. b; 54. a; 55. b; 56. d; 57. f; 58. f; 59. e; 60. f.

17.4. How the Early Embryo Takes Shape [pp.316–317]

17.5. The First Eight Weeks: Human Features Emerge [pp.318–319]

17.6. Development of the Fetus [pp.320–321]

1. F; 2. J; 3. G; 4. K; 5. A; 6. M; 7. I; 8. L; 9. B; 10. D; 11. H; 12. C; 13. E; 14. c; 15. b; 16. a; 17. d; 18. lanugo; 19. vernix caseosa; 20. second; 21. third; 22. distress; 23. umbilical; 24. umbilical; 25. lungs; 26. liver; 27. collapsed; 28. atrium; 29. foramen ovale; 30. birth; 31. heart; 32. increases; 33. foramen ovale; 34. pulmonary; 35. systemic; 36. venous; 37. yolk sac; 38. embryonic disk; 39. amniotic cavity; 40. chorionic cavity; 41. primitive streak; 42. neural tube; 43. future brain; 44. somites; 45. pharyngeal arches; 46. pharyngeal arches; 47. developing heart; 48. somites; 49. foot plates; 50. retinal pigment; 51. If an embryo has a Y chromosome, testes develop that produce hormones that will masculinize the embryo. In the absence of this hormone signal, the embryo will become a female.

17.7. Has the Age of Cloning Arrived? [p.322]

17.8. Birth and Beyond [pp.322–323]

17.9. Mother as Provider, Protector, and Potential Threat [pp.324–325]

17.10. Prenatal Diagnosis: Detecting Birth Defects [p.326]

17.11. The Path from Birth to Adulthood [p.327]

1. Therapeutic cloning enables tissues to be grown that are not rejected by a patient's immune system; 2. A very long time. In addition to technical obstacles, only single individuals would benefit at an incredibly high cost; 3. Parturition; 4. labor; 5. cervix; 6. amniotic; 7. birth; 8. two; 9. breech; 10. afterbirth; 11. umbilical cord; 12. carbon dioxide; 13. inhalation; 14. navel; 15. lungs; 16. colostrum; 17. lactation; 18. Oxytocin; 19. The developing individual is at the mercy of the mother's diet and health habits. The course of development is greatly influenced by these factors. Infection of the embryo prior to four months can result in malformations. The embryo is highly sensitive to drugs during the first trimester. Fetal alcohol syndrome (FAS) is a constellation of defects that can result from alcohol use by a pregnant woman. FAS is the third most common cause of mental retardation in the United States. Cigarette smoking harms fetal growth and development. Research shows that women smokers are at greater risk of miscarriage, stillbirth, and premature delivery; 20. CVS; 21. Amniocentesis; 22. preimplantation diagnosis; 23. Fetoscopy; 24. adolescence; 25. childhood; 26. neonate; 27. adult; 28. pubescent; 29. senescence.

17.12. Time's Toll: Everybody Ages [p.328]

17.13. Aging Skin, Muscle, Bones, and Reproductive Systems [p.329]

17.14. Age-Related Changes in Some Other Body Systems [pp.330–331]

1. G; 2. D; 3. F; 4. E; 5. H; 6. A; 7. C; 8. I; 9. B; 10. The shape of a collagen molecule is stabilized by bonds that form cross-links. More cross-links develop with age, making collagen more rigid. This can alter structure and function. This most likely occurs in other proteins besides collagen.

Self-Quiz

1. d; 2. b; 3. b; 4. a; 5. d; 6. d; 7. F; 8. B; 9. J; 10. I; 11. H; 12. L; 13. D; 14. M; 15. A; 16. N; 17. E; 18. C; 19. G; 20. K.

Chapter Objectives/Review Questions

1. organogenesis; 3. cell differentiation; 7. Capacitation; 8. blastocyst; 12. placenta; 13. neural tube; 19. senescence.

Media Menu Review Questions

1. sugar; 2. infertility; 3. ¾; 4. leukocytes; 5. three; 6. animal; 7. stem; 8. developmental; 9. Hox; 10. undernourishment.

Chapter 18 Life at Risk: Infectious Disease

Impacts, Issues: The Face of AIDS [p.335]

18.1. Viruses and Infectious Proteins [pp.336–337]

18.2. Bacteria—The Unseen Multitudes [pp.338–339]

1. The "bad news" is the genetic material of the virus that has the capacity to take over and destroy a cell; 2. is not; 3. reproduce; 4. T; 5. capsid; 6. T; 7. host cells; 8. E; 9. B; 10. F; 11. C; 12. A; 13. D; 14. receptors; 15. kills its host quickly; 16. reactivated; 17. epithelial; 18. Epstein-Barr; 19. retroviruses; 20. prions; 21. protein; 22. prokaryotic; 23. peptidoglycan; 24. coccus; 25. bacillus; 26. spirillum; 27. flagellum; 28. Pili; 29. prokaryotic fission; 30. one; 31. 20; 32. plasmids; 33. fertility; 34. making cheeses, therapeutic drugs, genetic engineering; 35. prescribing unnecessary antibiotics, not using all of a prescription of antibiotics; 36. 1940s; 37. bacteria and fungi; 38. bacteria; 39. interferons; 40. susceptible; 41. plasmids; 42. resistant.

18.3. Infectious Protozoa and Worms [p.340]

18.4. Infectious Foes Old and New [p.341]

18.5. Characteristics and Patterns of Infectious Diseases [pp.342–343]

1. B; 2. D; 3. C; 4. A; 5. E; 6. scratch through *Trypanosoma brucei*; 7. scratch through sexual intercourse; 8. scratch through *Cryptosporidium*; 9. b; 10. a; 11. d; 12. d; 13. a; 14. c; 15. b; 16. a; 17. c; 18. a; 19. d; 20. a; 21. b; 22. c; 23. b; 24. d; 25. a; 26. d; 27. d; 28. c; 29. d; 30. C; 31. G; 32. A; 33. B; 34. J; 35. F; 36. D; 37. E; 38. H; 39. I; 40. The immune response to gonorrhea may be so extreme that it results in sterility; 41. They secrete toxins that damage the nervous system and vital organs of the host; 42. There are more people on the planet, people are traveling more than ever, the misuse of antibiotics makes bacteria hard to control; 43. a. common cold, measles; b. zoonoses; c. animals are main reservoir; d. emerging; e. West Nile virus, Lyme disease; f. re-emerging; g. tuberculosis.

18.6. The Human Immunodeficiency Virus and AIDS [pp.344–345]

18.7. Treating and Preventing HIV Infection and AIDS
[p.346]
1. more than one; 2. immune; 3. T; 4. T; 5. sex with an
infected partner; 6. is not; 7. can; 8. are not; 9. AIDS;
10. T; 11. T; 12. helper T cells; 13. T; 14. are not; 15. d;
16. a; 17. c; 18. b; 19. e; 20. T; 21. assembly; 22. protease
inhibitor; 23. cells; 24. T; 25. T.

18.8. Protecting Yourself—and Others—From STDs
[p.347]
18.9. Common STDs Caused by Bacteria [pp.348–349]
18.10. A Rogue's Gallery of Viral STDs and Others
[pp.350–351]
1. a; 2. a; 3. a; 4. b; 5. b; 6. b; 7. a; 8. c; 9. d; 10. a; 11. a;
12. a; 13. a; 14. c; 15. g—but b for treatment; 16. a; 17. e;
18. f; 19. PID stands for pelvic inflammatory disease; it
may be caused by chlamydial or gonorrheal infections;
in both cases, the infection may be present for some time
without a woman knowing it; 20. primary—a small
chancre develops that is not painful; secondary—lesions
develop on mucous membranes and a rash breaks out,
then virus becomes latent; tertiary—lesions develop on
skin and damage occurs to internal organs, including the
brain, which possibly leads to insanity and paralysis.

Self-Quiz
1. a; 2. a; 3. f; 4. d; 5. b; 6. b; 7. c; 8. c; 9. c; 10. g; 11. g;
12. h; 13. e; 14. i.

Media Menu Review Questions
1. China, 200; 2. coronavirus; 3. genome; 4. mutates;
5. not closely related; 6. facilitated, animal; 7. papillomavirus; 8. genital herpes; 9. 70; 10. HPV.

Chapter 19 Cell Reproduction

Impacts, Issues: Henrietta's Immortal Cells [p.355]
19.1. Dividing Cells Bridge Generations [pp.356–357]
1. HeLa cells have been invaluable in the study of cell
division and of cancer; 2. E; 3. K; 4. A; 5. G; 6. H; 7. B;
8. J; 9. I; 10. D; 11. F; 12. C.

19.2. The Cell Cycle [p.358]
19.3. A Closer Look at Chromosomes [p.359]
1. interphase; 2. mitosis; 3. G_1; 4. S; 5. G_2; 6. prophase;
7. metaphase; 8. anaphase; 9. telophase; 10. cytokinesis;
11. 5; 12. 2; 13. 4; 14. 3; 15. 1; 16. 10; 17. 1; 18. histones;
19. nucleosome; 20. loosens; 21. condenses; 22. chromatid; 23. centromere, microtubules; 24. spindle;
25. microtubules.

19.4. The Four Stages of Mitosis [pp.360–361]
19.5. How the Cytoplasm Divides [p.362]
19.6. Concerns and Controversies over Irradiation [p.363]
1. interphase—daughter cells (F); 2. anaphase (A); 3. late
prophase (G); 4. metaphase (D); 5. interphase—parent
cell (E); 6. early prophase (C); 7. prometaphase (B);
8. telophase (H); 9. Prophase; 10. interphase; 11. sister
chromatids; 12. centromere; 13. chromosome; 14. condensed; 15. microtubules; 16. centrioles; 17. prophase;
18. Prometaphase; 19. chromosomes; 20. poles (sides);
21. chromatids; 22. metaphase; 23. anaphase; 24. chromatid; 25. Telophase; 26. nuclear envelope; 27. number;
28. telophase; 29. mitosis; 30. B; 31. D; 32. A; 33. C;

34. Natural sources include cosmic rays from outer
space and radioactive radon gas in rocks and soil;
35. Ionizing radiation damages cells by breaking apart
chromosomes and altering genes. If the damage occurs
in germ cells, the resulting gametes may give rise to
infants with genetic defects. If somatic (body) cells are
damaged, the damage may include burns, miscarriages,
eye cataracts, and cancers of the bone, thyroid, breast,
skin, and lungs; 36. In medicine, X-rays, MRI, and PET
scanning are valuable uses of irradiation. Irradiation
therapy is useful in treating cancer. Food is irradiated
to kill harmful microorganisms and to prolong shelf
life by preventing vegetables like potatoes from
sprouting.

19.7. Meiosis—The Beginnings of Eggs and Sperm
[pp.364–365]
19.8. A Visual Tour of the Stages of Meiosis [pp.366–367]
1. T; 2. gametes; 3. T; 4. Diploid; 5. T; 6. 1; 7. 2; 8. 4; 9. n;
10. gamete production; 11. E ($2n$); 12. D ($2n$); 13. B ($2n$);
14. A (n); 15. C (n); 16. 2 ($2n$); 17. 5 (n); 18. 4 (n); 19. 1
($2n$); 20. 3 (n); 21. Spermatogenesis produces four viable
sperm, while oogenesis produces one viable egg along
with three polar bodies that are discarded; 22. anaphase
II (H); 23. metaphase II (F); 24. metaphase I (A);
25. prophase II (B); 26. telophase II (C); 27. telophase
I (G); 28. prophase I (E); 29. anaphase I (D).

**19.9. The Second Stage of Meiosis—New Combinations
of Parents' Traits** [pp.368–369]
19.10. Meiosis and Mitosis Compared [pp.370–371]
1. B; 2. D; 3. E; 4. F; 5. A; 6. G; 7. C; 8. a. mitosis; b. mitosis; c. meiosis; d. meiosis; e. meiosis; f. meiosis; g. mitosis; h. meiosis; i. meiosis; 9. C; 10. F; 11. D; 12. A; 13. B;
14. E; 15. four; 16. eight; 17. four; 18. eight; 19. two.

Self-Quiz
1. a; 2. d; 3. a; 4. d; 5. c; 6. e; 7. a; 8. a; 9. d; 10. c; 11. b;
12. c; 13. b.

Chapter Objectives/Review Questions
1. DNA; 3. chromosomes; 5. diploid; 8. DNA, histones;
9. chromatids; 11. replicated, chromatids; 13. 23; 14.
meiosis I and II; 17. crossing over; 18. independent
assortment.

Media Menu Review Questions
1. telomerase, telomeres; 2. DNA; 3. stop;
4. replenish; 5. finite; 6. cancer; 7. aging; 8. genes;
9. cancerous; 10. transplants, aging.

Chapter 20 Observable Patterns of Inheritance

Impacts, Issues: Designer Genes? [p.375]
20.1. Basic Concepts of Heredity [p.376]
20.2. One Chromosome, One Copy of a Gene [p.377]
20.3. Figuring Genetic Probabilities [pp.378–379]
1. G; 2. A; 3. C; 4. K; 5. E; 6. I; 7. D; 8. H; 9. J; 10. B; 11. F;
12. monohybrid; 13. chromosome; 14. segregation;
15. haploid; 16. F_1; 17. *Cc*; 18. Punnett; 19. probability;
20. don't have to; 21. doesn't; 22. D; 23. A; 24. B; 25. E;
26. C; 27. *C* and *c*; b. *c*; c. *C* and *c* go across the top of the
Punnett square; *c*'s go down the side of the square; d. 1

Cc: 1 *cc*; e. 1 chin fissure: 1 smooth chin; f. 1/2; the predicted ratio remains constant for each fertilization event; 28. a. ½ × 1 = ½; b. ½ × 1 = ½; c. 1/2 chin fissure and 1/2 smooth chin.

20.4. The Testcross: A Tool for Discovering Genotypes [p.380]

20.5. How Genes for Different Traits Are Sorted into Gametes [pp.380–381]

20.6. More Gene-Sorting Possibilities [p.382]

1. C; 2. B; 3. A; 4. D; 5. The taster parent is heterozygous, *Aa*. The nontaster child must be homozygous and had to have received one nontasting gene from each of his parents; 6. The man is *cc* and the woman is *Cc*. The probability that the child will have either a smooth chin or a chin fissure is 1/2 for each; 7. Albino is *aa*, normal pigmentation is *AA* or *Aa*. The woman of normal pigmentation with an albino mother is genotype *Aa*; the woman received her recessive gene (*a*) from her mother and her dominant gene (*A*) from her father. The woman's husband is *aa* and will give each of their offspring an *a* allele. It is likely that half of the couple's children will be albinos (*aa*) and half will have normal pigmentation but be heterozygous (*Aa*); 8. a. 9/16; b. 3/16; c. 3/16; d. 1/16 (note the following Punnett square).

	BR	Br	bR	br
BR	BBRR	BBRr	BbRR	BbRr
Br	BBRr	BBrr	BbRr	Bbrr
bR	BbRR	BbRr	bbRR	bbRr
br	BbRr	Bbrr	bbRr	bbrr

9. a. ¾; b. ¼; c. ¾; d. ¼; e. ¾ × ¾ = ⁹⁄₁₆; f. ¾ × ¼ = ³⁄₁₆; g. ¼ × ¾ = ³⁄₁₆; h. ¼ × ¼ = ¹⁄₁₆

20.7. Single Genes, Varying Effects [p.383]

20.8. Other Gene Impacts and Interactions [pp.384–385]

20.9. Custom Cures? [p.385]

1. a. pleiotropy; b. multiple alleles; c. codominance; d. polygenic inheritance; 2. There is evidence that butyrate can reactivate dormant genes for fetal hemoglobin, genes that are normally turned off after birth; 3. f; 4. d; 5. f; 6. g; 7. b; 8. a; 9. c; 10. g; 11. e; 12. a.

Self-Quiz

1. d; 2. b; 3. a; 4. d; 5. d; 6. c; 7. a; 8. a; 9. c; 10. d.

Chapter Objectives/Review Questions

2. Gregor Mendel; 3. Genes; 5. homozygous, heterozygous; 7. Genotype, phenotype; 8. monohybrid; 9. segregation; 12. independent assortment; 16. pleiotropy; 18. variable.

Media Menu Review Questions

1. Pharmacogenetics; 2. loss, funding; 3. parenting; 4. thinks, feels; 5. genes; 6. susceptibility; 7. Pharmacogenetics; 8. markers; 9. 17, 7, 2; 10. speaking, move.

Integrating and Applying Key Concepts (genetics problem answer)

The first mating could only produce one genotype (*HhPp*) and one phenotype, with all individuals having normal hair and normal growth. Matings between genotypes like those of their children (*HhPp* × *HhPp*) would produce offspring with the following probability ratios and phenotypes: 9 normal hair, normal height; 3 normal hair, dwarf; 3 hypotrichotic, normal height; 1 hypotrichotic, dwarf.

Chapter 21 Chromosomes and Human Genetics

Impacts, Issues: Menacing Mucus [p.389]

21.1. Chromosomes and Inheritance [p.390]

21.2. Picturing Chromosomes with Karyotypes [p.391]

1. J; 2. E; 3. B; 4. F; 5. C; 6. G; 7. I; 8. K; 9. D; 10. M; 11. L; 12. N; 13. A; 14. H; 15. F; 16. A; 17. C; 18. D; 19. B; 20. G; 21. E.

21.3. How Sex Is Determined [pp.392–393]

21.4. Linked Genes [p.393]

21.5. Human Genetic Analysis [pp.394–395]

1. Two blocks of the Punnett square should be XX and two blocks should be XY; 2. sons; 3. mothers; 4. daughters; 5. nonsexual; 6. D; 7. I; 8. C; 9. J; 10. F; 11. A; 12. H; 13. B; 14. E; 15. G; 16. The farther apart two genes are on a chromosome, the more likely it is that crossing over and recombination between them will occur.

21.6. Inheritance of Genes on Autosomes [pp.396–397]

21.7. Inheritance of Genes on the X Chromosome [pp.398–399]

21.8. Sex-Influenced Inheritance [p.400]

1. b; 2. a; 3. c; 4. a; 5. b; 6. b; 7. a; 8. e; 9. c; 10. c; 11. a; 12. e; 13. c; 14. d; 15. d; 16. The woman's father is homozygous recessive, *aa*; the woman is heterozygous normal, *Aa*. The albino man, *aa*, has two heterozygous normal parents, *Aa*. The two normal children are heterozygous normal, *Aa*; the albino child is *aa*; 17. Assuming the father is heterozygous with Huntington disorder and the mother is normal, the chances are 1/2 (50%) that the son will develop the disease; 18. If only male offspring are considered, the probability is 1/2 (50%) that the couple will have a color-blind son; 19. The probability is that half of the sons will have hemophilia; the probability is 0 that a daughter will express hemophilia; the probability is that half of the daughters will be carriers; 20. If the woman marries a normal male, the chance that her son would be color blind is 1/2 (50%); if she marries a color-blind male, the chance that her son would also be color blind is also 1/2 (50%); 21. The nonbald man is homozygous *b1b1*, and his wife must be heterozygous *b1b* because her mother was bald. The first child can be bald only if it is a male (probability of 1/2) and also heterozygous for this gene (probability of 1/2). The chance of having a bald son is (1/2 × 1/2) = 1/4; 22. a. autosomal dominant; b. X-linked recessive; c. sex influenced; d. autosomal dominant; e. X-linked recessive; f. X-linked dominant; g. X-linked recessive; h. autosomal recessive;

i. autosomal recessive; j. autosomal recessive; k. autosomal dominant.

21.9. How a Chromosome's Structure Can Change [pp.400–401]

21.10. Changes in Chromosome Number [pp.402–403]
1. duplication (B); 2. deletion (A); 3. translocation (C); 4. A mutation is a change in one or more of the nucleotides that make up a particular gene; a mutation may have no phenotypic effect, may produce a positive change in phenotype, or may cause a genetic disorder; 5. a. aneuploidy; b. polyploidy; c. nondisjunction; d. trisomy; e. monosomy; 6. Most changes in the number of chromosomes (aneuploidy) arise through nondisjunction during gamete formation; 7. c; 8. b; 9. c; 10. d; 11. a; 12. b; 13. d; 14. c; 15. a.

Self-Quiz
1. d; 2. d; 3. b; 4. c; 5. b; 6. a; 7. c; 8. b; 9. b; 10. c.

Chapter Objective/Review Questions
1. genes; 2. homologous; 3. Alleles; 11. *SRY*; 13. linked; 15. pedigree; 16. abnormality, disorder; 17. sex-influenced; 18. aneuploidy; 19. polyploidy; 20. Nondisjunction; 21. Down, XO, XXX, Klinefelter; XYY.

Media Menu Review Questions
1. dozen; 2. sperm; 3. *SRY*; 4. 33; 5. learning; 6. predisposition; 7. nature, nurture, eugenics; 8. patterns; 9. environment; 10. 1993, blood; 11. 50 percent; 12. brain; 13. childbearing; 14. cure; 15. genetically engineered.

Chapter 22 DNA, Genes, and Biotechnology

Impacts, Issues: Ricin and Your Ribosomes [p.407]

22.1. DNA: A Double Helix [pp.408–409]

22.2. Passing on Genetic Instructions [pp.410–411]
1. castor beans; damages ribosomes, which stops protein synthesis; no; 2. a five-carbon sugar (deoxyribose), a phosphate group, and one of the four nitrogen-containing bases; 3. a. adenine; b. guanine; c. thymine; d. cytosine; e. double ring; f. double ring; g. single ring; h. single ring; i. thymine; j. cytosine; k. adenine; l. guanine; 4. four; 5. T; 6. five-carbon sugar; 7. T; 8. T; 9. helix; 10. T; 11. hydrogen; 12. deoxyribose; 13. phosphate group; 14. double; 15. single; 16. double; 17. single; 18. nucleotide; 19. covalent; 20. hydrogen; 21. a sequence of DNA nucleotides coding for a specific polypeptide chain;

22.
T - A	T - A
G - C	G - C
A - T	A - T
C - G	C - G
C - G	C - G
C - G	C - G

23. bases; 24. nucleotide; 25. semiconservative; 26. DNA polymerases; 27. repair; 28. two million; 29. mutation; 30. ultraviolet radiation; 31. thymine dimer; 32. gene mutations; 33. base-pair substitution; 34. inserted; 35. deleted; 36. expansion; 37. transposable elements; 38. proteins; 39. gonads; 40. beneficial.

22.3. DNA into RNA—The First Step in Making Proteins [pp.412–413]

22.4. Reading the Genetic Code [pp.414–415]

22.5. Translating the Genetic Code into Protein [pp.416–417]
1. a. deoxyribose; b. adenine, thymine, guanine, cytosine; c. ribose; d. adenine, uracil, guanine, cytosine; 2. transcription; 3. translation; 4. transcription; 5. translation; 6. polypeptide; 7. protein; 8. a. rRNA: nucleic acid chain that combines with certain proteins to form ribosomes, which are involved in assembly of polypeptide chains; b. mRNA: linear sequence of nucleic acids that deliver protein-building instructions to ribosomes for translation into polypeptide chains; c. tRNA: nucleic acid chain that picks up a specific amino acid and delivers it to the ribosome where it will pair with a specific mRNA code for that particular amino acid; 9. In transcription, only the gene segment serves as the template—not the whole DNA strand (as in DNA replication). Enzymes called RNA polymerases are involved instead of DNA polymerases. Transcription results in only a single-stranded molecule—not one with two strands (as in DNA replication); 10. E; 11. B; 12. F; 13. A; 14. C; 15. D; 16. AUG UUC UAU UGU AAU AAA GGA UGG CAG UAG; 17. DNA gene (E); 18. introns (B); 19. cap (F); 20. exons (A); 21. tail (D); 22. mature RNA transcript (C); 23. speed up, halt, or trigger transcription; 24. F; 25. G; 26. H; 27. A; 28. E; 29. D; 30. C; 31. B; 32. a. initiation; b. elongation; c. termination; 33. tRNA anticodon sequence: UAC AAG AUA ACA UUA UUU CCU ACC GUC AUC; 34. amino acids: met phe tyr cys asn lys gly try gln stop; 35. transcription (E); 36. messenger RNA (A); 37. ribosome subunits (D); 38. transfer RNA (G); 39. translation (B); 40. ribosome (F); 41. polypeptide (C).

22.6. Tools for "Engineering" Genes [pp.418–419]

22.7. "Sequencing" DNA [p.420]
1. The bacterial chromosome, a circular DNA molecule, contains all the genes necessary for normal growth and development. Plasmids are small, circular molecules of "extra" DNA that carry only a few genes and are self-replicating; 2. mutation; 3. recombinant DNA; 4. species; 5. replicate; 6. genetic engineering; 7. plasmids; 8. restriction enzyme; 9. "sticky"; 10. clone; 11. amplify; 12. d; 13. g; 14. h; 15. e; 16. f; 17. b; 18. a; 19. c; 20. PCR is used to amplify DNA fragments in a test tube, using primers instead of using bacteria as cloning vectors; 21. These short nucleotide chains base-pair with any complementary DNA sequences. DNA polymerases recognize primers as "start" tags; 22. DNA sequencing; 23. probe; 24. gene library.

22.8. Mapping the Human Genome [pp.420–421]

22.9. Some Applications of Biotechnology [pp.422–423]

22.10. Issues for a Biotechnological Society [p.424]

22.11. Engineering Bacteria, Animals, and Plants [p.425]

22.12. Mr. Jefferson's Genes [p.426]
1. nucleotides; 2. 30,000; 3. one and a half; 4. T; 5. T; 6. T; 7. has; 8. gene therapy; 9. viruses; 10. transfection; 11. virus; 12. allele; 13. seven; 14. 5; 15. cancer; 16. interleukin; 17. attack; 18. Lipoplexes; 19. immune system;

20. anti-tumor; 21. DNA fragments; 22. tandem repeats; 23. gel electrophoresis; 24. restriction enzymes; 25. Transgenic bacteria or viruses could mutate and become pathogenic; Bioengineered plants might escape as "superweeds"; Pest-resistant engineered crop plants could select for more resistant pests; Transgenic fish could replace native species and disrupt ecosystems; Genetic screening could allow insurance companies to discriminate against individuals carrying problem alleles; 26. "Designer plants" will improve agriculture and enhance crop yields; Genetically enhanced "oil-eating" bacteria help with oil spills; 27. transgenic; 28. Plasmids; 29. proteins; 30. insulin; 31. micro-injected; 32. diseases; 33. resistance.

Self-Quiz

1. d; 2. d; 3. d; 4. b; 5. c; 6. b; 7. c; 8. a; 9. c; 10. a; 11. d; 12. b; 13. a; 14. b. 15. b; 16. c; 17. c; 18. a; 19. d; 20. b.

Chapter Objectives/Review Questions

1. Watson, Crick; 3. TAAGCG; 8. ribosomal, messenger, transfer; 10. AUG-GUA-CUC-UFF-CGG-UGA; 11. promoter; 12. exons, introns; 19. Plasmids; 20. restriction, sticky; 26. fragment, polymorphisms; 30. confirmed.

Media Menu Review Questions

1. DNA, sequencing; 2. Month, DNA, 3. restrictions; 4. disease; 5. pharmacogenetics, designer; 6. threat, environment; 7. transgenic; 8. scary; 9. tumor-suppressing; 10. suicide.

Chapter 23 Genes and Disease: Cancer

Impacts, Issues: The Body Betrayed [p.431]

23.1. Cancer: Cell Controls Go Awry [pp.432–433]

23.2. The Genetic Triggers for Cancer [pp.434–435]

23.3. Assessing the Cancer Risk from Environmental Chemicals [p.436]

1. F; 2. G; 3. H; 4. B; 5. E; 6. C; 7. A; 8. D; 9. carcinogenesis; 10. Proto-oncogenes; 11. oncogene; 12. does not; 13. does not; 14. tumor suppressor gene; 15. one; 16. division; 17. promote; 18. activated; 19. triggering; 20. normal cell; 21. proto-oncogenes; 22. tumor suppressor genes; 23. abnormal cell; 24. breakdown; 25. proliferates; 26. tumor; 27. metastasis; 28. activation; 29. destroyed; 30. d; 31. b; 32. a; 33. e; 34. c; 35. b; 36. c; 37. a; 38. e; 39. d; 40. half; 41. 40; 42. pesticides; 43. spraying; 44. industrial.

23.4. Diagnosing Cancer [p.437]

23.5. Treating and Preventing Cancer [p.438]

1. bowel, bladder; 2. sore; 3. bleeding, bloody; 4. lump; 5. Indigestion; 6. wart, mole; 7. cough, hoarseness; 8. E; 9. F; 10. I; 11. D; 12. G; 13. A; 14. H; 15. K; 16. C; 17. B; 18. J; 19. L.

23.6. Some Major Types of Cancer [p.439]

23.7. Cancers of the Breast and Reproductive System [pp.440–441]

23.8. A Survey of Other Common Cancers [pp.442–443]

1. sarcomas; 2. carcinomas; 3. Adenocarcinomas; 4. Lymphomas; 5. leukemias; 6. h; 7. g; 8. e; 9. d; 10. a; 11. d; 12. g; 13. e; 14. g; 15. g; 16. h; 17. e; 18. g; 19. d; 20. c; 21. a; 22. b; 23. c; 24. d; 25. d; 26. f; 27. h; 28. e; 29. a; 30. g; 31. a; 32. b; 33. g; 34. d; 35. g; 36. g.

Self-Quiz

1. b; 2. d; 3. a; 4. c; 5. b; 6. a; 7. e; 8. d; 9. b; 10. e; 11. b; 12. e; 13. d; 14. a; 15. d.

Chapter Objectives/Review Questions

1. tumor; 2. Dysplasia; 6. oncogene; 7. tumor suppressor; 14. Chemotherapy.

Media Menu Review Questions

1. active; 2. tobacco; 3. calorie; 4. estrogen, insulin; 5. lung, esophagus; 6. healthy; 7. colon (and possibly prostate), high; 8. growth; 9. fiber; 10. genes.

Chapter 24 Principles of Evolution

Impacts, Issues: Measuring Time [p.447]

24.1. A Little Evolutionary History [p.448]

24.2. A Key Evolutionary Idea: Individuals Vary [p.449]

1. a. John Henslow; b. Cambridge University; c. HMS Beagle; d. Thomas Malthus; e. natural selection; 2. B; 3. E; 4. H; 5. F; 6. A; 7. G; 8. C; 9. D.

24.3. Microevolution: How New Species Arise [pp.450–451]

1. mutations; 2. Natural selection; 3. *On the Origin of Species*; 4. principles; 5. adaptation; 6. the founder effect; 7. genetic drift; 8. gene flow; 9. gene flow; 10. fertile offspring; 11. isolating; 12. divergence; 13. gradualism; 14. punctuated equilibrium; 15. species; 16. punctuated equilibrium; 17. a. variations; b. survival of the fittest; c. natural selection; d. evolution; e. diversity; 18. a. A and B; b. B and C; c. D; 19. pace and timing; no; gradualism probably occurs when the environment is changing gradually, while punctuated equilibrium explains what occurs when the environment changes rapidly.

24.4. Looking at Fossils and Biogeography [pp.452–453]

24.5. Comparing the Form and Development of Body Parts [pp.454–455]

24.6. Comparing Biochemistry [p.456]

1. L; 2. D; 3. G; 4. A; 5. F; 6. E; 7. B; 8. H; 9. C; 10. J; 11. I; 12. K; 13. T; 14. T; 15. are not; 16. T; 17. does.

24.7. How Species Come and Go [pp.456–457]

24.8. Endangered Species [p.457]

24.9. Evolution from a Human Perspective [pp.458–459]

24.10. Emergence of Early Humans [pp.460–461]

1. D; 2. A; 3. C; 4. B; 5. E; 6. Bonobos; 7. Primates; 8. prehensile; 9. opposable; 10. forward-directed eyes; 11. trees; 12. bow-shaped, smaller teeth of about the same length; 13. insects, a mixed diet; 14. fewer; 15. longer; 16. brain; 17. Culture; 18. language; 19. bipedalism; 20. shorter; 21. bipedalism; 22. anthropoids; 23. hominoids; 24. hominids; 25. Lucy; 26. did; 27. *erectus*; 28. 30,000; 29. African emergence; 30. multiregional; 31. cultural.

24.11. Earth's History and the Origin of Life [pp.462–463]

1. did not contain; 2. T; 3. oxygen-free; 4. 3.8 billion; 5. sunlight, lightning, or heat escaping from the crust; 6. T; 7. T; 8. RNA; 9. T.

Self-Quiz

1. c; 2. a; 3. c; 4. d; 5. b; 6. d; 7. c; 8. b; 9. c; 10. b; 11. b; 12. a; 13. b; 14. d; 15. d.

Chapter Objectives/Review Questions
1. Evolution; 6. Mutations; 9. genetic drift; 10. species; 13. mass extinction; 14. adaptive radiation.
Media Menu Review Questions
1. Geographic; 2. Mutations; 3. carbon dioxide; 4. leaves; 5. 700; 6. Africa; 7. Neandertals; 8. out-of-Africa; 9. volcanic ash; 10. skulls.

Chapter 25 Ecology and Human Concerns

Impacts, Issues: The Human Touch [p.467]
25.1. Some Basic Principles of Ecology [pp.468–469]
25.2. Feeding Levels and Food Webs [pp.470–471]
25.3. How Energy Flows Through Ecosystems [p.472]
1. Earth is an island in space similar to the Easter Islands on Earth. Using resources without conservation or population control will lead to the collapse of the ecosystem and civilization; 2. H; 3. J; 4. E; 5. A; 6. C; 7. G; 8. F; 9. B; 10. I; 11. D; 12. a; 13. b; 14. b; 15. a; 16. sun; 17. autotrophs; 18. consumers; 19. heterotrophs; 20. herbivores; 21. carnivores; 22. omnivores; 23. decomposers; 24. energy; 25. decomposers; 26. a; 27. d; 28. e; 29. c; 30. a; 31. a; 32. b; 33. input; 34. sun; 35. are; 36. trophic; 37. chain; 38. plants; 39. lower; 40. producers, consumers; 41. biomass; 42. decreases.
25.4. Chemical Cycles—An Overview [p.473]
25.5. The Water Cycle [p.474]
25.6. Cycling Chemicals from the Earth's Crust [p.475]
1. geochemical cycle; 2. decomposers; 3. weathering of rocks; 4. runoff; 5. nutrient reservoir in environment; 6. oxygen and hydrogen; 7. gas; 8. Earth's crust; 9. G; 10. H; 11. E; 12. A; 13. C; 14. D; 15. I; 16. F; 17. B.
25.7. The Carbon Cycle [pp.476–477]
25.8. Global Warming [pp.478–479]
1. C; 2. E; 3. B; 4. F; 5. A; 6. D; 7. greenhouse effect; 8. heat; 9. global warming; 10. sea; 11. rainfall; 12. carbon dioxide; 13. climate; 14. reverse; 15. political; 16. fossil fuels.

25.9. The Nitrogen Cycle [p.480]
25.10. The Danger of Biological Magnification [p.481]
1. d; 2. a; 3. c; 4. a; 5. b; 6. c; 7. b; 8. e; 9. mosquitoes; 10. body lice; 11. tissues; 12. biological magnification; 13. food chains; 14. songbirds; 15. salmon; 16. pests; 17. predators; 18. food chains; 19. brittle shells; 20. extinction; 21. pesticide; 22. banned; 23. ecosystem.
25.11. Human Population Growth [pp.482–483]
25.12. Nature's Controls on Population Growth [p.484]
1. nine; 2. decreases; 3. birth, death; 4. India; 5. above; 6. industrialized; 7. sixty; 8. demographics; 9. density; 10. distribution; 11. base; 12. slower; 13. exponential; 14. S-shaped; 15. high-density; 16. E; 17. F; 18. B; 19. A; 20. D; 21. C; 22. G; 23. a. 1962–1963; b. 2025 or sooner; c. Depends on age and optimism of the reader.
25.13. Assaults on Our Air [p.485]
25.14. Water, Wastes, and Other Problems [pp.486–487]
1. c; 2. d; 3. a; 4. b; 5. d; 6. e; 7. b; 8. b; 9. e; 10. c; 11. a; 12. e; 13. salty; 14. irrigated; 15. mineral; 16. salinization; 17. irrigation; 18. aquifer; 19. saltwater; 20. Agricultural; 21. factories; 22. oceans; 23. D; 24. E; 25. F; 26. A; 27. B; 28. C.
25.15. Concerns About Energy [p.488]
25.16. A Planetary Emergency: Loss of Biodiversity [p.489]
1. Net; 2. solar; 3. developing; 4. fossil; 5. consumption; 6. air pollution; 7. more; 8. meltdown; 9. solar cells; 10. fusion power; 11. hour; 12. tropical deforestation; 13. half; 14. more; 15. less.
Self-Quiz
1. b; 2. b; 3. c; 4. c; 5. b; 6. a; 7. a; 8. d; 9. b; 10. d; 11. d; 12. a; 13. a; 14. c; 15. b; 16. a; 17. d; 18. a; 19. d; 20. c.
Chapter Objectives/Review Questions
4. weight, energy; 6. phosphorus.
Media Menu Review Questions
1. small; 2. pictorial; 3. giants; 4. walked; 5. 110; 6. earlier, ranges; 7. ecosystem; 8. global warming; 9. the coral reef; 10. tolerance.